Modifying the
AERODYNAMICS
of Your Road Car

Step-by-step instructions to improve the aerodynamics of road cars

More from Veloce:

SpeedPro Series
4-Cylinder Engine Short Block High-Performance Manual – New Updated & Revised Edition (Hammill)
Aerodynamics of Your Road Car, Modifying the (Edgar and Barnard)
Alfa Romeo DOHC High-performance Manual (Kartalamakis)
Alfa Romeo V6 Engine High-performance Manual (Kartalamakis)
BMC 998cc A-series Engine, How to Power Tune (Hammill)
1275cc A-series High-performance Manual (Hammill)
Camshafts – How to Choose & Time Them For Maximum Power (Hammill)
Competition Car Datalogging Manual, The (Templeman)
Custom Air Suspension – How to install air suspension in your road car – on a budget! (Edgar)
Cylinder Heads, How to Build, Modify & Power Tune – Updated & Revised Edition (Burgess & Gollan)
Distributor-type Ignition Systems, How to Build & Power Tune – New 3rd Edition (Hammill)
Fast Road Car, How to Plan and Build – Revised & Updated Colour New Edition (Stapleton)
Ford SOHC 'Pinto' & Sierra Cosworth DOHC Engines, How to Power Tune – Updated & Enlarged Edition (Hammill)
Ford V8, How to Power Tune Small Block Engines (Hammill)
Harley-Davidson Evolution Engines, How to Build & Power Tune (Hammill)
Holley Carburetors, How to Build & Power Tune – Revised & Updated Edition (Hammill)
Honda Civic Type R High-Performance Manual, The (Cowland & Clifford)
Jaguar XK Engines, How to Power Tune – Revised & Updated Colour Edition (Hammill)
Land Rover Discovery, Defender & Range Rover – How to Modify Coil Sprung Models for High Performance & Off-Road Action (Hosier)
MG Midget & Austin-Healey Sprite, How to Power Tune – Enlarged & updated 4th Edition (Stapleton)
MGB 4-cylinder Engine, How to Power Tune (Burgess)
MGB V8 Power, How to Give Your – Third Colour Edition (Williams)
MGB, MGC & MGB V8, How to Improve – New 2nd Edition (Williams)
Mini Engines, How to Power Tune On a Small Budget – Colour Edition (Hammill)
Motorcycle-engined Racing Cars, How to Build (Pashley)
Motorsport, Getting Started in (Collins)
Nissan GT-R High-performance Manual, The (Gorodji)
Nitrous Oxide High-performance Manual, The (Langfield)
Optimising Car Performance Modifications (Edgar)
Race & Trackday Driving Techniques (Hornsey)
Retro or classic car for high performance, How to modify your (Stapleton)
Rover V8 Engines, How to Power Tune (Hammill)
Secrets of Speed – Today's techniques for 4-stroke engine blueprinting & tuning (Swager)
Sportscar & Kitcar Suspension & Brakes, How to Build & Modify – Revised 3rd Edition (Hammill)
SU Carburettor High-performance Manual (Hammill)
Successful Low-Cost Rally Car, How to Build a (Young)
Suzuki 4x4, How to Modify For Serious Off-road Action (Richardson)
Tiger Avon Sportscar, How to Build Your Own – Updated & Revised 2nd Edition (Dudley)
Triumph TR2, 3 & TR4, How to Improve (Williams)
Triumph TR5, 250 & TR6, How to Improve (Williams)
Triumph TR7 & TR8, How to Improve (Williams)
V8 Engine, How to Build a Short Block For High Performance (Hammill)
Volkswagen Beetle Suspension, Brakes & Chassis, How to Modify For High Performance (Hale)
Volkswagen Bus Suspension, Brakes & Chassis for High Performance, How to Modify – Updated & Enlarged New Edition (Hale)
Weber DCOE, & Dellorto DHLA Carburetors, How to Build & Power Tune – 3rd Edition (Hammill)

Workshop Pro Series
Car electrical and electronic systems (Edgar)
Setting up a home car workshop (Edgar)

RAC Handbooks
Caring for your car – How to maintain & service your car (Fry)
Caring for your car's bodywork and interior (Nixon)
Efficient Driver's Handbook, The (Moss)
Electric Cars – The Future is Now! (Linde)
First aid for your car – Your expert guide to common problems & how to fix them (Collins)
How your car works (Linde)
Pass the MoT test! – How to check & prepare your car for the annual MoT test (Paxton)
Selling your car – How to make your car look great and how to sell it fast (Knight)
Simple fixes for your car – How to do small jobs for yourself and save money (Collins)

Enthusiast's Restoration Manual Series
Citroën 2CV Restore (Porter)
Classic Car Bodywork, How to Restore (Thaddeus)
Classic British Car Electrical Systems (Astley)
Classic Car Electrics (Thaddeus)
Classic Car Suspension, Steering & Wheels, How to Restore & Improve (Parish – translator)
Classic Cars, How to Paint (Thaddeus)
Jaguar E-type (Crespin)
Reliant Regal, How to Restore (Payne)
Triumph TR2, 3, 3A, 4 & 4A, How to Restore (Williams)
Triumph TR5/250 & 6, How to Restore (Williams)
Triumph TR7/8, How to Restore (Williams)
Ultimate Mini Restoration Manual, The (Ayre & Webber)
Volkswagen Beetle, How to Restore (Tyler)
VW Bay Window Bus (Paxton)

Expert Guides
Land Rover Series I-III – Your expert guide to common problems & how to fix them (Thurman)
MG Midget & A-H Sprite – Your expert guide to common problems & how to fix them (Horler)
Triumph TR2, TR3, TR3A & TR3B – Your expert guide to common problems & how to fix them (Hogan)
Triumph TR4 & TR4A – Your expert guide to common problems & how to fix them (Hogan)
Triumph TR6 – Your expert guide to common problems & how to fix them (Hogan)

www.veloce.co.uk

First published in October 2018, reprinted June 2023 by Veloce Publishing Limited, Veloce House, Parkway Farm Business Park, Middle Farm Way, Poundbury, Dorchester DT1 3AR, England. Tel +44 (0)1305 260068 / Fax 01305 250479 /e-mail info@veloce.co.uk / web www.veloce.co.uk or www.velocebooks.com.
ISBN: 978-1-787112-83-4 UPC: 6-36847-01283-0.
© 2018 Julian Edgar and Veloce Publishing. All rights reserved. With the exception of quoting brief passages for the purpose of review, no part of this publication may be recorded, reproduced or transmitted by any means, including photocopying, without the written permission of Veloce Publishing Ltd. Throughout this book logos, model names and designations, etc, have been used for the purposes of identification, illustration and decoration. Such names are the property of the trademark holder as this is not an official publication. Readers with ideas for automotive books, or books on other transport or related hobby subjects, are invited to write to the editorial director of Veloce Publishing at the above address. British Library Cataloguing in Publication Data – A catalogue record for this book is available from the British Library. Typesetting, design and page make-up all by Veloce Publishing Ltd on Apple Mac.
Printed and bound by CPI Group (UK) Ltd, Croydon, CR0 4YY.

Modifying the
AERODYNAMICS
of Your Road Car

Step-by-step instructions to improve the aerodynamics of road cars

Julian Edgar
Technical consultant: R H Barnard

THE PUBLISHER OF FINE AUTOMOTIVE BOOKS

Contents

Introduction 6
Further research 7
Acknowledgements 7

Chapter 1: Theory 9
Types of drag 9
Form drag 9
Viscous drag 11
Wake size and drag 11
Body pressures causing drag
 and lift 12
Interference drag 14
Attached and separated flows ... 14
Force vectors 15
Reynolds numbers 17
Induced drag and trailing
 vortices 17
Trailing vortices and lift – chicken
 or the egg? 19
Undercar flows 19
Drag and lift coefficients 20
Changes in drag with top speed . 21
Wings 22
Yaw 22

Dynamic and static pressures 22
Veracity of drag and lift numbers .. 23

**Chapter 2: Aerodynamically
 optimised vehicles 25**
1921 Rumpler Tropfenwagen 25
1928 Opel RAK 2 winged rocket
 car 28
1936 Tatra T87 28
1938 Porsche Type 64 32
1950 Volkswagen Transporter 32
1967 NSU Ro 80 33
1969 Dodge Charger Daytona &
 1970 Plymouth Roadrunner
 Superbird 36
1988 Holden Special Vehicles
 Group A Commodore
 ('Walkinshaw') 39
1989 Opel Calibra 41
1996 General Motors EV1 43
2011 General Motors Chevrolet
 Volt 45
2013 Volkswagen XL-1 47
Independent testing 49

Some more fascinating aero cars .. 49

**Chapter 3: Flow
 visualisation 52**
On-road flow testing 52
Visualisation techniques 53
Wool-tufting 58
Tufts on long lengths of tape 58
Smoke streams 68
Eroding clay 68
Comparing tufts and eroding clay . 68
Dust 70

**Chapter 4: Pressure
 measurement 71**
Pressure measurement aims 72
Measuring instruments 72
Magnehelic gauges 73
Digital manometers 74
Pressure probes 75
Measuring pressures with a
 Magnehelic gauge 76
Step-by-step pressure
 measurement 77

CONTENTS

Using pressure measurement data .80
Undercar pressure measurements 81
Using pitot tubes to measure airspeed 81

CASE STUDIES
Testing the Jaguar XE 84
Testing the Tesla Model S ... 92
Testing the Mercedes-AMG GT 96

Chapter 5: Measuring changes in drag and lift ... 100
Measuring drag 101
Required power testing 101
Measuring required power 102
Coast-down testing 102
Fuel economy testing 105
Comparing rear vision mirror drag 106
Measuring lift and downforce.... 108
Measuring lift and downforce on the road 108
Bumpy roads, and getting on and off the throttle 112

Chapter 6: Reducing drag .. 115
Reducing frontal area 115
Reducing rear vision mirror drag 117
Lenses and mirrors? 120
Reducing cooling drag 120
Implications of reducing cooling airflow 125
Ride height and rake 125
Reducing the area of the wake.. 127
Wake area reduction on a pick-up 134
A larger wake 136
Reducing the strength of the wake 136

Achieving clean separation 137
Reducing separation bubbles ... 140
The Excelle GT 141
Safety under the car 150
Skoda Roomster – vortex generators on a front undertray 150
Undertrays 151
Wheels, tyres and wheelarches 157
Front deflectors? 158
Trailing vortices 162
Add-on modifications 164
Nissan Leaf 164
Testing multiple drag-reducing modifications 165
Reducing drag – development of a new car 165
Reducing drag – modifying an existing car 169
Philip's truck 170

Chapter 7: Reducing lift and improving stability 171
Lift 171
Stability 173
Forces affecting stability 175
Determining the centre of pressure and centre of gravity. 176
Undertrays and diffusers 182
Honda Insight undertray and diffuser 188
Spoilers 192
Wings 198
Active aerodynamics 204
Downforce in a road car 205

Chapter 8: Improving airflow through heat exchangers 210
Pressure differentials 211
Improving underbonnet intercooler airflow 212
Starting points 213

Testing a Peugeot 405 diesel intercooler 218
Trying to improve flow through an Audi S4 intercooler 219
Honda Insight – alternator cooling duct 220
Turning vanes 221
More on bonnet vents and ducts 222
The overall approach 222

Chapter 9: Engine intakes . 224
Benefits 224
Measuring pressures 225
Measuring engine intakes 226
Building a new engine intake 227
Example new intake – Honda Insight 230
Other examples 232

Chapter 10: Reducing aerodynamic noise 233
Causes of aerodynamic noise ... 233
Testing and measurement 234
Decreasing noise production 235
Damping versus isolating 236
Decreasing noise transmission .. 236
Modifications to reduce noise ... 237

References 240
Books 240
Papers 240
Magazine 242

Index 247

Introduction

People have been modifying their road cars for well over a century – car modification goes back at least as far as the Ford Model T. But the widespread aerodynamic modification of cars is much younger. In fact, if you weren't racing, but instead were working on your road car, aero modification is arguably little more 20 years old. It's in only that period that car modifiers have started to wool-tuft their cars to see airflow patterns, to directly measure surface body pressures to find areas of lift and downforce, and to measure pressures when improving flows through heat exchangers like intercoolers and radiators.

And it's no surprise that aerodynamic modification has exploded over just the same period that car manufacturers have been focusing harder on aerodynamics than ever before. Over that time, the amount of aerodynamic lift that a typical new car generates has fallen by 80 or 90 per cent, while drag figures have dropped by over 30 per cent – sometimes more. Sophisticated and stimulating aerodynamic design can now be seen on the cheapest new cars being sold: just look around you in traffic to see thrilling aero stuff everywhere!

But what makes aerodynamic modification of cars so exciting is that it's all within the reach of someone working at home. Compare that with engine design, for example. How many of us can build a new DOHC, variable compression engine with five valves per cylinder, working just in our home workshop? But in aerodynamic terms, we can make changes as sophisticated as that engine – all using simple materials, and testing the results on the road.

For a handful of coins, you can have testing equipment that will show you airflow patterns over your car, allowing you to spot where improvements are needed. For just a little more money and time, you can accurately measure downforce and lift – and all on the road at normal speeds.

You want to lower drag? You can have it. You want to reduce lift? You can have it too. (And if you want to have a road car with downforce – that's possible as well.) You want to improve airflow through an oil cooler or intercooler? You can make changes and measure the results. And often those modifications can be trialled using just plastic sheet cut to size with a pair of scissors and temporarily stuck in place with gaffer tape. If you cannot measure a change, take off your temporary modification. If it works, make a proper one.

If you're an experienced car modifier, all the skills you already have can be used in aerodynamic modification – working with fibreglass or carbon fibre, welding, shaping plastics and even building electronics. And if you're not an experienced modifier, making aerodynamic changes – eg under the floor of the car – is still within your reach.

If you're a hyper-miler, reducing drag and watching fuel economy improve is exciting and fun. And if you're a performance driver, feeling the effects of real downforce on a road car is like going from humdrum tyres to the stickiest you can afford – it's simply magic.

INTRODUCTION

For anyone interested in car modification, making aerodynamic changes is one of the best games in town.

FURTHER RESEARCH

So what further research can you do when you've finished this book?

(Note: with all the books I am citing here, the later the edition, the better – but any edition is worth buying. I recommend www.abebooks.com in sourcing these books.)

The best next stepping stone is *Road Vehicle Aerodynamic Design: An Introduction* by R H Barnard. It is not serendipitous that I asked Dick Barnard to be technical consultant on the book you are now reading. Dick is both a superb writer and an academic expert in the field. You will find more maths in his book than this one, but you'll also find a beautifully understated writing style that is extremely clear.

One book that you will find rightfully referenced everywhere in the field of car aerodynamics is *Aerodynamics of Road Vehicles*, edited by W H Hucho. This is the absolute gold standard, with an enormous amount of information, much of it showing experimental results on real cars. However, it can at times be heavy going, and often you will have to think hard of how to apply the ideas to your own car. Buy it and use it as a reference textbook, dipping in and out as needed.

It's now quite an old book, but *Road Vehicle Aerodynamics* by A J Scibor-Rylski is an excellent read. Because of its age, you need to remember that the author is often talking about cars with drag coefficients around 0.40, and cars that have rough underbodies – but keep these points in mind and there is still plenty of interest. However, I would not start with this book. (As I have kept coming back to this book over the years, I have thought it better and better, but I still remember my initial puzzlement with it!)

There are two other aero textbooks that I would recommend – and both should be used with caution. The first is *Race Car Aerodynamics: Designing for Speed* by Joseph Katz. The other is *The Leading Edge: Aerodynamic Design of Ultra-streamlined Land Vehicles* by Goro Tamai. I learned a lot from both books, but they are more complex than the publications I've covered so far. Furthermore, since neither is about road cars, you need to be very careful in taking the principles they espouse, and then applying them uncritically to cars on the road. But both are excellent food for thought.

Two wonderful books on historic streamlined vehicles are *Stromlinienautos in Deutschland* and *Stromlinienbusse in Deutschland*, about streamlined cars and buses (respectively) in Germany, primarily before WWII. Both books are by Ralf J F Kieselbach, and are in German and English. I found these books thought-provoking, and in fact my interest in rear fins (used to move the centre of pressure rearwards) came from these two books.

To keep up to date, nothing beats the US-based Society of Automotive Engineers papers. The SAE publishes a large number of technical papers each year – and many of these are on car aerodynamics. Be warned though – first, the papers are expensive to buy, and second, they are written for fellow engineers, so can be very complex. The best are typically those that cover the aerodynamic development of a specific car.

Car aerodynamics is not standing still. There is plenty of technology still to be employed – from blown wakes to flapping spoilers to active car aerodynamics (*really* active, not just spoilers that pop up and down). So this is one area where it's wise to stay abreast of the changes being made to road cars – after all, many of the techniques can be employed by anyone studying the outcome.

Finally, visit car museums! I have been lucky enough to visit some of the most important car museums in the world, and I find each is a veritable treasure trove of aerodynamic ideas. In many cases, idiosyncratic makers of cars were many decades ahead of their contemporaries. They may have produced commercial flops, but we can learn a lot from their cars – even 100 years later (I think here of the Rumpler Tropfenwagen). My wife always makes gentle fun of the fact that I am the only museum visitor who apparently likes lying on the floor to look under these vehicles, but I encourage you to do the same. (It's OK, most of the floors are clean and in only one museum did an 'attendant' mysteriously appear to chaperone me for the rest of my visit!)

ACKNOWLEDGEMENTS

Lots of people have contributed to the outcome that you are holding your hands.

For supplying pics and descriptions of their modified cars, thanks to Darin Cosgrove, Wyatt Fisher, Earl Poon, Jim Michler, Jerry Lee, Pascal Dunning, Gerrelt Molhoek, Arto Viinanen and Philip Knox. For contributing the section on locating the centre of pressure, thanks to Eva Hakansson and Bill Dube'.

John Lindsay kindly made available his Tesla Model S for wool tuft testing, and Jaguar Australia did the same with an XE Jaguar.

SPEEDPRO SERIES

Paul Lawford not only supplied his AMG-Mercedes AMG GT but also photographed it being tested.

Wayne Medway let me use his workshop hoist when I needed to take undercar photographs. Car modifier (and Boeing 747 pilot) Bill Sherwood read and commented on some material. Various chapters of the book were reviewed by Willem Toet (F1 aerodynamicist), Joseph Katz (Professor of Aerospace Engineering at San Diego State University) and Adrian Gaylard (Jaguar Land Rover chief aerodynamicist). Adrian also pointed me towards some significant engineering papers that I'd missed. My wife Georgina was test driver for much of the aerodynamic testing photography, a role she performed with grace, skill and patience.

My thanks also to the companies that made available images for this book. Many of the wind tunnel photos are from the archives of Mercedes-Benz, a firm that makes available to journalists and authors the best technical photography of any car company in the world. Jaguar also kindly gave me permission to use technical images not previously widely published. (Photos not otherwise credited were taken by me.)

Finally, technical consultant on the book, Dick Barnard, provided a good-humoured, firm and experienced guiding hand, especially in the more technical sections.

Julian Edgar

The author attaching a surface pressure measuring disc when aero testing an XE Jaguar. (Courtesy Georgina Edgar)

Chapter 1
Theory

- ♦ **Form drag**
- ♦ **Viscous drag**
- ♦ **Body pressures causing drag and lift**
- ♦ **Interference drag**
- ♦ **Attached and separated flows**
- ♦ **Reynolds numbers**
- ♦ **Induced drag and trailing vortices**
- ♦ **Undercar flows**
- ♦ **Drag and lift coefficients**
- ♦ **Yaw**
- ♦ **Dynamic and static pressures**

To understand what is occurring as air passes over (and around and under) your car, you need to first understand some theory. I could start by discussing the physical characteristics of air, but I think it makes the subject more accessible if instead we begin by looking at a real-world aspect of car aerodynamics – drag. Through looking at how drag occurs, we can begin to understand airflow and the forces it develops. Then, when we have done that, we can build on those ideas to look at more complex aspects.

Drag is the force imposed by the air that retards the car's forward movement. Higher drag increases fuel consumption and slows acceleration. So what types of drag are cars subjected to, and why does it occur?

TYPES OF DRAG
Figure 1-1 shows the different components into which drag can be broken down.

FORM DRAG
In conventional road vehicles, *form drag* makes up the largest component of the different types of drag. Form drag relates to the shape of the vehicle – so that's easy to remember, it is dependent on the car's form. To understand what form

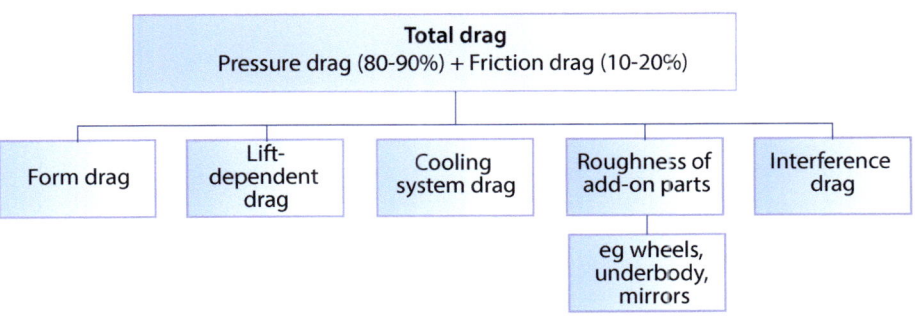

Figure 1-1: The breakdown of drag on a typical car.

SPEEDPRO SERIES

Figure 1-2: An older Mercedes being wind tunnel tested. Note how the airflow on the front two-thirds of the car follows the bodywork contours. This flow is attached. However, you can see that at the trailing edge of the roof the air flow separates – it goes its own way and no longer follows the body contours. (Courtesy Mercedes)

drag is, we need to split airflow over a car into two different types. These types of airflow are called 'attached' and 'separated.'

Look at Figure 1-2, which shows an older Mercedes being wind-tunnel tested. The smoke shows streamlines, and you can picture these streamlines as the paths the air is taking. You can see that the streamline directly above the bonnet (hood) is largely following the shape of the car. It travels along the bonnet, up the windscreen and then along the roof. In other words, the airflow is being guided by the surfaces over which it is flowing. This is called *attached flow*.

That's at the front of the car – but what about at the back? Looking again at Figure 1-2, you can see that at the end of the roof, the airflow *does not* follow the bodywork but instead goes its own way. That is, the streamline ceases to follow the shape of the car. For example, the streamline doesn't descend the rear window and then flow across the boot (trunk) lid. Where the airflow is no longer being guided by the car's surface, the airflow is said to have *separated*. In general, keeping the flow attached as far back as possible, both in side and plan views, reduces drag. (An exception to this is if the airflow then wraps around a rear-facing curved surface, where it can create a backward-pulling low pressure, and so increase drag. This is avoided by achieving *clean separation*, covered in more detail later.)

Look at Figure 1-3. We can see that, on this car, the airflow separates at the end of the roof. We can also see an area of disturbed air behind the car. This area is called the *wake*. In the wake the pressure is low, and so it is an area that is effectively pulling the car backward. The area of the car exposed to the wake is called the car's *base*.

Most cars are boat-tailed, ie the sides of the car get closer together over the last part of the body. In Figure 1-3 you can see that with this car, the roof also descends towards the rear, which also increases base pressure and so decreases drag. Even if we are dealing with a three-box sedan, it is possible with careful design to keep flow attached right to the trailing edge of the boot lid.

Figure 1-4 shows a computer model of airflow over a car. Here we can see not only the streamlines of the type we saw with smoke in the wind tunnel, we can also see the behaviour of air in the wake. You can see that in the wake immediately behind the car, the airflow is very

Figure 1-3: On this squareback car, the airflow separates at the end of the roof. There is also an area of disturbed air behind the car: this disturbed air is called the wake. (Courtesy Mercedes)

THEORY

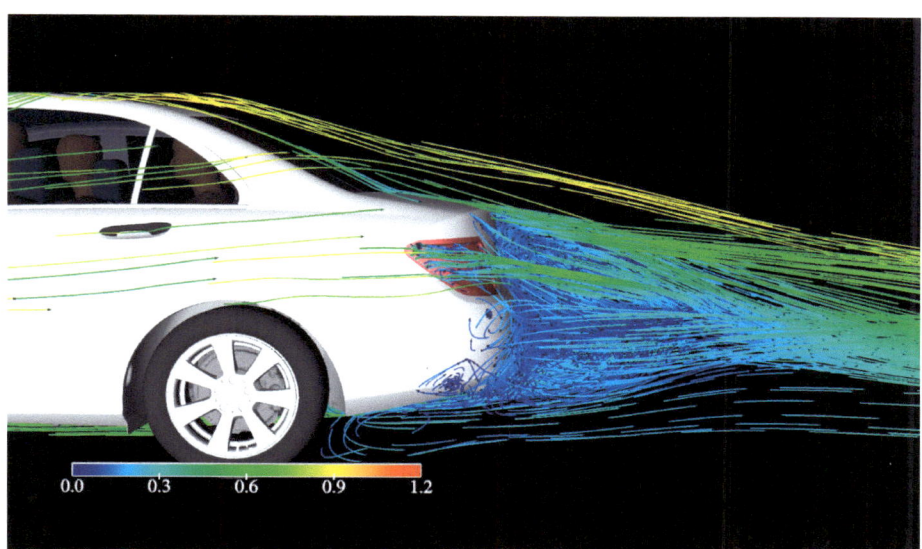

Figure 1-4: A computer model showing streamlines and the behaviour of the air in the wake. Note how because the flow is attached until the trailing edge of the boot lid, the height of the wake is much smaller than the height of the car. (Courtesy Mercedes)

WAKE SIZE AND DRAG

A wide wake does not necessarily imply a low rear end pressure; in fact, the reverse can be true. What is more important is the scale and intensity of *turbulence in the wake*, since high turbulence implies that a lot of energy has gone into the wake formation – and that energy has ultimately come from the engine.

In order to reduce drag, it is necessary to encourage the flow to return as closely as possible to its free-stream conditions of speed and pressure when it reaches the rear. In a well streamlined shape, the drag is small, and is mainly due to surface friction effects (see below). For normal road cars, it is not possible to produce a fully streamlined shape, but maintaining attached flow as far back as possible will minimise the drag by minimising the amount of energy that goes into producing turbulence.

However, it is the case that for normal car-shaped bodies, a small wake area normally indicates a low drag. Additionally, for a given car, reducing the size of the wake (eg by angled inward body extensions) typically reduces drag.

jumbled in the directions it is taking (this is called *turbulence*).

(Another contribution to drag arises from trailing vortices. In Figure 1-4 there are whirling swirls of air developing behind the car, these are examples of trailing vortices. These vortices may or may not be located in the wake. More on trailing vortices in a moment.)

VISCOUS DRAG

Air is a fluid, and like other fluids, has viscosity. *Viscosity is the ease with which particles of the fluid move past one another*. Honey, for example, is highly viscous – the particles don't want to slide past one another. Engine oil is less viscous, and air is much less viscous again. You can also think of air viscosity as being the 'stickiness' of the fluid. So why is the viscosity of air important? Because when the air passes over a car's body, it is being constantly sheared. Let's take a look at why this happens.

If you park a car on a dirt road, the car will soon be covered with a fine layer of dust. When you drive off, watch the surface of the bonnet. Despite it being covered in dust, the dust particles will largely remain in place. But why doesn't the dust get blown away by the airflow over the car? The answer is that the air in contact with the body isn't moving. That is, the tiny molecules of air closest to the body are not sliding over the paintwork. As you get further away from the car body surface, the airspeed gets greater and greater, until when you're far enough away from the body, the airspeed is the same as the car's forward speed (called freestream velocity).

Figure 1-5 shows the effect. Here the length of the arrows is proportional to the speed of air. You can see the lowest arrow, a little

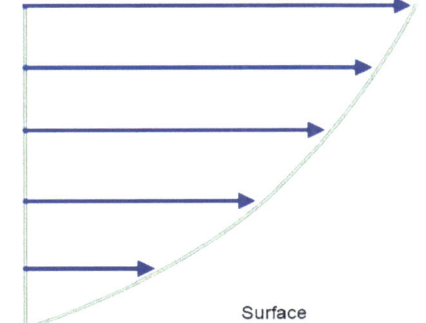

Figure 1-5: The length of the arrow shows air speed. At the surface the air is not moving relative to the car – it is being pulled along with the vehicle. The further away you get from the car's body, the faster the airflow, until, when you are far enough away, the airflow is at freestream speed. The different lengths of arrows show how the air is being constantly sheared as the car moves forward.

11

above the surface of the car, is much shorter than the top arrow, furthest from the car. Therefore, because the layers of air are travelling at different speeds, they are sliding over one another – or to put it another way, particles of the fluid are constantly moving past one another. This was our definition of viscosity – the ease with which particles of the fluid move past one another. Now you can see why the viscosity of the air has an impact.

The volume of low-speed flow attached to the surface of the car is called the boundary layer. An easy way to define the boundary layer is to think of it as comprising all the air volume where the velocity is less than 95 per cent of the freestream velocity. Therefore, it is the layer where the friction with the surface causes a significant change in air velocity over that found in the freestream. As you move towards the rear of the car – even one with attached flow from nose to tail – the boundary layer gets thicker. In other words, the number of air molecules getting dragged along with the car becomes greater.

Thinking about the two factors that we've discussed so far, you can see that the longer that flow stays attached to the body, the lower will be the form drag. So that's good! However, the longer the flow stays attached to the body, the greater the amount of viscous shearing that is occurring in the boundary layer around the body – and that's bad. However, overall, it is still much better to have as much attached flow as possible. Viscous drag makes up only a small component of drag on current road cars, but as form drag is reduced in the future with better designs, the importance of viscous drag will increase.

Incidentally, an aerodynamics lecturer once told me about a great idea some of his students (thought) they'd had. They were interested in the Land Speed Record for human-powered vehicles, and decided the best approach was to have a very long, streamlined vehicle. The length of the vehicle (say 10m, or 30ft) would also allow them to have plenty of riders to power the machine. All the riders would be semi-reclined, making for a very small frontal area. With almost no form drag, the vehicle was sure to break the record! The problem? With such a long vehicle, the viscous behaviour of the air would still result in lots of drag. Back to the drawing board.

BODY PRESSURES CAUSING DRAG AND LIFT

We have already looked at the lower pressure in the wake that then influences car base pressure. But what about other pressure variations over the car's body? These can create both drag and lift.

To understand these pressures, we need to be aware of some more ideas. The first is that while the car may be travelling forward at a constant speed, *the speed of airflow over different parts of the car varies*. In fact, the localised airflow speed might vary from zero to two or even three times the car's forward velocity.

So how can this occur – how, for example, can the speed of the air on the car's body be zero if the car is moving forward? Think of the airflow approaching the front of the car. Somewhere on the front of the car, the airflow splits to head off in two directions. Some of the airflow will pass over the car, and some will head downward and go under the car. The line at which the airflow splits to go in different directions is called the *stagnation* line, and at this precise location, the airflow speed will be zero. (You can remember this as the 'confused air' that doesn't know whether to go up or down, and so just stops!)

On the other hand, where the airflow is wrapping around a curved surface, the airflow speeds up. For example, the airflow passing from the windscreen around the curve onto the roof will travel more rapidly than the car's forward speed.

Figure 1-6 shows a computer simulation of air speeds, with blue indicating low speeds and red indicating high speeds. Note the blue streamlines at the front of the car, where the airspeed is slowed down. See also the red airstreams, where the airflow wraps around the curves at the leading edge of the bonnet, and where the airflow passes from the windscreen onto the roof.

Now, the second idea. *The higher the air speed, the lower the pressure*. (And of course, the converse is also the case: the lower the airspeed, the higher the pressure.) Look again at Figure 1-6, but this time instead of thinking of the colours showing air speed, think of the colours as showing pressures. Now the blue shows high pressures and the red, low pressures. Look at all the red over the curved roof of the car – this is an area of low pressure, and so lift. But what about at the back of the car? If the airflow remains attached onto the rear window, you can see that there will be a line of low pressure at this transition zone. Some of the 'pull' exerted by this low pressure will be backward, as well as upward. Therefore, it will be causing both drag and lift (see the breakout box on page 15 for more on how both of these forces occur).

Figure 1-7 shows pressure measurements made on a Porsche 924 in a wind tunnel. Here you can see the pattern of low and

THEORY

Figure 1-6: A computer simulation of air speeds, with blue indicating low speeds and red indicating high speeds. The higher the air speed, the lower the pressure (and vice versa), so you can also see areas of low and high pressure. (Courtesy Mercedes)

Figure 1-7: The numbers show the measured pressures on the upper surfaces of a Porsche 924, with the measurements made in a wind tunnel. The pressure is low at the front of the car (-20) as the airflow wraps around the curve from bumper to bonnet. At the base of the windscreen the pressure is high (+40), and then over the long curve from the top of the windscreen to the rear edge of the hatch the pressure is low. The low pressures over the hatch are causing both lift and drag. (Courtesy Porsche)

high pressures on the car's upper body. As the airflow wraps around the leading edge of the bonnet, there is a low pressure of -20 units. The pressure rises at the base of the windscreen (+40), but then decreases at the top of the windscreen (-150). However, look at the low pressures across the rear hatch (-40, -20, -20). If you imagine all these pressures acting on the surface, you can see that these low pressures would be creating both lift and drag. (That is, they have both upward and backward force components.) In Chapter 4 we cover how you can make pressure measurements like this on the road – you don't need to use a wind tunnel!

On any car body surface that has a vertical component, pressures above or below atmospheric will potentially be causing drag. For example, a high pressure on the front of the car (eg the grille, headlights, front spoiler, windscreen, etc) will cause drag. In Figure 1-8, the surface pressures over an older, bluff-fronted car are modelled. The 'hotter' the colour, the higher the pressure. You can see in this image the large amount of high pressure in the grille and headlights area of the car, creating drag. A low pressure on the rear of the car (eg the rear window) will also cause drag, and

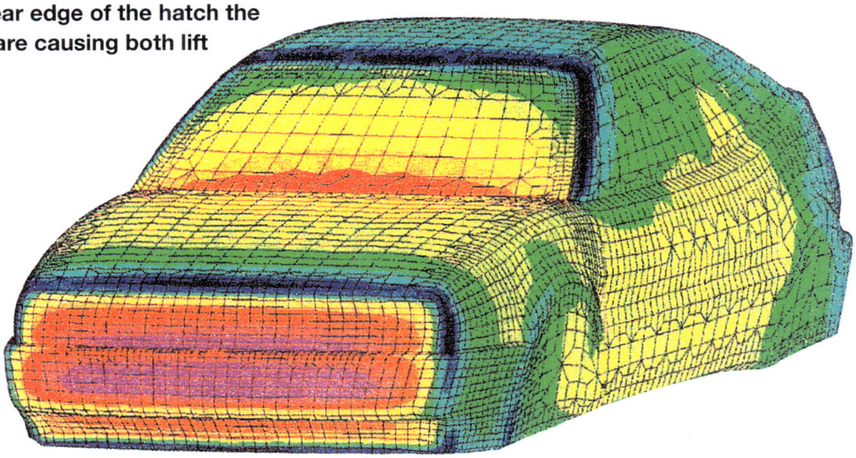

Figure 1-8: This simulation shows the surface pressures on an older car with an upright front. The 'hotter' the colour, the higher the pressure. (Courtesy Volkswagen)

13

– while we can't see it in this image – there are sure to be low pressure areas at the back of the car causing drag in this manner.

So why don't we avoid this sort of drag by creating earlier airflow separation? After all, if the air isn't following the body contours, it will be as if the body isn't there – and so we won't get this 'pressure problem' of drag creation. That's quite true, but what we will get instead is more form drag. In fact, the original meaning of the word 'spoiler' was to spoil the airflow – ie to cause separation. In later chapters we will look at how creating earlier separation can cause changes in lift and drag.

However, where the air finally *does* have to separate from the car's body, it is best that this occurs without wrapping around trailing curves that create suction peaks. This is usually achieved by *separation edges* – sharp changes in direction that the airflow cannot follow. Figure 1-9 shows the separation edge on the side of a rear light.

Figure 1-9: Where separation finally has to occur at the rear of the car, clean separation is desired. That is, we don't want airflow wrapping around curves and creating suction peaks causing drag. The arrow points to a separation edge moulded into the rear light.

INTERFERENCE DRAG

Interference drag occurs because vehicles are not single bodies without appendages. And unfortunately, when different shapes are combined to form a practical vehicle, the total drag is usually greater than the sum of its parts.

For example, an external rear vision mirror tested in a wind tunnel might have a certain drag value, but when it is attached to the car, it will increase the drag of the car body, as well as imposing its own drag. In effect, the flow over one surface interferes with the flow over another surface.

The rear vision mirror is a good example of this, because as we've already seen, when airflow wraps around a corner, the speed increases. In the case of a car, high airflow speeds occur around A-pillars, and so a mirror mounted in the conventional position, and subject to these high airflow speeds, creates much more drag than its size and shape would suggest. Figure 1-11 shows flow disturbance on a front door window caused by a poor A-pillar design, and the interaction of that and the rear vision mirror. Note how the wool tufts are pointing in almost every direction but towards the back of the car!

The less protuberances on the car, the lower will be the interference drag.

ATTACHED AND SEPARATED FLOWS

We've seen that attached flow follows the contours of the car's body, while separated flow does not. We've also seen that to reduce form drag, attached flow should be maintained as far rearward as possible. So what actually causes unwanted flow separation to occur?

Flow separation occurs when the air is subject to an *adverse pressure gradient*. Air can flow from a low pressure to a high pressure only by slowing down. For example, it does this when passing from the top of the windscreen (a lower-pressure area) to rear of the roof (an area with higher pressure). If the increase in pressure is gradual,

Figure 1-10: Here the airflow separates across a sharp upper edge at the trailing end of the boot to prevent a suction peak forming. (Courtesy Mercedes)

THEORY

FORCE VECTORS

On pages 12-14 it is discussed how high and low pressures act on the surface of the car's body, and how – for example – a low pressure can cause both drag and lift. To understand how these forces come about, it helps if you can picture force vectors, and especially the 'triangle of forces.'

The photo on the right shows an older Mercedes being tested in the Mercedes wind tunnel. Note how the flow, as indicated by the smoke streams, wraps around the windscreen header rail (the transition from windscreen to roof). Also note how here the streamlines are closer together, indicating a higher airflow speed, and so a lower pressure being developed on the car body.

In the picture below a red arrow has been added to show the created force. (This arrow is called a 'force vector.') Note how the arrow goes through the 'kink' in the streamlines.

The green and yellow arrows (below right) show the vertical and horizontal components of this force. You can see that a right-angled triangle is developed, thus explaining the 'triangle of forces' name. The green arrow represents the lift component of the force. So what is the yellow arrow showing? It's showing that there is actually a thrust component to this force; that is, the low pressure here is helping to suck the car along. (Thrust is the opposite of drag.) In this diagram, the length of the arrows is proportional to the forces involved. Therefore, in the case of the Mercedes, about three times as much lift as thrust is developed at this point. (However, note that the area the thrust acts on is very small, and this small amount of thrust will be rapidly offset by the viscous drag on the roof.)

When you are thinking about aerodynamic forces acting on car surfaces that are not horizontal or vertical (and that's most of them), always think about how those forces can be broken down into horizontal and vertical components. For example, making a change that increases pressure on a sloping rear window (eg by the addition of a rear spoiler) will reduce both lift and drag, although not necessarily by the same amount.

When drawing the triangles, remember that resolved drag and thrust forces are always parallel to the ground, and lift and downforce are at right-angles to the ground.

SPEEDPRO SERIES

the outer layers of the boundary layer will have a chance to mix with, and pull along, the air closer to the bodywork. However, if the increase in pressure is fast and so the airflow slows more quickly, there will not be time for the layers nearer the bodywork to be energised, and so this air will stop moving. Once the lower part of the layer is no longer moving, separation will occur. You can see that a thicker boundary layer is more likely to separate – it's harder to keep the lower layer moving.

It's this 'energising of the layers' that explains why a turbulent boundary layer will stay attached in a situation where a laminar boundary layer will separate. While it's common to think of attached flow as being laminar (ie comprising layers sliding over one another like a pack of playing cards), in fact only the very first part of attached flow on a car is ever like this. Most of the attached flow comprises a turbulent boundary layer with lots of exchanges of kinetic energy occurring between the levels. (Vortex generators, covered in Chapter 6, help with that energy exchange between the layers.)

Flows can separate and then reattach. For example, on many three-box sedans, the flow separates part way down the rear window and then reattaches on the boot lid. In this circumstance, the area of separated flow is called a *separation bubble*. Figure 1-12 shows a close-up of wool-tufts on the rear of a three-box sedan. The highlighted areas show a separation bubble at the base of the rear window and on the central front part of the boot lid. Note that the tufts before and after this area show attached flow – they are pointing in the direction of the main airstream. Figure 1-13 shows the separation bubble in which the occupants sit in this open car.

Figure 1-11: Different components on the car have aerodynamic effects on each other. The wool tufts show flow disturbance on a front door window caused by the A-pillar and rear vision mirror.

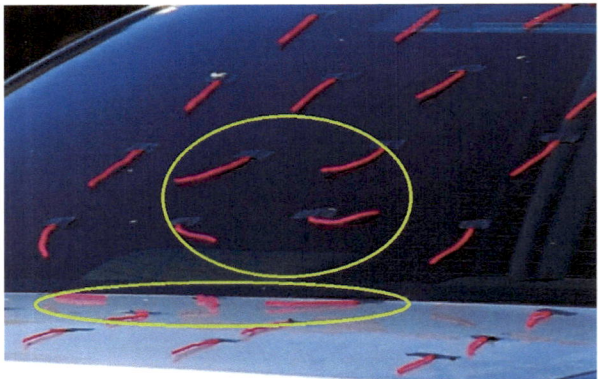

Figure 1-12: The separation bubble at the base of the rear window and on the central front part of the boot lid. Note that the tufts before and after this area show attached flow.

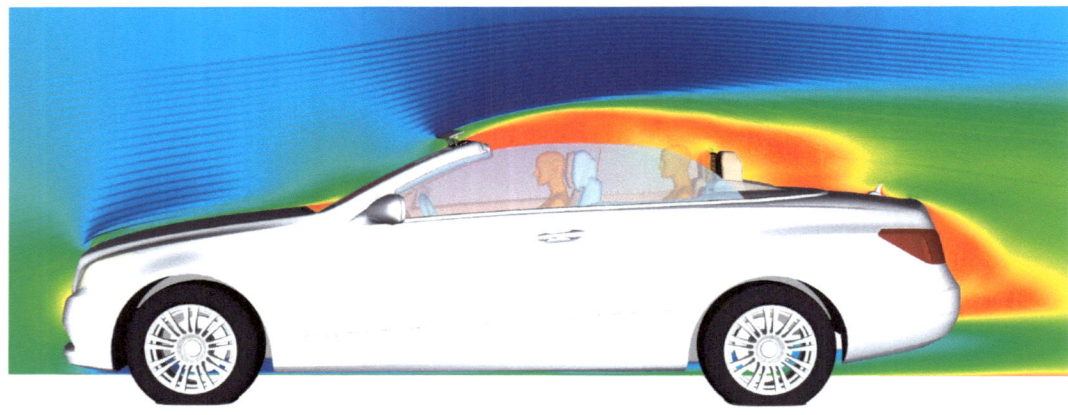

Figure 1-13: The occupants of this open car are positioned within a separation bubble. (Courtesy Mercedes)

THEORY

REYNOLDS NUMBERS

The characteristics of the boundary layer influence viscous drag and airflow separation. In turn, boundary layer characteristics are determined by speed, air density, air viscosity and vehicle size. The last aspect – size – is relevant because it leads us to another factor that is important in aerodynamic testing of models and in wing calculations – Reynolds numbers.

Let's assume that the laminar boundary layer extends for the first 300mm (~12in), and after that, the boundary layer is turbulent. On a small model tested in a wind tunnel, the boundary layer is therefore likely to be laminar for much of the model's length. As we've seen, laminar boundary layers separate more easily than turbulent boundary layers, so on the model, separation is likely to occur early. On the other hand, on a full-size car with the same 300mm length of laminar boundary layer, the presence of a turbulent boundary layer over most of the car means that separation will be delayed. We therefore need some way of quantifying the effect of size and the other factors that influence flow behaviour.

The dependence of flow patterns on speed, density, viscosity and length is expressed via the *Reynolds number*, which is calculated by:

$$\frac{\text{density} \times \text{speed} \times \text{length}}{\text{viscosity}}$$

Looking at the equation, you can see that if the size of the object is reduced (eg when testing a small model in a wind tunnel), then wind tunnel speed will need to be increased if the testing is to remain accurate. In reality, this may not always be practical. A one-fifth scale model would have to be

Figure 1-14: Reynolds numbers are important when working with models in wind tunnels as the 'size' dimension influences flow. However, when testing on the road, they're much less important. (Courtesy Mercedes)

tested at five times the road speed to give the same Reynolds number, so to simulate 150km/h (93mph), the model would need to be run at 750km/h (465mph)! Even if we had a suitable tunnel, at this speed we would start to incur compressibility effects (ie the way that air behaves would begin to alter). This is an inherent problem with testing scale models in wind tunnels, and partly explains why full scale and model results do not always tie up well.

When testing vehicles on the road, the variation in Reynolds number is not a major concern for us. However, if you are looking at drag and lift specifications of wings taken from wind-tunnel data, the Reynolds number will need to be known. If you know the dimensions of the wing, and the road speeds you will be using, it is easy enough to calculate the Reynolds number. However, this might not necessarily help, as wind tunnel data is not usually conducted over a large range of Reynolds numbers.

INDUCED DRAG AND TRAILING VORTICES

Any shape that creates lift or downforce develops drag. This type of drag is referred to as 'induced drag.' A by-product of induced drag is the creation of trailing vortices (also see the breakout box on page 19). Trailing vortices are like longitudinal whirlwinds appearing behind the vehicle. The less lift being developed by the car, the weaker these trailing vortices. Trailing vortices are a *symptom* of induced drag, rather than a creator of drag in their own right.

Trailing vortices are present within the car's wake and are also shed by angled, upright edges – for example, the C-pillars. The strength and location of the generated vortices is heavily dependent on the rear shape of the car:

- **Squareback (or wagon)** – There will be quite a bit of random turbulence in the wake, but even a brick-shaped vehicle will generate some lift if close to the ground. So, in addition to the random turbulence, there will also be a pair of trailing vortices whose strength will depend on the amount of lift generated.
- **Fastback** – A pair of strong vortices will usually appear, often emanating initially from the inclined rear pillars. There may also be a separation bubble on the rear window.
- **Sedan (notchback or three-box)** – A pair of vortices will be

apparent behind the rear of the vehicle. There is often a flow separation bubble on the rear window.

In addition, vortices can form off the A-pillars – and other body shape details – and then interact with the vortices generated by the rear pillars.

The direction of rotation of the large vortices depends on whether the body is developing lift or downforce. Picture a car viewed from behind. Where lift is occurring, the vortex trailing behind the car on the left-hand side of the car rotates clockwise, and the one on the right-hand side of the car rotates anti-clockwise. Figure 1-15 shows this flow. Note how this diagram shows that there are sometimes 'vortices within vortices.'

When downforce is occurring, the direction of vortex rotation is the opposite – that is, the vortex trailing behind the car on the left-hand side of the car rotates anti-clockwise, and the one on the right-hand side of the car rotates clockwise.

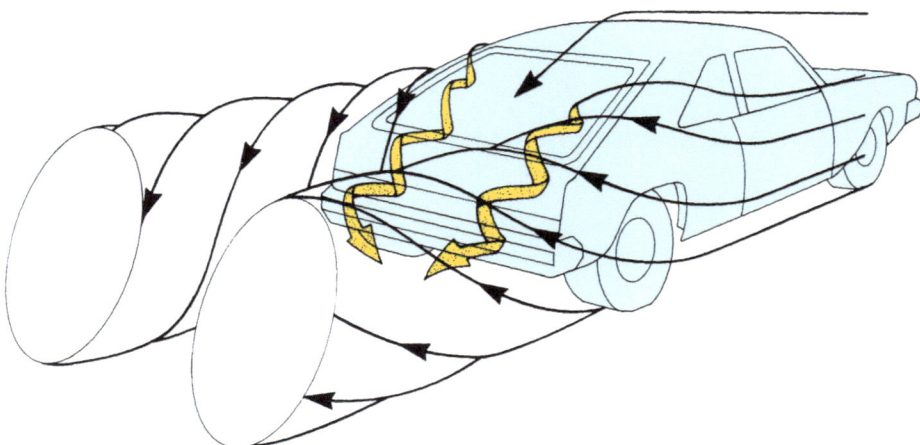

Figure 1-15: Sedan and fastback body shapes are associated with trailing vortices of air. Where lift is occurring, the vortex trailing behind the car on the left-hand side of the car rotates clockwise, and the one on the right-hand side of the car rotates anti-clockwise. (Based on Allen)

Just occasionally, conditions are appropriate to see some aspects of these vortices in road-driven vehicles. For example, if you are following a car on a wet road after a rain shower has cleared, you can sometimes see rotational movement of the fine drops behind the vehicle.

When viewing the behaviour of wool tufts (see Chapter 3), sometimes you will notice that over a C pillar, the tufts are angled diagonally as the airflow wraps around the curved pillar, passing from the higher-pressure sides to the lower-pressure rear window. This airflow

Figure 1-16: An example of a wool tuft pattern on a C pillar where a trailing vortex is being formed. Note the angled position of the tufts, indicative of flow wrapping around the pillar.

THEORY

TRAILING VORTICES AND LIFT – CHICKEN OR THE EGG?

In the description of trailing vortices used in this section, I have associated trailing vortex generation with lift. However, if you read many Society of Automotive Engineers (SAE) papers on aerodynamics, you will often find that the engineers state that they are chasing reduced trailing vortex generation *per se* – rather than trying to reduce lift that leads to vortex generation.

For example, take the rearmost pillars on a sedan. In all modern cars, they're quite shallow in angle and curved in cross-section. Air flows around them and, because of their angle, creates lift and drag. There will also be trailing vortices starting from these pillars. If the pillars are changed in shape, the strength of the trailing vortices may be reduced, and so will the lift and drag forces caused by these vortices. But why actually did drag decrease?

A car aerodynamicist working in a wind tunnel will be confident it was the reduction in strength of trailing vortices that did it. The academic aerodynamicist will be equally confident that, as the vortices were caused by the lift properties of the body, lower body lift will have reduced vortex intensity – and that's what caused the change in drag.

Does knowing which came first – the lift or the trailing vortices – matter to us working on our cars? In most cases – no! In the same way as reducing the area of the wake on a given car will almost always reduce drag, reducing vortex generation on a given car will usually reduce drag.

However, there *is* one important implication. Because it is very hard to see trailing vortices and so assess changes in their behaviour, sometimes you will need to work with a 'rule of thumb.' That rule of thumb is this: f you can reduce lift without at the same time doing anything that will greatly increase drag, you are likely to reduce the strength of trailing vortices, and so reduce overall drag. For example, fitting an undertray with rear diffuser (see Chapter 7) potentially can achieve this. And, as covered in Chapter 5, measuring lift/downforce is straightforward and can be done quite accurately.

pattern is likely to be the start of a trailing vortex. Figure 1-16 shows an example of this tuft pattern.

As you'd expect given its name, lift-induced drag is reduced by decreasing lift. As a result of this reduction in overall lift, the strength of trailing vortices is diminished. Note, though, that reducing lift does *not* mean that in all cases drag will be reduced – this is because the method used for lift reduction may cause drag in itself. Identifying the location, size and strength of trailing vortices is difficult to achieve on the road.

UNDERCAR FLOWS

So far, we've been concentrating on airflow that occurs on the upper surfaces of the car. But more than ever before, it is what is happening under the car that's important.

In the past, the undersides of most cars were very rough and caused lots of drag. As described in Chapters 6 and 7, the solution was to use a low front air dam (spoiler)

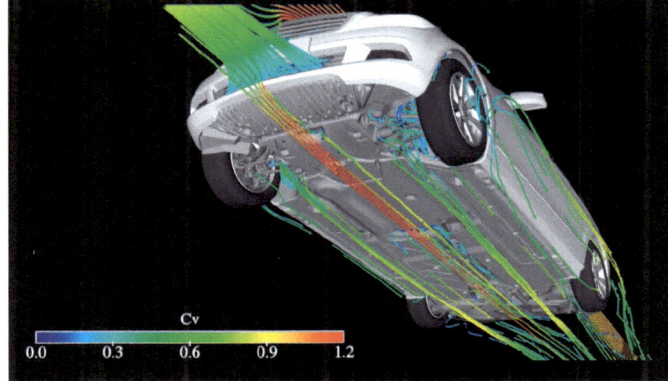

Figure 1-17: The underside of the car is important to aerodynamic performance. This graphic shows airflow speeds, and as can be seen, on this smooth underside car they mostly remain high. (Courtesy Mercedes)

Figure 1-18: Good airflow under the car can reduce the size of the wake. Note, though, that here the car has been elevated, making it easier to get more undercar flow. (Courtesy Mercedes)

19

that blocked the underside from the airflow. However, taking this approach had two disadvantages. First, frontal area increased – so the benefit in decreasing drag had to more than offset the penalty in increasing frontal area. Second, and more importantly, use could not be made of that undercar airflow. Two uses are available.

One is to use this air to reduce the size of the wake. If no airflow is able to pass under the car, the wake extends further downward. Conversely, if more air can pass under the car, the wake will not extend so far downward. The other use for this undercar airflow is to reduce lift by creating a lower pressure beneath the car. As mentioned earlier, faster speeds equal lower pressures, and if we can gain high-speed airflow under the car, then the pressure will drop. Because the area under the car is so large, even a slight lowering of pressure below atmospheric can create a significant downward force. (Note: it still might not be providing the car with downforce *per se*, but it will help offset the lift being developed by the upper surfaces.) This reduction in lift also results in weakened trailing vortices.

DRAG AND LIFT COEFFICIENTS

So how are drag and lift quantified? (Note: The following calculations are given in just metric units, as showing imperial conversions would make it hard to follow.)

The following equation shows how drag is calculated:

$$\text{Drag force} = 0.5 \times \rho \times V^2 \times C_D \times A$$

where drag is in Newtons, ρ (rho) is air density in kg/m³, V is velocity in metres per second, C_D

Figure 1-19: The projected frontal area of the car is all that you can see facing you here. It is measured in ft² or m². (Courtesy BMW)

is the coefficient of drag, and A is the projected frontal area in square metres.

Before we go any further, let's divide this equation up for further study. The $(0.5 \times \rho \times V^2)$ part of the equation is the air's 'dynamic pressure' – more on this in a moment. However, you can see it is dependent only on air density and air speed. It's the other part of the equation I want you to look at – $(C_D \times A)$, that is, drag coefficient multiplied by the projected frontal area. (This is usually shown as $C_D A$ and is known as 'drag area.')

On a given day with fixed atmospheric conditions, drag goes up proportionally with C_D and frontal area. This means you can increase C_D by 10 per cent and decrease frontal area by 10 per cent – *and you get the same total drag*. Or you can have a bigger car with a smaller C_D, or a smaller car with a larger C_D – and total drag is much the same. The seemingly inexorable move to larger and larger cars means that to maintain the total drag at the same level, manufacturers must achieve lower C_D values.

Anyway, back to using the equation to calculate drag. You can see that to use the equation, we need the following data:

- air density
- speed
- projected frontal area
- drag coefficient

Air density is dependent on temperature, atmospheric pressure and humidity. For example, with a temperature of 20°C, an atmospheric pressure of 1013hPa and a relative humidity of 55 per cent, the calculated air density is 1.20kg/m³. You can easily find an on-line calculator to do this – search under 'air density calculator.' (However, just using 1.2kg/m³ as a nominal figure is fine in most applications.)

The projected frontal area is the area of the car that you can see if you were to stand directly in front of the car, but a long way away, and view the car front-on through binoculars. In most cars it is about 80 per cent of the height multiplied by the width. If you want to make a more accurate measurement, take a photograph of the car front-on using a telephoto lens, then overlay the photo with a grid of suitable scale. Count the grid squares that are on top of the car and then, using your scaling, convert this to area in square metres. Another way to do this is to reduce the resolution of the photo using an image processing program, and then enlarge it massively. The individual pixels of the front of the car will be easy to count, and you can work out their scale by measuring how many pixels there are across the width of the car and then comparing that number with the real thing.

The drag coefficient in the equation will typically use the manufacturer's quoted figure.

Let's do the calculation for a Gen I Honda Insight. The quoted height and width dimensions are 1.355 x 1.695m, giving an area of 2.3m². Multiply that by 0.8 to get

THEORY

CHANGES IN DRAG WITH TOP SPEED

As described in Chapter 5, it is very hard to measure drag without having access to a full-size, five-belt wind tunnel run by experts. However, if you know the car's standard C_D and top speed, and you know the new top speed when the car has been equipped with modifications that change drag, you can get an indication of the new C_D. (It is an 'indication' only because it doesn't take into account that the gearing has not been altered to cater for the lower or higher drag, and so the engine may be developing slightly different power at the new top speed. However, unless you are radically changing the top speed, this is not a major issue.)

Let's do the figures for a vehicle with these specs:

- standard C_D 0.44 (as stated by the manufacturer)
- frontal area: 2.7m²
- standard top speed: 155km/h (43.1m/s)
- top speed with low drag modifications: 169km/h (47m/s)

We will use an air density figure of 1.2kg/m³.

The calculation for the standard vehicle is as follows:

Drag force = 0.5 x ρ x V² x C_D x A
Drag force = 0.5 x 1.2 x 43.1 x 43.1 x 0.44 x 2.7
Drag force = 1324 Newtons

Because this is the total force required to be overcome at maximum speed, we know the engine was capable of providing this – *and no more*. To convert this to engine power, we would need to know the engine rpm and the ratios of all the gearing (ie transmission, differential and tyre diameter), but for our purposes, we can just leave it in Newtons force.

Let's now rearrange the equation to put C_D on the left.

$$C_D = \frac{\text{Drag force}}{0.5 \times \rho \times V^2 \times A}$$

We can now find out what the new C_D is.

$$C_D = \frac{1324}{0.5 \times 1.2 \times 47 \times 47 \times 2.7}$$

C_D = about 0.37

It's unfortunate that in most countries you can't legally achieve top speed – it's one the most accurate ways of measuring changes in drag.

the actual frontal area of the car and we have a figure of 1.84m². I want to use a speed of 100km/h (which we can convert to metres/sec by multiplying by 0.278), so giving us 27.8m/s. The quoted C_D of the Insight is 0.25, and I'll assume the air density is 1.2kg/m³.

Therefore, we have the following data:

- air density = 1.2
- velocity = 27.8
- C_D = 0.25
- area = 1.84

Drag force = 0.5 x 1.2 x 27.8 x 27.8 x 0.25 x 1.84
= 213 Newtons

Divide Newtons by 9.81 to get kilograms-force, which gives us 21.8 kilograms-force required to push the car forward against the air at 100km/h. (In addition to air resistance, there is also rolling resistance that would need to be added to this figure to gain the total required force.)

Note there are on-line calculators that can solve this equation for you – that's especially useful if you want to use different units.

Now what about the coefficient of lift? The equation is identical but for the changing of the coefficient.

$$\text{Lift} = 0.5 \times \rho \times V^2 \times C_L \times A$$

where lift is in Newtons, ρ is air density in kg/m³, V is velocity in metres per second, C_L is the coefficient of lift, and A is the projected frontal area in square metres. (Note that some car data will list the lift coefficients for the front and rear axles separately – these are shown as C_{Lf} and C_{Lr} respectively.) For example, in air with a density of 1.2kg/m³, a car with a frontal area of 1.7m², travelling at a speed of 120km/h (120km/h x 0.278 = 33.4m/s), with a coefficient of lift at the front axle (C_{Lf}) of 0.118, will give:

Lift force = 0.5 x 1.2 x 33.4 x 33.4 x 0.118 x 1.7
= 134 Newtons, or 13.7kg

Note that a positive coefficient of lift indicates an upward force, and a negative coefficient of lift indicates a downward force (ie downforce). Measuring drag force is almost impossible without having access to a wind tunnel, but as described in Chapter 5, it *is* possible to accurately measure lift and downforce on the road. So can the coefficient of lift be calculated from just on-road data? Yes it can – let's take a look.

In Chapter 7 I describe my Honda Insight (fitted with an effective undertray and rear diffuser, and running a low ride height) as developing, at 160km/h, 110kg front downforce and 45kg rear

SPEEDPRO SERIES

> **WINGS**
> When assessing the lift/drag characteristics of a wing, you should know that the way in which lift and drag coefficients are calculated for wings differs from the approach taken for cars.
>
> As we have seen, for car coefficients, the projected frontal area of the car is used. However, for wings, it is the *wing plan area* that is used when the coefficients are being calculated. To convert wing data into the more familiar car data, the wing data needs to be multiplied by (wing area divided by the car frontal area).

Figure 1-20: Note here the excellent undercar flow – but also note that the wheels are not turning, which would influence the resulting flow pattern. (Courtesy Mercedes)

downforce. Let's convert these into the units we're using: at 44.5m/s, front downforce was 1078 Newtons and rear downforce was 441 Newtons. Frontal area of the car is 1.84m².

The formula for calculating the coefficient of lift is:

$$C_L = \frac{\text{Lift}}{0.5 \times \rho \times V^2 \times A}$$

$$C_{Lf} = \frac{1078}{0.5 \times 1.2 \times 44.5 \times 44.5 \times 1.84}$$

$$C_{Lf} = -0.49$$

$$C_{Lr} = \frac{441}{0.5 \times 1.2 \times 44.5 \times 44.5 \times 1.84}$$

$$C_{Lr} = -0.20$$

I also have measurements for downforce at 100km/h (60mph), and do these coefficients support those measurements? At 100km/h, the calculated front downforce with a C_{Lf} of -0.49 is 23kg, and the calculated rear downforce with a C_{Lr} of -0.20 is 9.5kg – and both figures stack up against the actual measurements based on suspension deflection.

YAW

The C_D and C_L figures that you see most commonly quoted are those for when the airflow is directly approaching the car. That is, the car is placed in the wind tunnel so that it is facing straight into the airflow. However, on the road, there is often a cross-wind component to the airflow. This is called a *yaw*, and the angle of yaw indicates the degree to which the airflow is not directly front-on. Nissan recently calculated that in the US, a yaw angle of 4° is the median actual yaw to which vehicles are subjected.

The concept of yaw is important for two reasons. First, in most cars, both drag and lift rise with increasing yaw. So a car might have a quoted C_D of 0.30 at 0° yaw, but this rises to 0.34 at 5° yaw. Or, the C_{Lf} might be 0.1 at 0° yaw but 0.2 at 12° yaw – that is, front lift doubles in a strong crosswind! To cater for this real-world variation in flow direction, many cars today are being developed that have less sensitivity in drag and lift to small yaw angles.

Note that the technical possibility exists for airflow at high yaw angles to create thrust by acting on aerodynamic surfaces; after all, that's how a sailboat can head windward.

Second, the idea that the airflow is often angled to the forward direction of the car is significant when aerodynamic stability is considered. I talked earlier about airflow separating at the rear of the vehicle, and so creating a low-pressure wake, but in crosswinds, this separation can occur on the leeward (downwind) side of the car. When looking at car shapes, it is therefore good to think of airflow separation resulting from a yawed airflow. The location of the lateral centre of pressure versus the centre of gravity is also important for stability – more on this in Chapter 7.

DYNAMIC AND STATIC PRESSURES

When we are dealing with pressures acting on a car, two different types of pressures need to be understood. These are static

THEORY

VERACITY OF DRAG AND LIFT NUMBERS

Before we leave lift and drag coefficients, it's important to realise that these figures are not like many other car measurements with which we're familiar.

What does this mean? Well, if we want to measure the diameter of a piston gudgeon (wrist) pin, we get out a micrometer. We can measure the pin's diameter, and be confident that we have a measurement that is very accurate. (A standard one-inch micrometer has readout divisions of 0.001in and a rated accuracy of +/-0.0001in.)

But lift and drag coefficients are nothing like this! Typically, they are derived from measurements made in wind tunnels, and these figures are only as accurate as the wind tunnel allows. When a single car is measured in different wind tunnels (even well-respected, large wind tunnels run by major manufacturers and institutions) the results vary. Other lift and drag figures that you will see used are derived from wind tunnel measurements made of models (not full-size cars), Computational Fluid Dynamics (CFD) modelling or coast-down testing.

This means we end up with a lot of figures, many of which may not have much credibility. I am *not* saying that all this data is wrong, but what I am saying is that you shouldn't take every C_D and C_L figure you see as gospel.

For example, the wonderful Tatra T87 (covered in the next chapter) had a drag coefficient, when measured full-size in the highly-respected Volkswagen wind tunnel, of 0.36. However, models of the car were being measured at the time of the car's development at 0.24!

The ground-breaking pre-WWII research vehicles, created by the German FKFS institution under the leadership of Wunibald Kamm, were dubbed the K1, K2 and K3 models. One reference states that K1 had a C_D of 0.23 (and a model measured 0.196), K2 apparently had a C_D of 0.23 – but K3 had the much higher C_D of 0.37! So what happened to K3? – after all, it looked much the same as the two earlier cars. Well, K3 was tested full-size in the Volkswagen wind tunnel in 1979 …

In fact, it's very likely that K1 and K2 had drag coefficients that were similar to K3. (That's still a lot better than nearly all their contemporaries, so they were indeed low drag cars for the time – just not as much as claimed!)

To look at current (mis)information, one amateur website did extensive CFD modelling of a car and gained a C_D of 0.22 – considering the car being modelled, I'd suggest that is out by at least 25 per cent. As described in the next chapter, drag figures achieved through an independent test failed to match the manufacturer's figures for most of the tested cars. I have seen data that suggests dramatic decreases in C_D achieved through modifications, and yet top speed changed relatively little.

Drag and lift figures are certainly very interesting, and where a car is being aerodynamically developed in a full-size wind tunnel, the *relative* changes are likely to be accurate. Professional aerodynamicists working for major manufacturers and using complex, sophisticated and expensive CFD software are also likely to be producing accurate data.

When reading this section, Jaguar chief aerodynamicist, Adrian Gaylard, made an interesting comment. "Measured values are also highly dependent on the car's configuration. For example, different manufacturers' tyres can change drag significantly – even for the same nominal specifications. Different rim choices of the same diameter can shift vehicle drag by 10 per cent or more."

Always keep in mind that the fact that a figure is given to three decimal places doesn't mean it is accurate. At minimum, look for some type of corroborative evidence. For example, a car with much reduced lift (or having increased downforce) should feel quite different on the road. A car with a dramatically lowered drag coefficient should get better fuel economy (mileage) and have a higher top speed. (And a media release or technical paper that talks only of percentage changes, and never states the absolute figures, is probably trying to hide unimpressive coefficients!)

This book is full of C_D and C_L figures, but always remember that they're not like using that micrometer …

---

pressure and dynamic pressure. Rather as we did at the beginning of this chapter, where we used drag to introduce some characteristics of airflow, let's initially explore these pressures by looking at a practical application.

A pitot tube measures air speed, and it does so by measuring these two different types of pressure. (Pitot tubes are used to measure airspeed around the car body in Chapter 4. Note that 'pitot' is pronounced 'pee-toe'.)

Figure 1-21 (on the next page) shows the construction of a pitot tube. It has one opening that faces directly into the oncoming airstream, and one or more openings that are at right-angles to it. These openings connect to hoses that run to our measuring instrument. Let's start off by looking at the openings that are not facing into the air steam – they are at right-angles to it. These openings measure static pressure.

Static pressure is caused by the constant random motion of molecules in a gas. When these molecules strike a surface, they

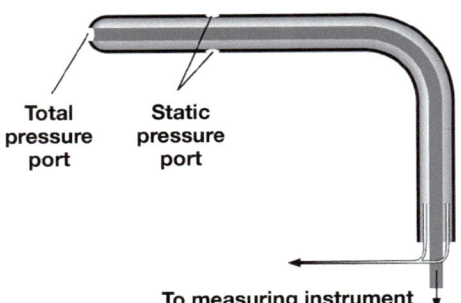

Figure 1-21: The internal construction of pitot tube used for measuring airspeed.

produce a force, just as you could produce a force on a wall by throwing a handful of pebbles against it. If you compress the gas into a smaller volume (as you do when you inflate a tyre), there are more moving molecules trapped in the same volume, and they impact the walls more often and so create a greater force – that is, the measured pressure is higher.

We are thinking of static pressure when we think of atmospheric pressure (the 'highs' and 'lows' on a weather map), turbo boost pressure, tyre pressures, or the pressure of oxygen in a welding cylinder.

Let's now think about the pressure measured by the other opening – the one that faces into the airflow. This air is moving, and because air has mass, it has kinetic energy – the ability to do work. We already know that moving air can do work – that's how a wind turbine works. But what are some units that we can use to describe this? Perhaps confusingly, we again use units of pressure. Remember, though, that this dynamic pressure is actually proportional to the kinetic energy of a moving volume (eg $1m^3$) of air.

The pressure of air that is measured by the probe facing into the airflow is the dynamic pressure (caused by the movement of the air) plus the static pressure (that caused by the random movement of molecules). Therefore, this probe port measures total pressure.

total pressure = static pressure + dynamic pressure

By subtracting the static from the dynamic pressure, we can work out how fast the air must be flowing – and so use the pitot tube to measure air speed.

Now let's look in more detail at static and dynamic pressures.

Earlier, we looked at the equation for dynamic pressure. It was: $0.5 \times \rho \times V^2$. That is, the dynamic pressure is dependent on the density of the air, and heavily dependent on the speed with which it is moving. (And that makes sense when we think of it in terms of kinetic energy.)

If we call the static pressure P_S, we then get the total pressure by:

total pressure = $P_S + 0.5 \times \rho \times V^2$

So what happens if we reduce the speed of the airflow (that is, we reduce V)? To maintain the same total pressure, if V decreases, P_S must rise. That's why if airflow speed over the car decreases, the static pressure measured at that point rises, and why if the airflow speed increases, the static pressure falls. In Figure 1-7 (page 13) the pressures we are showing on the surface of the car are the static pressures. And, just as with the pitot tube, in wind tunnels this pressure is measured via sensing tubes connected underneath to small holes drilled in the panels of the car.

(So why then do we have a pressure lower than atmospheric in the wake – and yet there's no apparent airflow direction? The answer is that the theory we are using here applies only to steady flows – turbulent flow is another story.)

It's already been noted that the local airflows over the car vary in speed, and so therefore cause changes in static pressure. But there's an additional point to think about. Let's take the example of the airflow approaching the car. As described above, at some area on the front of the car, the airflow will come to a complete stop – the stagnation zone.

Stopping the air at the front of the pitot tube means that the kinetic energy produces a rise in the pressure at that point equal to $0.5 \times \rho \times V^2$, so the pressure there is now $P_S + 0.5 \times \rho \times V^2$. This is known as the stagnation pressure, and is the maximum aerodynamic pressure available on the car.

Chapter 2
Aerodynamically optimised vehicles

- **Rumpler Tropfenwagen**
- **Opel RAK 2 winged rocket car**
- **Tatra T87**
- **Porsche Type 64**
- **Volkswagen Transporter**
- **NSU Ro 80**
- **Dodge Charger Daytona & Plymouth Roadrunner Superbird**
- **Holden Special Vehicles Group A Commodore**
- **Opel Calibra**
- **General Motors EV1**
- **General Motors Chevrolet Volt**
- **Volkswagen XL-1**

In this chapter I want to look at some of the most important vehicles that have been produced over the years. Irrespective of their age or what their original design purpose was, each of the featured cars in this chapter can potentially teach you about car aerodynamics. The selection is idiosyncratic: these are cars that I find interesting, and for most part, cars I have been lucky enough to see in the metal. Irrespective of their aerodynamics, most of these cars are also of an exciting technical make-up, so I also touch on these aspects. As this book concentrates on road cars, so will this section – apart from an occasional diversion!

1921 RUMPLER TROPFENWAGEN

The world's first aerodynamically optimised production car was the Tropfenwagen. Developed by Edmund Rumpler in 1921, it not only incorporated extreme low-drag design, but also used an innovative six-cylinder, W-configuration mid-mounted engine, and had independent rear suspension. So what is the story of this car – and the man who designed it?

Edmund Rumpler was born in 1872. He was an Austrian automobile and aircraft designer who carried out most of his work in Germany. After studying engineering, Rumpler joined Daimler in 1898 as a car designer, then in 1902 moved to Adler becoming, at the age of just 30, technical director of that car company. Here he designed the first German car where the engine and gearbox were mounted together as a unit. In 1903, he patented a swing axle rear suspension design.

After the Wright Brothers invented the aircraft, Rumpler's attention turned to aircraft design. He left Adler in 1907, and in 1910 Rumpler became the first aircraft manufacturer in Germany. The most famous aircraft produced by the Rumpler company was the Taube, based on a design by Igo Etrich. Subsequently, the company went on to produce a range of aircraft of its own design.

Following World War I, Edmund Rumpler returned to automotive design. The result was the

SPEEDPRO SERIES

Figure 2-1: The extraordinary 1921 Rumpler Tropfenwagen was like no other car. It had curved front glass, independent rear suspension, and used a 6-cylinder, mid-mounted engine.

Figure 2-2: The extreme boat-tailing used at the rear of the car for both the main bodywork and the cabin can be seen here.

Tropfenwagen, released at the 1921 Berlin Auto Show. The Tropfenwagen (German for 'drop car' – as in a drop of water) had the extraordinary low C_D of 0.28, as measured in 1979 in the Volkswagen wind tunnel. (Frontal area was 2.57m^2, or 27.7ft^2.)

It achieved this low drag by a number of measures, all carefully tested beforehand by Rumpler working with models in a wind tunnel.

- The windscreen and side windows were formed from glass panels curved in a single plane – another world first for a car. In fact, the degree of curvature of the windscreen was radical.
- Disc wheels were used on most of the cars, rather than the exposed spoke wheels then universal.
- Mudguards, instead of being curved to match the wheels, were straight and mounted horizontally, giving a very small frontal area.
- The lights were integrated into the aerodynamic shape, rather than being drag-inducing add-ons.

However, the most radical aerodynamic innovation was the 'teardrop' plan-view shape. The boat-tailing was so extreme that the bodywork reduced in width to form a trailing vertical edge, where the exhaust pipe was placed. Mirroring this boat-tailing of the main body was the cabin, that also boat-tailed to a trailing vertical edge. In addition to the extreme teardrop shape when viewed from above, the Tropfenwagen also tapered a little in height towards the rear. A further radical change over all other cars of the time was the use of a smooth undertray and a rear diffuser.

The engine, in unit with the gearbox (really a transaxle), was placed ahead of the rear wheels. The radiator was mounted behind the engine and had louvred exits for the cooling air. The louvres were angled so that the cooling air didn't disrupt the airflow over the bodywork. The engine was built by the Siemens and Halske company. It used a W-design with three banks of paired cylinders

Figure 2-3: Drag-reducing disc wheels were fitted to most of the Tropfenwagens. At the time, the use of spoked wheels was almost universal.

displacing 2580cc. Power output was 26kW (35hp). The independent, swing-arm rear suspension used long, trailing cantilever leaf springs that were angled outwards and joined the axles near the hubs. Further rear wheel location was provided by angled rods mounted behind the axles. Along with the wheels, the rear suspension was fully exposed to the

AERODYNAMICALLY OPTIMISED VEHICLES

Figure 2-4: The front springs were mounted within the bodywork, and the upper and lower frontal surfaces used compound curves to keep the flow attached.

Figure 2-5: Note the small frontal area of the mudguards (fenders). The integrated lights are an elegant touch.

Figure 2-6: Engine cooling air exhaust louvres directed the air along the bodywork.

airflow. A recent reference suggests that the drag penalty of not enclosing the wheels may have been as high as 50 per cent. The front suspension used a non-independent beam axle located by leading cantilever leaf springs mounted parallel with the body. These springs attached to the axle well inboard from the wheels, and so must have provided very little front roll stiffness.

The Tropfenwagen was a genuine production car – but it was not a commercial success. The W-engine had problems, the steering could shimmy, and the first versions of the car did not have a boot (trunk). Roads on which cars could travel at higher speeds had not yet been developed, and so the aerodynamic advantages weren't reflected in improved fuel consumption. One hundred of the cars were built, of which only two now remain.

The Tropfenwagen was available in at least four different versions. One version was a five-seat, long-wheelbase, open car that used a more conventional 2614cc in-line four-cylinder engine, developing 37kW (50hp). It had wire wheels and the different side air intake. All cars used a single, central seat for the driver. In the photos shown here, the lighter blue car is the long-wheelbase model and the dark blue car, the short wheelbase.

After the Tropfenwagen failed in the marketplace, Rumpler returned to aircraft. Rumpler was Jewish, and he was imprisoned after Adolf Hitler

Figure 2-7: When tested in the Volkswagen wind tunnel in 1979, the Tropfenwagen had the extraordinary low C_D of 0.28. Frontal area was 2.57m^2 (27.7ft^2). (Courtesy Volkswagen)

SPEEDPRO SERIES

Figure 2-8: A view unlike any car before or since!

came to power in 1933. He was soon released, but his career was ruined. He died in 1940.

It is only when you see the Tropfenwagen next to a conventional car of the time that you can fully appreciate the radical nature of the Rumpler machine. The Tropfenwagen may have been a commercial failure, but its swing axle rear suspension, mid-mount engine and transaxle, and stunning low-drag design helped create the path followed by the later Auto Union racing cars of the 1930s.

The two remaining Tropfenwagens can be seen at two museums in Germany – Deutsches Museum Verkehrszentrum, Munich and Deutsches Technikmuseum, Berlin.

1928 OPEL RAK 2 WINGED ROCKET CAR

On May 23, 1928, at Berlin's AVUS race track, Fritz von Opel set a speed record for rocket-propelled cars, hitting an estimated 238km/h (148mph) in his RAK 2.

Von Opel, grandson of Opel Motor Car Company founder Adam Opel, and his partners Friedrich Sander and Max Valier, used 24 solid-fuel rockets packed with 120kg (260lb) of explosives to propel RAK 2 forward. Each time von Opel pressed on the gas pedal he ignited two rockets, increasing power until the propulsion system reached full strength, and he gained his world record.

"I stopped thinking. I was acting on instinct alone, with uncontrollable forces raging behind me," von Opel said when he stepped from the car.

The car was fitted with wings – probably the first car in the world so-equipped. But what was the purpose of the wings? That's contentious. Most references suggest that they were to provide downforce, but the camber of the wings looks much more like ones designed to create lift. And why could that have been required? The rocket centre looks to be above the centre of gravity, resulting in a downward push on the front wheels. Perhaps the wings were designed to counteract this? But then on the other hand, the wings' angle of attack looks correct for providing downforce – so perhaps the wing camber was simply a mistake.

A replica of Opel Rak II can be seen in the Technik Museum Speyer, Germany.

Figure 2-9: The Tropfenwagen was a genuine production car – but it was not a commercial success. This is the long wheelbase version.

1936 TATRA T87

The Tatras from Czechoslovakia are among the most extraordinary looking cars ever produced … and yet people know little or nothing about them. And not only do the cars look amazing, but their underskin technology is equally as interesting.

Tatra has existed in various forms since 1850 – along with Peugeot and Mercedes, it is among the oldest car companies in the world. It produced its first car in 1897, when the company was known as Nesselsdorfer Wagenbau-Fabriksgesellschaft. Famed automotive designer Hans Ledwinka then joined the concern, with his first design produced in 1898. By

AERODYNAMICALLY OPTIMISED VEHICLES

Figure 2-10: Tatra T26-30 truck of 1927. It had six forward and two reverse speeds, and the four independently suspended rear driving wheels could have their axle differentials locked. Its innovation was to be reflected in the aerodynamics of later Tatras.

Figure 2-11: The Tatra T-87 is one of the most extraordinary cars ever produced. The all-steel monocoque body shape is curved and elegant. The rear wheels are covered, and the door handles recessed. The body uses a central rear fin for aerodynamic stability.

the engine included a double-walled intake manifold where oil flowed between the walls, both cooling the oil and warming the intake air.

One of the company's early products is the Tatra T26-30 truck of 1927. It has six forward and two reverse speeds, and the four independently suspended rear driving wheels can have their axle differentials locked. At the front is positioned the air-cooled twin cylinder engine, and independent suspension sprung by a transverse leaf and friction dampers. The Tatra truck was so advanced for its time that the vehicle shown here was anonymously purchased by Studebaker in the United States, which was manufacturing trucks for the US Army and wanted to investigate Tatra technology.

But while the underskin technology was state of the art, the body shapes of the early Tatras looked much the same as other contemporary manufacturers. However, in 1934, all that changed, with the release of the T-77. This extraordinary car not only used a 2.6-litre air-cooled V8, but the V8 was placed at the rear of the car. Despite still being wood-framed, the bodywork was startlingly low in drag, and all-round independent suspension was again used.

Next was the T87 – replacing the T-77 after only three years. The T-87 was one of the most extraordinary cars ever produced. The all-steel monocoque body shape was curved and elegant. The rear wheels were covered and the door handles recessed. The body used a central rear fin – a real fin, for aerodynamic stability. Scoops each side of the car provided cooling air for the rear-mounted V8, which was now a new OHC alloy unit featuring hemispherical combustion chambers and upsized to 3 litres. Power was

1922, the innovative engineering approach had been set: the T-12 model used a large central tube to form the frame and had rear swing axles. At the front was an aluminium, horizontally-opposed, air-cooled twin cylinder engine using three valves per cylinder. Engineering niceties of

SPEEDPRO SERIES

Figure 2-12: Scoops each side of the car provide cooling air for the rear-mounted V8, which was an OHC alloy unit featuring hemispherical combustion chambers and a 3-litre capacity.

Figure 2-13: The underside of the T-87, viewed from the rear. Note the almost completely flat undertray, which in part is responsible for the car's C_D of 0.36 (as measured in Volkswagen's wind tunnel in 1979).

quoted at only 53kW (71hp) at 3600rpm, but even so, this large car had a top speed of 155km/h (96mph).

When tested in 1979, in Volkswagen's wind tunnel, the T87 was found to have a C_D figure of 0.36 – extraordinarily good for the 1930s. Making that C_D figure even more impressive is that the windscreen used flat glass. To achieve a curved airflow around the A pillar, inset extra glass panels were fitted. Under the car, a full-length flat undertray was provided. Rear suspension was by quarter-length semi-elliptic springs that provided trailing-arm location of the swing axles, while at the front, transverse leaf springs were used.

Three headlights were fitted – the T87 could not be mistaken for any other car of the 1930s! Under the front cover sat a 55-litre (12-gallon) fuel tank, two spare wheels (one with tyre and the other bare), two 6V batteries and an engine oil cooler.

Inside, the car was roomy, with front bucket seats that could be fully reclined to form a bed. The walnut dashboard contained a mixture of white and black-faced dials, with a large speedometer in front of the driver and an equally large clock in front of the passenger. An oil temperature gauge was fitted. The steering wheel was white-rimmed Bakelite and a long gearlever sprouted from the floor. Vision was good – except towards the rear, where a view could be obtained only through the engine cover's louvres.

So what was the T87 like on the road? With its swing rear axles and rear-mounted engine, it is said that the car could easily leave the road tail-first. In fact, during World War II, when the car was kept in limited production by the Nazi occupiers, German officers were apparently banned from driving the Tatra!

Writing immediately after the war, an English motoring journalist said of the T87:

"There is little or no noise from the engine … and all the gears seem equally silent. There is an excellent view out, both forwards and sideways, because the rear squab [seat] is not only well forward of the axle, but actually in front of the leading edge of the wheels. Steering is remarkably positive and has that smooth, turn-table feeling always provided by a good rack and pinion mechanism. [But] as we had expected, the big Tatra is a great oversteerer."

Production of the T87 finished in 1950; just 3056 were produced.

The T87 was always going to be a large, luxurious and fast car. But what of those people who didn't require such a machine? For them Tatra produced the smaller T97. Similar in design layout to the T87, the T97 used a rear-mounted, flat-four, air-cooled engine. The monocoque construction body used independent suspension on all four wheels, and hydraulic brakes and rack and pinion steering were fitted. Manufacture of the T97 started in 1938 but was soon stopped: Czechoslovakia was invaded by Germany, and, on the orders of Hitler, production of the T97 ceased. Why? It was too close in design to the Volkswagen, a car about to be released.

Post World War II, the Tatraplan

AERODYNAMICALLY OPTIMISED VEHICLES

Figure 2-14: The T97 was built in 1938, but, on Hitler's order, production soon stopped. Why? The car was too close in design to the Volkswagen, a car about to be released.

Figure 2-15: The Tatraplan was built between 1946 and 1951. From this view, the aerodynamic styling looks more restrained than on the previous cars …

was released. Compared to the previous cars, this model was made in relatively large numbers – 6342 being produced between 1946 and 1951. Body styling became a little more sedate, with the ducts for the engine cooling intakes integrated into the upper part of the engine cover. Rear windows were now also fitted, and the rear fin became much smaller. Aerodynamic testing of the body design was undertaken on the road using wool tufting, and in model form in smoke tunnels. Some references claim a C_D figure of 0.33; however, this has not been independently verified. Top speed was 130km/h (81mph).

The Tatraplan had a four-cylinder, horizontally opposed engine of 1952cc. Overhead valves were used and the quoted output was 39kW (52hp) at 4000rpm. Early engines used a vertical axis cooling fan and later engines had a horizontal fan. Front springing was by two transverse leaf springs, and the rear used torsion bars – both front and rear systems were independent.

Then followed other Tatra models, but the aerodynamic innovation had been lost. Production of Tatra cars stopped entirely in 1999, although the company still makes trucks. But pre-World War II, its nearest design competitor was Ferdinand Porsche and his products; it's not stretching credulity too far to suggest that if history had turned out just a little differently, the name of Tatra could now be as revered as Porsche in automotive circles …

A good range of Tatras can be seen at the Tampa Bay Automobile Museum in Florida, USA. A T87 can be seen at the Deutches Museum, Germany. A Tatraplan can be found at the Gosford Classic Car Museum, north of Sydney, Australia.

Figure 2-16: … however, from the rear, the Tatraplan could not be anything other than a Tatra! Some sources suggest a C_D of 0.33, but this figure has not been independently verified.

1938 PORSCHE TYPE 64

Regarded as the forefather of all Porsches, the Type 64 was originally designed for the Berlin–Rome long-distance race. The race was scheduled for mid-September 1939, but the advent of WWII caused its cancellation.

The Type 64 dates back to not only the beginning of Porsche cars, but also the beginning of the Volkswagen – the Type 64 was built on a very early Volkswagen chassis. Three of the curvaceous cars were produced, with the hand-formed aluminium body made by the custom coachwork firm, Reutter. With just 24kW (32hp) at its disposal, the 1131cc flat-four could still propel the slippery body to 130km/h (81mph) – indicative of a very low drag value indeed.

And what happened to the three cars? One was crashed and destroyed in the early years of World War II, one car survived the war but was driven to destruction by triumphant American forces, and the other survives to this day.

A replica of the Type 64 body can be seen in the Porsche museum, Stuttgart, Germany.

1950 VOLKSWAGEN TRANSPORTER

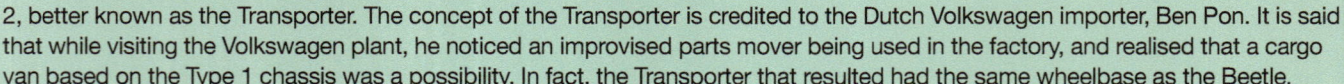

The Volkswagen Type 1 (the Beetle) was the first commercially successful car that used streamlining. That statement might look a little odd (the Beetle streamlined?!), but it needs to remembered that in the 1930s, when what became the Beetle was being developed by Ferdinand Porsche, other cars had appalling drag coefficients. The Beetle (then the KdF) had, in initial testing, a C_D of 0.40, while the production model of 1938 was said to have a C_D of 0.385 – which seems hugely optimistic, as other sources suggest 0.48! However, photos exist of the prototype being tested in the full-size wind tunnel of the DVL (Deutsche Versuchsanstalt für Luftfahrt or German Aerospace Centre) in Berlin, so the car certainly did undergo specific drag testing.

However, here I want to talk about not the Beetle but the next model – the Type 2, better known as the Transporter. The concept of the Transporter is credited to the Dutch Volkswagen importer, Ben Pon. It is said that while visiting the Volkswagen plant, he noticed an improvised parts mover being used in the factory, and realised that a cargo van based on the Type 1 chassis was a possibility. In fact, the Transporter that resulted had the same wheelbase as the Beetle.

But this book is about aerodynamics. By 1930 it was realised that rounding the leading-edge radii on rectangular bodies would result in a reduction in drag. However, this knowledge was not widely implemented, although some long-distance German buses of the late 1930s show these characteristics. When the Transporter was developed, wind tunnel testing of models was carried out. The first models had the now-familiar nose profile, but the vertical front corners were sharp-edged. This caused flow separation down the sides of the model, resulting in a C_D of 0.76. Rounded front corners were substituted (in fact, much more rounded than strictly necessary) and the wool-tufts on the photographed model then show attached flow down the length of the sides. The drag coefficient? It dropped by a whopping 45 per cent to 0.42! Unfortunately, the styling was also symmetrical in that the tail end of the Transporter was also rounded; it would have been better if sharp edges had been used at the rear to promote clean separation.

The next time you see a Volkswagen Transporter of any vintage, consider that it's probably the longest running model series in the world that has been continuously developed in wind tunnels.

The prototype Volkswagen Transporter can be seen in the AutoMuseum Volkswagen, Wolfsburg, Germany.

AERODYNAMICALLY OPTIMISED VEHICLES

1967 NSU RO 80

When the NSU Ro 80 was released in 1967, the car was greeted with universal acclaim. It immediately won Car of the Year awards, and road tests in Europe, the US and Australia were glowing in their praise. Here was a car with not only a brand new design of rotary engine, but also a stylish and elegant body that had an aerodynamic drag figure superior to nearly everything then available. Add to that front-wheel drive, four-wheel disc brakes, class-leading power steering and a well-developed suspension. And all from a company, NSU, previously known only for motorcycles and small rear-engine cars!

The Ro 80 was a tour-de-force – but it sent NSU broke. So how did one of the most advanced and welcomed cars ever made prove such a disaster? It's a fascinating technical story.

Let's start with the engine. Mounted longitudinally at the very front of the car, the twin rotor Wankel used peripheral intake and exhaust ports and displaced 1 litre (497.5cc x 2). While one of the very first production cars in the world to use a rotary engine, the Wankel was certainly not an unknown design to NSU: at the time of release of the Ro 80, the company had built 3000 small rotary outboards and 2500 single-rotor NSU Spider cars.

The Ro 80 engine used aluminium alloy rotor housings and a 9:1 compression ratio. The rotor tip seals were cast iron and the wearing surface of the rotor housings was a nickel and silicon carbide coating that was electrically deposited. Two sparkplugs per chamber were fitted (in later engines this was reduced to one plug per chamber) and the two plugs were fired simultaneously. Ignition on early engines was by traditional coil; this was upgraded to electronic ignition on later cars.

Figure 2-17: The NSU Ro 80 was a radical car indeed. Not only were its aerodynamics advanced but it also featured a rotary engine, front-wheel drive and a semi-automatic transmission. (Courtesy Audi)

Figure 2-18: The low drag coefficient gave a high top speed (188km/h,117mph) for the Ro 80's size and power. Ride height is high – the car had long-travel suspension that made it a superb long-distance tourer. (Courtesy Audi)

Carburetion was by a twin-choke Solex, the unusual aspect being the use of one very small choke (18mm) and very large choke (32mm). These chokes fed long intake runners that merged only at the inlet ports. The exhausts of each rotor were kept completely separate. Power was listed at 84kW (113hp) at 5500rpm and peak torque at 159Nm (117ft-lb) at a high 4500rpm. The peripheral porting helped give a strong top-end, but low-rpm torque was poor. The Wankel was said to weigh only two-thirds that of a reciprocating engine developing the same peak power, and furthermore, NSU said the cost of producing the engine would be reduced to well below that of a conventional engine – the target was 60 per cent of the cost of a six-cylinder of equivalent power.

At the time of first release, NSU engineers were said to have overcome the development problems of the rotary: chatter marks on the epitrochoidal surface, apex seal wear and high oil consumption. The car's warranty was two years or 40,000km (24,000mi) far better than the then prevailing average. Road testers and owners universally commented on the engine's smoothness and ability to rev. Later, they also commented on the incredibly short time the engines lived …

Bolted to the back of the rotary was an interesting transmission. Similar to the Sportmatic design used in some Porsches of the time, the Fichtel and Sachs semi-automatic used a vacuum servo-controlled clutch, a torque converter and a three-speed all-synchro transmission. Incorporated in the

SPEEDPRO SERIES

Figure 2-19: This view shows what made the NSU Ro 80 body so radical. Note the high boot-lid and shallow angle of the rear glass – both unheard-of at the time. You can also see the sharp separation line on the trailing edge of the boot.

gearlever knob was a micro-switch that triggered the clutch. In normal town use the gearbox could be left in second gear, being driven as a full automatic. However, for better performance, the gearbox was manually changed. This semi-auto trans was the only transmission available; it was fitted after NSU engineers experienced snatching in the driveline on throttle lifts. The presence of the torque converter overcame this problem. Some drivers loved the transmission, while others hated it. It appears that the system had to be finely adjusted if jerkiness – especially on down-changes – and noise were to be avoided. In any case, off-the-line acceleration was always leisurely. Drive was to the front wheels – the whole engine and transmission being mounted ahead of the axle line.

Just the refined and sophisticated driveline would have made the Ro 80 a startlingly innovative car, but there was far more.

The front suspension was not ground-breaking in the mould of the contemporary Citroën hydro-pneumatic systems, and the hydraulic Hydrolastic system fitted to some Austin models, but it was still extraordinarily well developed. Front suspension used MacPherson struts, something then still relatively rare on a front-wheel drive car. The front coils were offset from the axis of the strut to reduce stiction and side-loadings, and the front wheels had no less than 188mm (7.4in) of travel. An anti-roll bar was used – another then unusual aspect of the design was that the anti-roll bar did not help locate the wheels; instead a lower triangulated wishbone was used. (Like many aspects of the Ro 80, this is something we now take for granted in many MacPherson strut designs.)

Rear suspension took more than a passing glance at BMW and Mercedes-Benz practice, and used semi-trailing arms (angled at only 10 degrees) and mounted on a subframe. Like the front suspension, spring and dampers were concentric. No rear anti-roll bar was fitted, something that certainly raises modern eyebrows. Rear wheel travel was an astonishing 250mm (10in). Road tests and owners alike commented on the extraordinarily comfortable ride, one that had little pitch. Initial tests were also glowing in their praise of the car's handling, it being said to have only a touch of understeer. However, later tests – perhaps within the context of improving front-wheel drives – talked about a lot of body roll and understeer that could not be throttle-controlled. The tail, for example, stayed resolutely planted, no matter what was done with the accelerator pedal or steering! Tyres were to modern eyes tiny – just 175mm width on 14 inch rims (alloy on later cars). The specified tyres were, however, very high-performance – Michelin XAS radials.

The ZF steering used a geared rack and pinion system, aided as standard by a hydraulic ram. Steering feel was said to be excellent at all speeds, however, as time passed and power steering became more common, the 3.75 turns lock-to-lock attracted some criticism. Very unusually for the time, the steering column did not extend past the front bulkhead, improving safety in the event of a front-end crash. In contrast, most cars then had steering columns that extended well forward; the steering wheel being pushed into the driver as the front crumpled.

Even in the braking system the Ro 80 was unusual for the time. Front discs were large and mounted inboard either side of the gearbox. Rear brakes were also disc, and incorporated separate drums for the handbrake. The braking system used a tandem master cylinder with two independent circuits – one circuit operating all four wheels and the other, just the front wheels through the second pair of cylinders on the twin piston brakes. The rear brakes

AERODYNAMICALLY OPTIMISED VEHICLES

used a load-sensitive proportioning valve.

But the reason that the NSU Ro 80 is featured in this book is its body – the shape of the Ro 80 was at the time, quite startling. A low nose, high boot, concealed rain gutters and lack of ornamental chrome gave rise to plenty of comments that the body was shaped by engineers, not stylists! (The Ro 80 doesn't look startling today because most of the styling cues were picked up in the later Audi 100, a car that set the aerodynamic sedan styling agenda for the next 20 years.)

In fact, it is interesting to compare the body shape of the Ro 80 with that of a contemporary – the Mercedes W114. Windscreen angles are where the differences between the cars begin – the windscreen angle is about 42° from horizontal for the Mercedes versus a more laid-back 35° for the Ro 80. At the back of the car it's a similar story – the Mercedes has rear window angle of 40°, while the Ro 80 is a much flatter 33°. If we measure the angle formed from a line drawn from the top of the rear window to the trailing edge of the boot, the Mercedes is 24° and the Ro 80 just 20°. Furthermore, the transition from windscreen to roof, and from roof to rear window, are much more abrupt in the Mercedes than the Ro 80, with the Ro 80 smoothing these transitions by having a more cambered roof. In short, the Ro 80 introduced to the world the idea that there should be a continuous flow line from the leading edge of the bonnet (hood) to the trailing edge of the boot (trunk), all achievable in a three-box sedan.

The claimed drag coefficient was 0.355, a figure then at least 25 per cent better than achieved by a typical car of the same class. Wind tunnel wool tuft testing shows amazingly good flow, especially in the reattachment on the boot-lid. (I have never seen any evidence that any other three-box cars of the era achieved attached flow on the final third of the car. Typically, any wool tuft photos of such cars are careful not to show this area!) The Ro 80 doesn't achieve attached flow on the rear window, but the reattachment on the boot-lid would have much reduced the wake size over its sedan-based contemporaries.

The low drag coefficient (and relatively small frontal area – the car was quite low) were responsible for the ability of the Ro 80 to run to 188km/h on only 84kW (that's 117mph on only 113hp). Even today, that's very good for a car that can easily carry four adults in comfort. Interestingly, a US magazine measured aerodynamic lift at 160km/h (100 mph) at a high 68kg (150lb) front and 64kg (140lb) rear. At least the lift was largely the same front and rear!

Initially, the demand for the Ro 80 was so high that there was a waiting list. Owners raved about the cars, particularly their ability to travel long distance in silent stability. Fuel economy, especially away from city confines, was within the ballpark of other cars of the same performance, and if performance was a bit sluggish off the line, well the mooted triple rotor Ro 80 would soon fix that.

And then, all at once, the whole thing came crashing down.

Simply, the innovative rotary engine proved to have a fearsomely short life. Engines worked perfectly for, in some cases, *just days*. Few cars did more than 50,000km (30,000mi) on the original engine. The standing joke was that if you were driving a Ro 80 and saw

Figure 2-20: This wool tuft photo of the Ro 80 is worth a long look. Note how the flow reattaches on the boot – unheard-of in this era of three-box sedans. (Courtesy Audi)

another, you'd not wave but instead hold up the number of fingers that corresponded to how many engines you'd gone through …

The eccentric shaft was upgraded with massively increased lubrication, tip seals were modified, an over-rev dashboard buzzer added, the ignition system and number of sparkplugs changed – but still engines failed. And not only did they fail catastrophically, but owners also complained about oil-fouled plugs and difficult starting. One enthusiast driver recounted that he became adept at changing plugs, removing carburettor jets and blowing through them, resetting the clutch servo and adjusting idle

SPEEDPRO SERIES

Figure 2-21: When Audi absorbed NSU, many of the Ro 80's engineers came with the purchase. The Audi 100, that had the extremely low C_D of 0.30 (base model), was the result.

speed! And even then, that particular car's engine died at just over 30,000km (19,000miles) …

At least maintenance oil costs were reduced – the engines used so much oil that the factory decided to drop oil changes, instead using a 'total loss' approach. To be fair, oil was always intended to be injected into the combustion chambers, but the mind boggles at the thought of an oil consumption so high the complete sump contents were changed frequently enough to never need draining. When engines failed, NSU replaced them at a nominal charge (and often free), but the costs of doing so sent the company into a downward spiral that was arrested only when it was taken over by Audi/Volkswagen.

The Ro 80 remains one of the few really brave cars launched by any car company in the world. (Arguably, it took until 1998 for Toyota to match the scale of innovation with the Prius.) The Ro 80 incorporated the very best clean-sheet design aspects that could then be integrated into a safe and stylish car. From a modern-as-tomorrow engine (rotaries were then seen as the future of automotive powerplants) to aerodynamics that established new standards for passenger cars – and then it all came to nothing …

An NSU Ro 80 can be seen in the Technik Museum Speyer, Germany.

1969 DODGE CHARGER DAYTONA & 1970 PLYMOUTH ROADRUNNER SUPERBIRD

The 1969 Dodge Charger Daytona and 1970 Plymouth Roadrunner Superbird are probably the best-known aero specials ever built – and for good reason. Featuring an extended droop snout, fastback window and enormous, high-mounted wing, the cars look like nothing built before or since. (Note that the two models are not identical, but they are very similar in approach. This section draws on the SAE paper on the Charger Daytona for technical information, but is illustrated with pictures of the Plymouth Roadrunner Superbird.)

The cars were built for NASCAR homologation purposes, with the rules requiring that at least 500 street cars be produced with appropriate bodies and motors. These were large cars – the Superbird weighed over 1700kg (3700lb) and was 5.6 metres (221in) long – but they also had the power to match their dimensions. The street Superbird was available with 440 and 426 cubic inch V8 engines developing about 320kW (425hp). Performance was as quick as a 14.3 second quarter mile and the estimated 0-60 mph (97km/h) time was just 5.5 seconds. Top speed? Well, on the racetrack these were genuine 320km/h (200mph) machines, while the road-going versions were variably credited with maximum speeds of 210-260km/h (130-160mph). (And talking of top speed, the engineers responsible for the racing version suggest that it would have topped an astonishing 350km/h (220mph) in a straight line.)

Racing on high-speed banked ovals meant that the aerodynamic development of the cars needed to achieve both low drag *and* low lift, unusual in cars designed primarily for racing where downforce typically results in a faster car around the track – even at the expense of increased drag.

SAE paper 700036 (*The Aerodynamic Development of the Charger Daytona for Stock*

Figure 2-22: The 1969 Dodge Charger Daytona and its 1970 Plymouth Roadrunner Superbird brother (shown here) are amongst the best aero specials ever produced.

AERODYNAMICALLY OPTIMISED VEHICLES

Car Competition) is detailed and interesting. The comparison car for the engineers was the 1969 Charger, a car that had already been produced in modified form for racing – but one that had much less development than the Superbird was to gain.

Said the engineers:

"Preliminary calculations showed that a 15% drag reduction from the level of the 1969 race Charger was required to gain a 5mph lap speed increase. The 1969 race car was, to begin with, a better car aerodynamically than the standard 1969 Charger; the grille and backlight areas were significantly improved from a drag standpoint, and a front undernose spoiler was used to reduce front end lift. Although the front end treatment of the 1969 car was improved by a different grille treatment, this area was still considered a high drag region capable of being significantly improved. In addition, the abrupt windshield to side glass treatment, the vehicle ground clearance, and the relatively large vehicle cross-sectional area were considered areas which could be improved to reduce drag."

The engineers listed these changes to the 1970 car:

"The Charger Daytona has been modified in the following way to improve performance and handling through aerodynamics:
1. Extended streamlined front end – An 18in low form drag extension has been mounted on the 'loop' bumper mounts of the standard 1970 Charger.
2. Front undernose spoiler – A fixed 5in chord, 51in span spoiler has been mounted 13in aft of the nose leading edge at a 45° angle to the ground.
3. Backlight modification – The rear window or backlight slope has been changed from approximately 45° to 22° from the horizontal, and the side window to backlight junction has been faired. [This was a carry-over from the 1969 race car.]
4. Rear deck vertical stabilizers – Two fixed symmetrical vertical stabilizers have been mounted at the rear of the rear fenders.
5. Rear deck horizontal stabilizer – a 58in span, 7½in chord aerodynamic control surface has been mounted 23½in above the rear deck between the vertical stabilizers. An adjustment range of +2° to -10° is provided."

So what were the effects of this package? At zero degrees yaw, the C_D of the Charger Daytona was just over 0.30. This rose asymmetrically, reaching a little over 0.32 at -5° yaw, and 0.34 at +5° yaw. By +10° yaw, C_D was about 0.39 – see Figure 2-23. The C_{Lf} was largely unchanging over the ±5° range, varying little from 0.07. C_{Lr} was -0.07 at zero yaw, but then altered asymmetrically at greater positive and negative yaw values – see Figure 2-24.

When compared with the previous race car, these figures represented a drop in C_D of about 18 per cent. The rear downforce of the winged car was the same as the previous unwinged model at zero yaw. However, as yaw increased, the unwinged car decreased in downforce to a greater extent than the winged car.

Development occurred primarily on ⅜ scale models made from fibreglass reinforced with wood and metal. The front-end of the models was made removeable so that alternative shapes could be trialled, and the rear window was prefitted with clay and fibreglass shapes. Unusually for the time, the models had detailed underbodies, including the petrol (gas) tank, suspension, exhaust system and structural members. The grille, wheel wells, and engine compartment were also modelled. Even a scale fan (driven by a small motor) was used, together with a scale radiator.

Front-end development was aimed at reducing drag and lift. The blunt front end was streamlined, the cooling opening was reduced in size, and the rough underbody shielded by the use of a front spoiler. Multiple

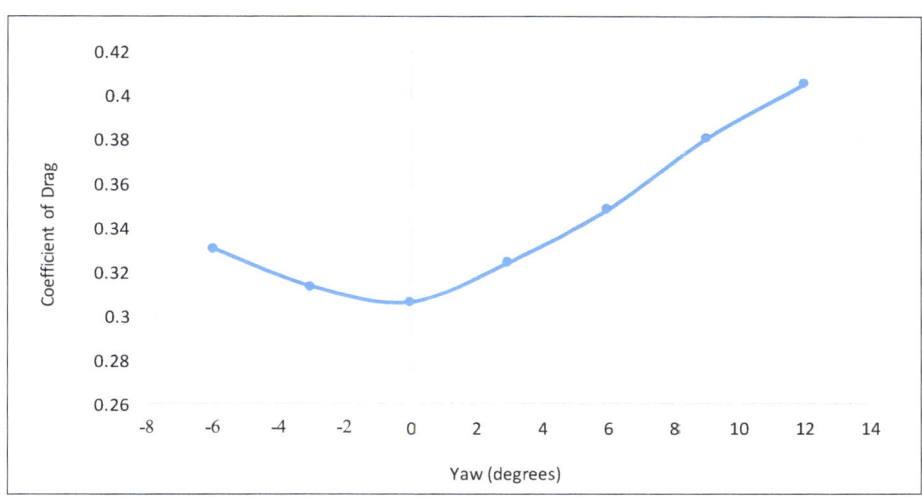

Figure 2-23: The Charger Daytona's drag coefficient was low at zero yaw, being just over 0.30. It then rose asymmetrically at higher yaw angles, reaching over 0.40 at +12° of yaw. (Data points from SAE 700036)

37

SPEEDPRO SERIES

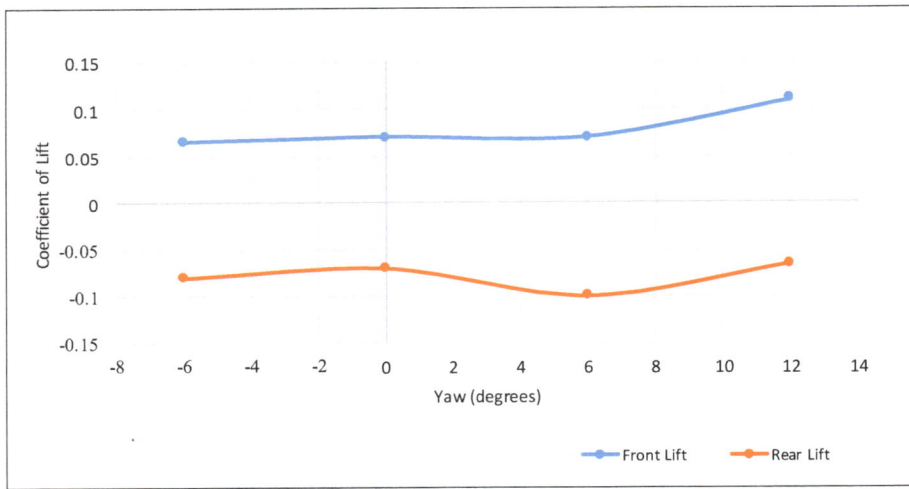

Figure 2-24: The Daytona showed a positive front lift coefficient and a negative lift coefficient (ie rear downforce), with both changing only a relatively small amount with yaw. (Data points from SAE 700036)

Figure 2-25: An extended nose smoothed air past what otherwise was a bluff front, and on the sister Daytona a chin spoiler decreased flow under the rough underside of the car. Drag and lift were reduced.

nose configurations were trialled, with the nose extensions falling into two size categories – 230mm (9in) and (460mm) 18in. Testing showed that when no spoiler was fitted, the 230mm extension was superior to the 460mm extension in producing low drag. However, with the spoiler fitted, the longer extension worked better. A 460mm extension with spoiler was thus chosen – it also gave room for pop-up headlights on the street version.

The location of the spoiler on the underside of the nose was also varied in tests. The further forward the spoiler was mounted under the nose, the less pressure that occurred on the surface immediately above it. Therefore, a forward spoiler location gave less lift. In comparison, drag changed relatively little for different front spoiler positions – that is, it blocked undercar flow equally in all positions.

Drag and lift were found to be influenced greatly by vehicle rake. The minimum drag was achieved with a negative rake (ie nose-down) angle of -0.5°. However, minimum C_{Lf} occurred at about -2.5°. In fact, the C_{Lf} went from nearly +0.1 to -0.1 with a rake change from 0 to -2.5°! The final chosen rake was -0.5°. (Note that C_{Lr} was relatively insensitive to rake changes.)

Not covered in the technical paper but visible on the final cars are two rear-facing scoops above the front wheels. After track testing was carried out, it was found that greater clearance was needed for the front tyres. Some references state that this was because of the front downforce compressing the suspension, but that is odd since the SAE paper shows only front lift (albeit small) occurring! Also fitted to the cars were A-pillar fairings that smoothed the flow past the otherwise exposed gutters.

While the eye is immediately drawn to the rear wing on these cars, what should not be overlooked is that the wing is supported on large stabilisers – ie aerodynamic fins. The cross-section of these fins was based on the NACA 0012 symmetrical profile. They had a geometric aspect ratio of 2.34 and each had an area of $0.11m^2$ ($1.17ft^2$). In the paper on the Daytona, the centre of pressure is located closer to the centre of gravity than in the standard Charger, but, significantly, is still ahead of it. (Note that the later Superbird (pictured here) had vertical fins about 40 per cent bigger in area than the Daytona – perhaps this change moved the centre of pressure aft of the centre of gravity.)

The wing itself was $0.28m^2$ ($3ft^2$) in area, had a geometric aspect ratio of 7.75, and in profile comprised an inverted Clark Y aerofoil. The fins were made from poured aluminium and the wing from aluminium extrusion. These parts are strong – there are photos of two race car drivers sitting on the wing!

A wide range of rear lift coefficients was available with changes in the wing's angle of

AERODYNAMICALLY OPTIMISED VEHICLES

Figure 2-26: The massive rear stabilisers (fins) supported an inverted Clark Y aerofoil that was adjustable for angle of attack. Rear lift, especially at yaw angles away from zero, was reduced.

attack. At -10° wing angle, the C_{Lr} was about -0.25. Incidentally, note that the C_{Lr} shown in Figure 2-24 is much lower than this – it's about -0.07. Obviously, the engineers chose to run much less rear downforce than was actually available. One reason they did this is that drag increased with higher wing attack angles. The final chosen wing angle appears to have been -5°, where drag over the unwinged car was increased by 7 per cent at zero yaw angle. The lift to drag ratio of the wing was 19.

The rear window angle was changed from the standard Charger's 45° to 22° from the horizontal. (This was a carryover from the previous homologation car.) In the standard car with the 45° window, flow was separated over the entire rear window region. The 22° car had attached flow. To help retain this attached flow, the pillars were also rounded to promote flow from the side of the car onto the window. Making the rear window change decreased C_D by a startling 0.021, and also decreased C_{Lr} by 0.065 (both in zero yaw conditions). Note that the 22° final window angle was chosen after extensive testing, that showed at angles greater *or* lesser than this, both drag and lift increased.

And how did the engineers conclude their paper?
"1. The Daytona Aerodynamic Package meets the goals of the program; that is, lap speed increase of 5mph and improved handling qualities.
2. The streamlined front end-front spoiler combination is a highly interactive aerodynamic system which is an effective reducer of front axle lift and vehicle axial force [drag].
3. A backlight angle somewhat less than a full fastback represents an optimum from an axial force and rear lift stand point.
4. Changing the position of the front undernose spoiler changes front axle lift significantly with a minimal axial force penalty.
5. Small changes in vehicle attitude affect the vehicle aerodynamics significantly.
6. The front undernose spoiler is much more effective on streamlined front ends than on blunt front ends.
7. The Daytona rear deck stabilizers are of significant benefit in controlling vehicle lift distribution and improving directional stability."

About 550 street Daytonas were produced, with over 300 surviving. The street Plymouth Superbird was produced in larger quantities, with about 2000 manufactured and about half that number surviving today.

Note: You can see these cars in a number of museums around the world, including the Simeone Foundation Automotive Museum in Philadelphia, Pennsylvania, USA.

1988 HOLDEN SPECIAL VEHICLES GROUP A COMMODORE ('WALKINSHAW')

Outside of Australia, very few people will have heard of the 1988 Holden Special Vehicles Group A Commodore, better known as the 'Walkinshaw.' However, along with the 1969 Dodge Charger Daytona and the 1970 Plymouth Roadrunner Superbird, the Commodore rates as one of the best aerodynamic makeovers of a standard car for homologation purposes that has ever occurred. The story is a little involved – politics of motor racing and car sales made the car much more adventurous than it would otherwise have been.

Throughout much of the latter part of last century, Holden (the Australian arm of General Motors) was heavily involved in the local racing scene. It also had a favourite racing car driver – Peter Brock – who in 1980 started to develop sportier models of the Holden best-seller, the Commodore. These cars, called Holden Dealer Team (or HDT) cars, were sold through selected dealers and were very successful. Often, they were simply known as 'Brock Commodores.'

When the international Group A racing arrived in Australia in 1985, the HDT cars also allowed the homologation of parts that could then be used in the racing cars. In November 1986, HDT produced the Commodore SS Group A – but that's *not* the car pictured here! The HDT car featured the then-traditional Peter Brock style of bodykit, comprising a low front spoiler and rear 'ducktail' spoiler. An odd NACA-inspired scoop sat on the rear edge of the bonnet and fed air to the carburettor V8.

But then suddenly the relationship between Peter Brock and Holden soured. Brock had started fitting all HDT-produced road cars with a device called the Energy Polariser, a small box containing crystals and magnets in epoxy resin that, he claimed, improved the performance and handling of vehicles through "aligning the

SPEEDPRO SERIES

Figure 2-27: The Holden Special Vehicles Group A Commodore is one of the most radical production car specials ever released.

Figure 2-28: Here is a standard Holden Commodore for comparison. Note that this is the Calais model, that has a slightly different nose treatment to the car on which the Group A was based. (Courtesy Holden)

molecules." Not surprisingly, testing by Holden engineers could find no benefits from the Polariser, but Brock refused to drop the device. Holden had no option but to sever the relationship with Brock. But that left it with a major hole in its product line, with the most sporty, revered cars suddenly no longer available.

In response, Holden formed its own Holden Special Vehicles operation, and Tom Walkinshaw Racing in the UK was contracted to develop a new Group A VL Commodore road car. (That's right, there were two different special versions of the same model then going to be made available in sequence!) It was clear that the instructions to TWR were to 'go for broke' – not only was the new car going to be fitted with a twin-throttle, fuel injected V8 (replacing the rather anaemic former carburettor V8), but the bodykit for the new model was going to be developed scientifically, rather than by rule of thumb.

The development of the car was in the UK's MIRA full-size wind tunnel, and resulted in the most radical bodykit ever fitted to a production car. Unfortunately, no technical information on lift and drag was ever released.

At the front, a low spoiler (it replaced the whole front bumper) was fitted. It had an even lower lip, that was detachable to allow sufficient ground clearance to let the cars be driven off the delivery trucks at dealers. Large brake ducts were positioned either side of the radiator cooling flow intake. The intake, normally positioned between the headlights, was blanked-off. A large exit air vent was provided in the bonnet immediately behind the radiator. Louvres in this vent directed air out onto a bonnet bulge that was needed to clear the twin-throttle plenum of the new engine. At the time it was said that this vent prevented pressure build up in the engine bay that was causing front lift. Deep side skirts were fitted. These had three vents – two were for access to the normal Commodore jacking points, but the third was said to allow a 'controlled leakage' of air.

But it was at the rear of the car where the most radical changes occurred. Combining some educated guessing with the results of the wool-tuft testing on my own car (a standard VL Commodore – alas, I *just* couldn't afford a Walkinshaw at the time!), a huge increase in rear boot-lid height was required to regain attached flow. (In the standard car, the whole of the rear window and boot-lid was a mass of turbulence; flow was separating at the end of the roof and not reattaching.) Flow onto this rear area on the modified car was also enhanced by a revised C-pillar moulding – initially filling-in the window in this pillar, and when that conflicted with Group A rules, reinstating most of the opening. With flow reattached, a large ducktail spoiler could then be used to increase rear pressures. Notably, this spoiler was placed as far rearwards as possible – its trailing edge lined-up with the bumper below. Completing the kit was a rear bumper cover replacement that visually integrated with the rest of the additions.

The drag coefficient was said to be 25 per cent lower than that of the standard car, but as I said, no numerical data was officially released. (One suggestion is that

AERODYNAMICALLY OPTIMISED VEHICLES

Figure 2-29: The Group A features a very deep front spoiler with a removable lip, large bonnet vent with louvres directing flow, a bulge to cover the engine's taller plenum, side skirts, and an enormous increase in boot height carrying a large duck-tail spoiler.

Figure 2-30: Note how far rearwards the engineers have placed the spoiler. The standard car did not have flow reattachment after the trailing edge of the roof, but on this car, with the much higher rear height, the spoiler would have been effective.

the standard VL Commodore had a C_D of 0.43, and the Holden Special Vehicles Group A Commodore was 0.32 – ie 26 per cent lower.) Top speed of the new car was of course higher than the Brock VL Group A – but then the Walkinshaw also had more power.

However, very much like the public response to the Dodge Charger Daytona and the Plymouth Roadrunner Superbird, car buyers greeted the appearance of the 1988 Holden Special Vehicles Group A Commodore with amazed bemusement. Road tests of the time were complimentary (here was a car with a lot more power as well as better aerodynamics) and at least one test described the high-speed stability of the car as being like a 'giant hand' holding it in place on the road. I haven't driven a Holden Special Vehicles Group A Commodore, but the standard VL Commodore was certainly very wandery at high speed. Holden made 750 of the Group A Commodores and struggled to sell them; they are now collectors' items that change hands at high prices.

The Holden Special Vehicles Group A Commodore was one of those very rare cars where the requirements of racing, politics and technology resulted in a radical aerodynamic special, one where the changes are so massive that we can look at them and learn.

A Holden Special Vehicles Group A Commodore can be found at the Gosford Classic Car Museum, north of Sydney, Australia.

1989 OPEL CALIBRA

The Opel Calibra three-door hatchback coupé was released in 1989. Based on the underpinnings of the Vectra sedan, it was sold initially with 2-litre 8-valve or 16-valve engines. Later, a turbocharged version was added, along with all-wheel drive, and then in 1993 a 2.5-litre V6 version was introduced. Both the turbo and V6 versions sold in relatively small numbers; the car as a whole sold about 240,000 vehicles. Reading that description, you could think the Calibra was just a humdrum personal coupé of no great interest – except for one thing.

When it was released, the Calibra had the lowest drag coefficient of any production car in the world, with a C_D of just 0.26. It would be a decade before that figure was bettered. Making the Calibra of additional interest is that GM released a fascinating SAE paper on the aerodynamic development of the car – SAE 900317.

Not fortuitously, the aerodynamic development of the Calibra was intense. Wind tunnels throughout Europe were used, including one that had a rolling road. Initial development occurred on a 1:5 scale model developed from styling sketches. Unlike many cars of the era, where aerodynamic development attempted to fix the problems created by stylists, the Calibra had good low-drag essentials from the beginning. These included a front-end with smooth, rounded transitions; a boat-tailed rear in plan-view; sharp flow

41

Figure 2-31: When it was released in 1989, the Opel Calibra had the lowest drag coefficient of any production car in the world – just 0.26. (Courtesy Opel)

separation edges around the rear; and tapering of the C pillar.

The scale models were used to ascertain proportions that would be later difficult and expensive to change on full-size cars. This included establishing an optimal rear boot-lid height. Using an 890mm (35in) rear height as the baseline, C_D decreased as the rear height grew. At 940mm (37in) height (the optimum), the reduction in C_D was just over 0.020.

The influence of rear boat-tailing was also explored in the scale model. The test results showed a gradually decreasing drag coefficient as the amount of boat-tailing increased. Compared with no boat-tailing, having a final rear width 180mm (7in) narrower than the rest of the car body (ie 90mm – 3½in – each side) gave a reduction in drag of 0.011.

The front spoiler lip of the Calibra was first developed on the scale model, but was then changed substantially when full-size testing was undertaken. The front spoiler was developed on the philosophy that "impingement of high speed airflow onto underbody parts is considerably reduced." (This is important to note: it contrasts with the more current trend towards the use of smooth underbodies under which airflow is encouraged.) From a starting point of 210mm (8.3in) clearance between the front spoiler and the road, the spoiler was progressively lowered, and drag measurements made. At 150mm (6in) clearance, the change in C_D was 0.022. However, for ground clearance reasons, final clearance was set at 180mm (7in), resulting in a change in C_D of about 0.018.

As work was performed on the full-size car in the wind tunnel, major changes were then made to the front spoiler. Flow investigations carried out behind the car showed that the wake was wide, especially near the road. Testing showed that the front wheels caused the airflow to diverge outwards by about 30° to the centreline of the car. To try to reduce this divergence, and so make the flow more parallel to the centreline of the car, the engineers decided to reduce the pressure differential between the underside and external flows. This was achieved by cutting away 40mm (1½in) of the centre portion of the front spoiler, with maximum improvement in rear wake measurements found with a 965mm (38in) wide front spoiler cut-out. The corners of the spoiler were then moved 60mm (just under 2½in) outwards to better cover the front wheels; this resulted in a C_D reduction of 0.006. The final front spoiler complexion gave a C_D reduction over having no spoiler of 0.012.

With the increased flow now gained under the car (and remember, without any underbody streamlining at this stage), there were concerns that the increased underbody pressure at the rear would lead to lateral flows, so resulting in angled flow occurring at the rear wheels (and presumably a wider wake). Rocker panel extension (side skirts) were then fitted to inhibit this lateral flow; their addition reduced C_D by 0.005. Only one aerodynamic feature was fitted to the underbody of the base car: a panel that placed between the fuel tank and rear bumper cover. This decreased C_D by 0.003.

The Calibra body styling allowed for two air cooling inlets at the front of the car – one above the bumper between the headlights, and the other beneath the numberplate. In the base vehicle, it was possible to

Figure 2-32: The rear of the car was boat-tailed a total of 90mm (3½in) each side and the rear deck height was carefully developed to give low drag. (Courtesy Opel)

AERODYNAMICALLY OPTIMISED VEHICLES

Figure 2-33: Gaps in the front end were sealed, and, in the base model, the upper cooling airflow opening was blanked-off. However, in this V6 model, it was open – contributing to higher drag. (Courtesy Opel)

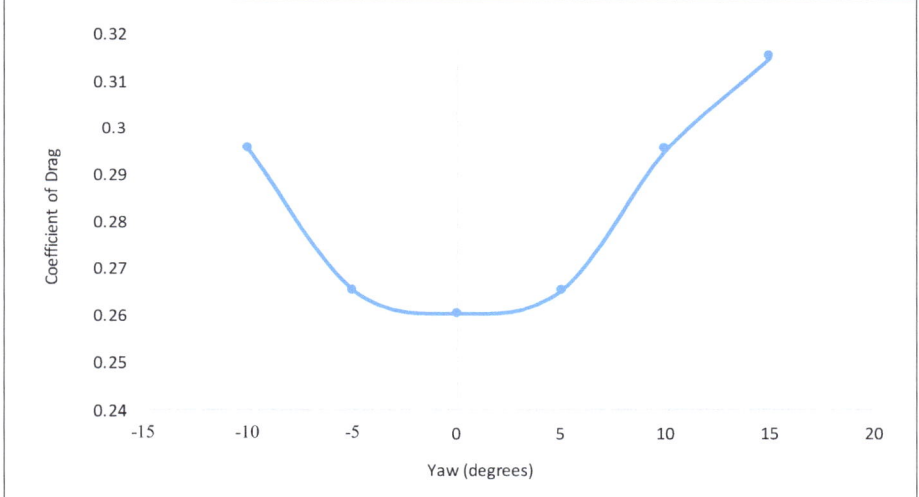

Figure 2-34: At zero yaw the Calibra had a C_D of 0.26. This rose symmetrically as yaw angles increased, rising to 0.295 at a yaw angle of ±10°. (Data points from SAE 900317)

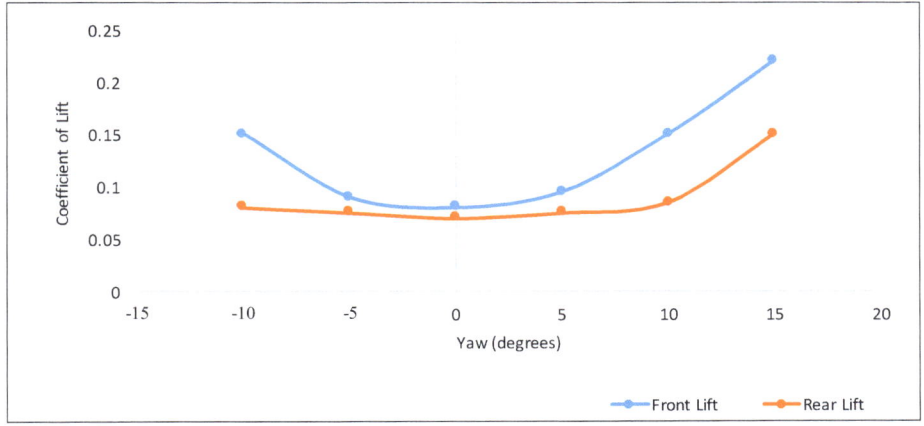

Figure 2-35: The Calibra had lift both front and back, but at low yaw angles, the lift coefficients were also low. (Data points from SAE 900317)

close-off the upper opening while still gaining sufficient cooling airflow. However, when the vehicle was used for towing or was fitted with an automatic transmission, this upper cover was removed. (Note that the exemplary C_D of 0.26 applied only to the base vehicle.) Attention was paid to front-end sealing, with seals positioned between the bonnet and headlights and grille, and between the headlights and bumper. Further seals were placed to prevent cooling airflow from passing below the radiator. The paper states that with these approaches reduced C_D by 0.014.

Attention was also paid to the drag and lift values experienced during yaw (cross-wind) conditions. As mentioned, at 0° yaw, the C_D was 0.26; this rose symmetrically, reaching 0.295 at a yaw angle of ±10° (see Figure 2-34). The front lift coefficient (C_{Lf}) was about 0.08 at 0° yaw, and reached 0.15 at ±10° of yaw. The rear lift coefficient (C_{Lr}) was about 0.07 at 0° yaw, rising only a little to 0.08 at ±10° yaw (see Figure 2-35).

The Calibra is a fascinating car not only because of its historical significance as a car having the then-best drag coefficient, but also because Opel released such extensive data on how it had been achieved. You can be sure that every aerodynamicist in the world working on cars in the 1990s read the paper ...

I am not aware of any museums featuring the Calibra; at the time of writing, they can be purchased for nearly nothing as they haven't achieved collector status.

1996 GENERAL MOTORS EV1

The General Motors EV1 was an electric car produced by GM between 1996 and 1999. The EV1

SPEEDPRO SERIES

Figure 2-36: The General Motors EV1 had a C_D of just 0.19. It was made in very small numbers between 1996-1999, and leased to customers for a limited time. (Courtesy GM)

Figure 2-37: Regarded by many as the car that could have changed the automotive landscape in the way the later Toyota Prius did, the EV1 still impresses in its design.

had the lowest coefficient of drag of any commercially available car from a major manufacturer. Designed from the ground up as an electric car, the C_D was just 0.19.

The EV1 has become famous (or perhaps infamous) for its demise. The car was never sold by GM; instead it was leased to customers. In February 2002, after 1117 cars had been produced, GM notified lessees that all cars were to be returned to GM. Despite a significant number of lessees requesting that they be allowed to buy the cars, and committing to covering maintenance and any other costs, GM refused. The cars were duly returned to GM – and then all but 40 were crushed! Those that were preserved were deactivated and the museums and other institutions to which they were donated had to agree that they would never be driven. Pictures circulated of otherwise perfectly driveable EV1s crushed and stacked in piles. The reasons that GM chose to take this approach are still open to conjecture, but few would argue that it was, in retrospect, a wise move.

And the car itself? Two versions were produced, with the first having lead-acid batteries and the later car using nickel-metal hydride batteries. Other improvements in the second generation included weight reduction and quieter operation. The initial car had a mass of 1350kg (2970lb) and was a compact two-door, with a length of 4309mm (169.7in) and a width of 1766mm (69.5in). Height was just 1281mm (50.5in) and projected frontal area was $1.89m^2$ ($20.3ft^2$).

Range in this initial form was stated as being 113km (70mi) in city conditions, and 145km (90mi) on the highway. The electric motor was rated at 102kW (137hp) and the performance included a top speed of 129km/h (80mph) and a 0-60mph (97km/h) time of "under 9 seconds." In our time now of Tesla and electric cars from other manufacturers, these specifications do not sound anything special, but at the time they were excellent.

Figure 2-38: The EV1 used a narrower rear track and had interesting shaped rear quarters, with wrap-around pillars and a heavily curved rear window. A flat undertray was used.

AERODYNAMICALLY OPTIMISED VEHICLES

The EV1 used a bonded and welded aluminium spaceframe. At 132kg (290lb), it was said to be 40 per cent lighter than a steel equivalent. Cladding the frame were three different types of plastic – reinforced fibreglass moulding was used for the roof, doors, and bonnet. Where more flexibility was needed, such as for the bumpers and interior facias, reaction injection moulded (RIM) polyurethane was used. To improve aerodynamics under the car, a structural reinforced injection moulding (SRIM) was used to form the undertray (belly pan). Aluminium and plastics were extensively used in the suspension.

Aerodynamic optimisation included using a 226mm (8.9in) narrower rear track with the wheels partly skirted. A flush undertray was fitted from front to back. No grille was fitted; instead a small air inlet under the front fascia cooled the radiator for the propulsion motor and power electronics bay. The cooling system used a small (40W) pump and a 75W fan that operated only when the air-conditioning was running. The radio antenna comprised a flat sheet located inside the fibreglass roof panel. Longitudinal fences were fitted between the front and rear wheels, down each underside of the car.

The electric motor was a 3-phase, AC design that delivered power to the front wheels through a single-speed reduction drive. Braking used a mix of regenerative braking, electro-hydraulic front brakes and electric rear brakes. Many design aspects that were to become common in later hybrid, electric and conventionally-powered vehicles were included in the EV1, including electric power steering, self-sealing tyres, aluminium body construction, low rolling resistance tyres, and cast magnesium components.

At the time of its release, John F Smith, the Chairman of GM, said:

"The EV1 is the world's most advanced vehicle platform. It is the world's most energy efficient vehicle. It is the world's most aerodynamic production vehicle. The car carries 23 new patents in a variety of critical new technologies. When auto industry historians look back, they will see this car as the first in the new generation of vehicles. And they will note that GM made it."

And all that is true – but unfortunately, auto industry historians are more likely to look back and see it as an opportunity that was missed.

The GM1 can be seen at a number of US museums, including The Henry Ford in Detroit, Michigan, and the Peterson Automotive Museum, Los Angeles, California.

2011 GENERAL MOTORS CHEVROLET VOLT

The Chevrolet Volt was first shown as a concept car at the North American International Auto Show in January 2007. The concept was radically different to the earlier EV1 (also from General Motors) with the Volt featuring four doors, seating for four passengers and a rear lift-up hatch. A series plug-in hybrid, the concept had a 64km (40mi) all-electric range. It was proposed that the production car would use a lithium ion battery pack – but the concept car got by with just two 12V lead-acid batteries!

Public response to the concept car was positive and so GM decided to put the car into production. However, an enormous amount of work was then needed – not only were the batteries not yet available, but the aerodynamics of the concept car were poor indeed. In fact, almost every aspect of the car's design underwent change between the concept and the production car, not least the dropping of the pure series hybrid approach and the adoption of four different propulsion modes from the electric motor and internal combustion engine.

The first stage in the aerodynamic optimisation of the car was to rescale the concept car styling

Figure 2-39: The Chevrolet Volt concept car, first shown at the North American International Auto Show in January 2007. The concept had a dreadful C_D of about 0.42. (Courtesy GM)

SPEEDPRO SERIES

Figure 2-40: Major improvements in drag over the concept car were achieved in the production Volt – it has a C_D of 0.28. Visible here are the sharp rear separation edges and the slightly kicked-up rear tail surface. (Courtesy GM)

cues onto an appropriate GM small car platform. This was done at one-third scale using clay models that were extensively wind tunnel tested. At this stage, it was determined that both underbody panels and a front air dam would be needed to meet drag targets. After testing of one-third scale models was complete, a full-size car was produced, again in clay. More wind tunnel work followed, backed-up by CFD.

The final key aerodynamics design features were:

- a reduction in the front grille open area, with much of the upper grille blocked. This resulted in a decrease in C_D of 0.010
- a kick-up of the rear boot-lid trailing edge. This 5mm high rise dropped C_D by 0.005.
- sharp separation edges on the back of the car. This reduced C_D by about 0.010
- door-mounted mirrors (as opposed to being mounted on the sails), of a specific low-drag design. These decreased C_D by another 0.008
- the use of a front air dam that not only reduced drag but also increased flow through the cooling system. The reduction in C_D was 0.025
- four underbody panels, that reduce C_D by 0.013

Ride height was lowered to "the minimum allowable by good engineering judgement and corporate standards" – clearance is about 90mm (3½ in) under the front air dam. (Most cars have a minimum ground clearance of about 150mm or about 6in.) The lowered ride height decreased C_D by 0.010.

Vented panels ahead of the wheelarches and behind the air dam reduced C_D by 0.004.

Note that the Volt is unusual in running an air dam while also using underside smoothing panels. (Normally, as described in Chapter 6, an air dam is effective in reducing drag only if the underside is rough ie not panelled.) However, the Volt uses relatively few underbody panels – the centre portion of the engine/transmission is completely unpanelled, panels are present under the middle of the car, the rear suspension is unpanelled, and the panelling behind the rear axle is minimal.

The final C_D of the Volt was 0.28. The technical paper on the car (SAE 2011-01-0168) suggests that this

Figure 2-41: The very low front spoiler (air dam) of the Volt gave a ground clearance of only 90mm (3½in). Many owners complained of its incessant scraping on the road, but the flexible rubber component proved to be durable and could be replaced when worn. (Courtesy GM)

AERODYNAMICALLY OPTIMISED VEHICLES

represents a 33 per cent decrease over the starting point, implying that the concept car had a C_D of about 0.42! The Volt is a good example of aerodynamicists and stylists working miracles on a concept car that, aerodynamically, was a very poor starting point.

2013 VOLKSWAGEN XL-1

The Volkswagen XL1 is a good example of what a major car manufacturer can achieve if it wishes to produce a very expensive, highly impressive car with the world's best fuel economy. At the time of writing, the car is in very limited production, with it expected that 250 units will be made.

Motive power is via a 0.8-litre, two-cylinder diesel engine developing 35kW (47hp), and a 20kW electric motor powered by a lithium-ion battery. Power is delivered to the rear wheels via a 7-speed, dual clutch gearbox. The production car has a C_D of just 0.189 (up a fraction over the earlier prototypes) and has a frontal area of 1.50m².

At 100km/h (62mph), the XL1 needs just 6.2kW (8.3hp) to propel it. The hybrid system is of the plug-in design, with pure electric propulsion able to take the XL1 up to 50km (31mi). Fuel consumption is 0.9 litres/100km (261 US mpg) in the New European Driving Cycle (NEDC). In driving mode, and maintaining the battery level without a plug-in boost, consumption is 2.1l/100km (112 US mpg) and emissions 70g/km. Top speed is electronically limited to 160km/h (100mph) – although 200km/h (124mph) is possible without the limiting – and the car can accelerate to 100km/h (62mph) in 12.7 seconds.

The XL1 is 3888mm long, 1665mm wide and just 1153mm tall (153.1 x 65.6 x 45.4in). The two-

Figure 2-42: With a C_D of just 0.189 and a frontal area of 1.50m² (16.15ft²), the XL1 needs only 6.2kW (8.3hp) to propel it at 100km/h (62mph). (Courtesy Volkswagen)

Figure 2-43: Cameras replace rear vision mirrors. Tyres are skinny and the ground clearance is low. (Courtesy Volkswagen

seater is also low in mass at 795kg (1749lb). The monocoque and exterior parts are made from carbon fibre. The body's torsional stiffness is more than 30,000Nm (22,100lb-ft) per degree. Carbon fibre isn't limited to just the body, with the anti-roll bars also made from this material. The weight breakdown is: 227kg (499lb) for the entire drive unit including the battery, 153kg (337lb) for the running gear, 80kg (176lb) for the equipment and 105kg (231lb) for the electrical system.

The hybrid powertrain is housed above the vehicle's driven rear axle, with the actual hybrid module with electric motor and clutch positioned between the engine and gearbox. The electric drive system can be decoupled from the diesel engine via a third clutch. The direct injected, turbo diesel was derived from

SPEEDPRO SERIES

Figure 2-44: The cabin is elegant and well-finished. The two seats are slightly offset to give better shoulder room. (Courtesy Volkswagen)

Figure 2-45: Note the attached flow from nose to tail, and the strong undercar airflow. The front wheels retain some openings around their peripheries. (Courtesy Volkswagen)

Figure 2-46: Top, side and bottom views of the Volkswagen XL1. (Courtesy Volkswagen)

Volkswagen's four-cylinder 1.6-litre diesel, and so shares the same 88mm cylinder spacing. A balance shaft has been added to the smaller engine. The lithium-ion battery (capacity: 5.5 kWh) operates at 220V. The body electrical system of the XL1 is supplied with 12V via a DC/DC converter and a small auxiliary battery.

The engine is cooled by a mechanical water pump that is activated by the engine management only as required. The air intake system at the front of the vehicle is regulated in opening to reduce cooling drag. An electric water pump, which is also used only as needed, circulates a separate lower-temperature coolant loop to cool the starter/generator and power electronics.

Weight is reduced by the use of aluminium parts (including suspension components, brake calipers, dampers, steering gear housing), ceramics (brake discs) and plastics (steering wheel). Friction-optimised wheel bearings and drive shafts, as well as optimised low rolling resistance tyres from Michelin (front: 115/80 R 15; rear: 145/55 R 16) are used. Magnesium wheels are fitted; these each weigh 3.16kg (6.97lb) at the front, and 3.71kg (8.18lb) at the rear.

And the aerodynamics? From the use of rear vision cameras to the completely flat underfloor, from the small frontal area to the boat-tailing, the XL1 takes every aerodynamic lesson and applies it in spades to achieve the world's slipperiest road car.

AERODYNAMICALLY OPTIMISED VEHICLES

INDEPENDENT TESTING

In June 2014, US magazine *Car and Driver* did a fascinating test. In an unnamed wind tunnel, and with unnamed experts advising, it tested five of the slipperiest cars then available. The cars were:
- Chevrolet Volt
- Mercedes-Benz CLA250
- Nissan Leaf
- Tesla Model S P85
- Toyota Prius

The results gained are shown in the following table.

	Chevrolet Volt	Mercedes-Benz CLA250	Nissan Leaf	Tesla Model S P85	Toyota Prius
C_D	0.28	0.30	0.32	0.24	0.26
A (ft²)	6.7	7.0	7.8	6.2	6.2
Drag force at 70mph (lb)	84	88	97	77	78
Aero power at 70mph (hp)	16	16	18	14	14
Aero power at 100mph (hp)	45	48	53	42	42
Front axle lift at 70mph (lb)	-15	46	-12	23	-4
Rear axle lift at 70mph (lb)	26	44	11	17	17

(Despite the rest of this book giving metric measurements first and imperial second, these figures are in their original form. It makes it much easier to understand them. Multiply mph by 1.61 to get km/h; divide lb by 2.2 to get kilograms.)

The C_D figures measured by *Car and Driver* differ from the manufacturer's figures, sometimes by quite a lot. For example, Mercedes claims a C_D of 0.22 for the CLA250 – *or does it?* In fact, that figure is only for the slipperiest of CLA250 models produced – not the US model. The Nissan Leaf is said by the manufacturer to have a C_D of 0.29. The magazine also mentions that it took along a 2001 Honda Insight as a reference – and it recorded a C_D of 0.30, way off the Honda figure of 0.25.

So, what to make of these figures? I think that by and large, they're probably valid. As the owner of a Gen I Insight, it pains me to say so, but considering the standard state of the underside of the Insight, the 0.25 C_D has always seemed to me to be a touch optimistic. (But 0.30? – I don't think so!)

Two other points. For all the cars, the aero power figures at 70mph (113km/h) are really quite superb. 14hp (10.4kW) to overcome aero drag at this speed, in a vehicle comfortably carrying four people? Wow! The other aspect to look at is the lift figures. With the exception of the Mercedes, these cars have very low lift.

SOME MORE FASCINATING AERO CARS

The MG EX181 of 1957 was a Land Speed Record car in the class for cars with engines between 1.1 and 1.5 litres. Fuelled with methanol (and with some other chemicals in the mix!), the supercharged 1.5-litre twin-cam, four-cylinder developed 216kW (290hp) at 7000rpm. Later, it produced even more power. At Bonneville it did 395.31km/h (245.64mph), and then the next year it went even better at 410.23km/h (254.91mph). The vehicle's C_D was just 0.12. Note how much it looks like the solar racing cars of 30 and 40 years later – the MG was extraordinarily advanced for its time.

SPEEDPRO SERIES

SOME MORE FASCINATING AERO CARS (continued)

The 1948 Dynavia was an experimental car built in France by Panhard, a well-established manufacturer. Power came from an OHV two-cylinder boxer engine of just over 600cc displacement. The engine drove the front wheels through a four-speed gearbox. Independent suspension and rack and pinion steering were fitted. The body panels were made from an aluminium/magnesium alloy. Vehicle mass was 650kg (1433lb). Despite having only 20.9kW (28hp), the vehicle could reach a top speed of 130km/h (81mph). Testing of models revealed drag coefficients of 0.26 and 0.24, while the full-sized car recorded 0.26.

Introduced in May 1972, the BMW 3.0CSL was an homologation special, built to make the car eligible for the European Touring Car Championship. Earlier versions of the CSL were available without the full aero kit shown here, but the July 1973 kit added a deep front spoiler, front fender strakes, a flow-directing spoiler above the rear window, and a wing mounted on wide-based fins. The car was powered by a 3.2-litre six – not the 3003cc unit as found in the normal CSL – and it produced 154kW (206hp) at 5600rpm. No aerodynamic data is available on the car, but its track performance speaks volumes – it won five consecutive Group 2 racing seasons.

The Mercedes Type 80 was prepared in the late 1930s for an attempt on the world Land Speed Record. However, World War II intervened, and the car never ran. Josef Mikcl did the aero design that was superbly streamlined, with a claimed C_D of just 0.18. In addition to the all-enveloping low-drag body, thin horizontal wing extensions each side of the body were used to reduce lift. The vehicle was to use a cutting-edge aircraft engine – the Daimler Benz DB600 series V12. Power was estimated at being between 2100-2600kW (2830-3500hp). A realistic potential speed for the T80 was 650km/h (400mph) – in 1947, John Cobb took the record, achieving 595km/h (370mph) on about 1870kW (2500hp) in a very streamlined, four-wheel drive machine.

AERODYNAMICALLY OPTIMISED VEHICLES

In 2002, Volkswagen released a working concept car aimed at achieving absolute best fuel economy, while still achieving a safe, usable and roadworthy vehicle. With a width of just 1.25m (49in), the car was extraordinarily narrow, the driver and passenger sitting one behind the other. The engine was a 0.3-litre diesel with DOHC and direct injection. It had an output of 6.3kW (8.5hp) at 4000rpm. Drag coefficient was just 0.159, and – for a two-seater – projected frontal area was tiny. Top speed was 120km/h (75mph) and fuel economy reflected the car's name – 1 litre/100km, or 235 US mpg.

The 1996 Honda Dream had among the lowest of drag coefficients of any car that has ever travelled on a public road. Following the successful 1990 and 1993 cars, the 1996 racing machine had a C_D of 0.101. With a frontal area of 0.999m², the C_DA was 0.101. C_{Lf} at zero yaw was 0.14 (enough to cause an increase in drag at speed through rising front ride height) and C_{Lr} was about -0.08 but rose quickly with yaw, reaching 0.2 at 9° yaw. The two-seater won the 1996 World Solar Challenge race across Australia. (Courtesy Honda)

The Ford Prove IV was a concept car created by Ford in 1983. In the history of car aerodynamics, the Probe IV has been largely forgotten, but it was an extraordinary innovative car that used drag-reduction technologies that were literally three or more decades ahead of their time. The Probe, a four-seater, had a C_D of 0.153 and a projected frontal area of 1.904m² (20.5ft²). A flat floor, flush side glass, shaped deflectors before the front tyres (and before and after the rear tyres), and an optimised rear fastback shape were all used. The front spoiler was actively height controlled, as was the air suspension. That these are all techniques used on current low drag cars speaks volumes – but the Probe IV was the first. However, the Probe IV also used some techniques not currently seen. The engine cooling radiator and air-conditioning condenser were mounted each side at the rear of the car, with coolant pipes connecting to the front-mounted engine. Air was drawn through side-mounted grilles by fans and then exhausted out of the rear, giving a blown wake. The front of the car was sealed and the exhaust pipe – carried above the undertray – was wrapped in insulation. Wheels were enclosed by the bodywork, with the front wheel skirts flexible to allow steering. The Probe developed front downforce (C_{Lf} -0.281) but had rear lift (C_{Lr} 0.211), giving an overall C_L of 0.070. Especially if you are thinking of developing a low-drag car from scratch, the Probe deserves a long look – see SAE paper 831000. (Courtesy Ford)

Chapter 3
Flow visualisation

- **On-road flow testing**
- **Visualisation techniques**
- **Wool-tufting**
- **Smoke streams**
- **Eroding clay**
- **Dust**

Along with pressure testing (covered in the next chapter), flow visualisation is one of the most important aerodynamic testing techniques. It allows you to see both the direction of airflow and the aerodynamic behaviour of that flow – as an example of the latter, you can see whether the flow is attached or separated. But before I describe the different types of flow visualisation techniques that are available to you, I want to spend a few moments talking about on-road testing.

ON-ROAD FLOW TESTING
Having the ability to aerodynamically test cars on normal roads is of huge benefit to anyone wanting to modify or assess the aerodynamics of their car. However, people unused to this type of testing are often sceptical at first of its practicality and what can be achieved, so let's look at some typical questions.

Speed?
Much on-road aerodynamic testing can be done at quite low speeds. For example, the wool-tuft testing shown in this chapter was completed at speeds of 60-80km/h (35-50mph). This is fast enough that the global airflow pattern becomes visible, and is also fast

Figure 3-1: Many people think of airflow past a car as looking like this stylised graphic, but that is definitely not the case! The flow visualisation techniques covered in this chapter will allow you to see what is actually happening. (Courtesy Honda)

FLOW VISUALISATION

enough that minor crosswinds will have little impact on the patterns that you can see. Clay testing is done more quickly at slightly higher speeds, but they are not necessary to get good results. For some on-road aerodynamic testing (eg coast-down testing and downforce/lift measurements), it can be an advantage to do testing at much higher speeds, but for flow visualisation, those high speeds are not needed.

Location?

Any road on which you can drive your car can be used for flow visualisation testing. For example, if you want to see the flow pattern over the rear window of your car, wool-tuft it and then go for a drive, occasionally glancing at the tufts in the rear vision mirror. If you want to see the airflow pattern on the boot (trunk) lid, wool-tuft it and then get a friend or your partner to drive by you as you stand by the side of the road, observing the flow pattern. Use a video or still camera to record the vehicle (more on this in a moment) and then you can examine the flow patterns at your leisure. If you want to use a chase car from which to observe or photograph the test car, you will need to have access to a (preferably quiet) multi-lane road.

Legality?

If you use common sense, you're unlikely to encounter any problems with the police. It's easy to think of approaches that *would* be illegal (like densely wool-tufting the windscreen in the driver's line of sight, or blocking other traffic as two cars drive slowly down a multi-lane road, one tracking the wool-tufted other), but in normal flow visualisation testing on public roads, you shouldn't have an issue. Also, don't forget that you need to test only the part of the car that you're interested in – and that can often be done quite unobtrusively.

Validity?

Flow visualisation (eg by wool tufting) has been used for almost a century in both wind tunnel and on-road use. The most revered of car companies have used tufting to reveal flow patterns – from Lamborghini to Mercedes to Volkswagen. (And that's not even mentioning how tufting has been used in aircraft development.) Flow visualisation techniques are powerful and effective. And if you're wondering how flow visualisation on the road compares with the wind tunnel, road testing is actually *better*! Why? Well it takes into account real-world atmospheric turbulence, the influence of other cars, and doesn't have problems caused by wind tunnel blockage factors or wind tunnel boundary layers.

Cost?

One of the best aspects of on-road testing is that it costs so little. In fact, much testing will cost you only the fuel you are using and your time. Tuft testing requires that you buy a ball of wool and some masking tape (and these are likely to last you years), while clay testing can also be done very cheaply. Pressure testing on the road (covered in the next chapter) requires that you buy a gauge or electronic instrument, but again this will last you forever and if purchased secondhand, won't cost much.

Downsides?

And the downsides of on-road aerodynamic visualisation? Most testing can be carried out only in fine weather with good visibility, and when wind speeds are low (eg less than about 15km/h, or 9mph). In addition, a lot of on-road testing requires two people – one to drive and one to observe or photograph. If you are using a tracking car, you will also need an additional car and driver.

VISUALISATION TECHNIQUES

As its name suggests, flow visualisation techniques allow you to see the airflow. Common flow visualisation approaches are:

Tufting

This where small pieces of yarn (eg wool) are temporarily stuck to the surface of the car body. These tufts move under the influence of the air passing over them. In areas of attached flow, they line up in neat rows, their ends only wriggling a little. In areas of separated flow, the turbulence makes the tufts whirl around, wriggle or even stand vertically – a random, high-speed motion. Tufting is easily done at low cost by anyone who wants to be able to see the airflow characteristics of their car.

Figure 3-2: A wool tuft testing kit! Not much to it, is there? However, the use of tufts can tell you a huge amount about the airflow pattern over your car.

SPEEDPRO SERIES

Figure 3-3: Mercedes factory picture of flow visualisation by tuft testing in a wind tunnel. Note the use of red tufts for the body and white tufts for the glass area: an approach that works well on most cars. (Courtesy Mercedes)

Smoke streams
This is most often seen in wind tunnel photographs. Smoke is produced and emitted from a wand that is placed so that a thin stream of smoke flows over the areas of interest. Smoke testing can be used to show separated and attached flows, and smoke can also be injected into the wake to show its size. Smoke testing on the road is difficult. Professional aerodynamicists have often suggested to me that while wind tunnel smoke testing looks good, it actually is not very effective as a development tool.

Figure 3-4: Mercedes factory picture of the big W126 S-class – this body shape was first released in 1979. Drag coefficient was 0.36. In this picture disturbed flow can be seen most distinctly before and after the front wheelarch, and behind the A-pillar. Note the car is not fitted with a right-hand rear vision mirror. (Courtesy Mercedes)

Figure 3-5: This view of the W126 shows substantially worse flow on the front door glass than in the photo above – the result of the door mirror. Flow separates at the end of the roof but reattaches earlier on the boot lid than on the smaller W123 Mercedes also shown in this chapter. Note how the flow reattaches sooner on each side of the boot lid than along the centreline of the car. (Courtesy Mercedes)

FLOW VISUALISATION

Figure 3-6: To an extent, smoke-stream pictures taken along the centreline of the car tend to make all cars look good! However, note how the smoke trace that flows up the windscreen is just visible leaving the trailing edge of the roof, meaning the rear glass is not in attached flow – something more clearly able to be seen on the wool tuft pictures. (Courtesy Mercedes)

Figure 3-7: Smoke flow around the A pillar of this Chevrolet Cruze shows that some separation would be occurring on the front door glass. (Courtesy GM)

Figure 3-8: Smoke being used to fill the low-pressure wake behind a car in the wind tunnel. (Courtesy Mercedes)

Figure 3-10: Using a smoke machine, mains power generator, plastic pipe and operator in a chase vehicle to inject smoke into the wake of an R32 Skyline GT-R. This testing was done on a hired racetrack. (Courtesy David Bryant)

Figure 3-9: This is how the smoke trails are created in a wind tunnel but doing so on the road is difficult, if not impossible. (Courtesy Mercedes)

Clay mixes

A clay mix uses clay particles suspended in a fluid. The mixture is sprayed or sponged on the car and then the car is driven, with the airflow shown by the resulting patterns formed in the mix.

There are two important aspects to this. The first is the use of clay. Clay comprises very small particles of rock, and these stay suspended in the fluid. The particles are deposited on the surface only when the fluid evaporates. Larger particles of rock (such as found in common soil) do not stay suspended in the fluid. Don't be

55

tempted to use other powdered substances (like flour); clay washes off easily, but other materials may not.

The second important aspect is the type of fluid used. Often in wind tunnels the carrier fluid is kerosene (paraffin), or some other type of light oil. When using kerosene, a few drops of oleic acid is added as a wetting agent, and either titanium oxide powder (white) or florescent dies of various colours are added to leave evidence of airflow visible at the end of the experiment. However, not many people want to spray a smelly solvent such as kerosene on their car, so for road testing, water is used instead. Clay mixes are particularly good in showing separation and reattachment areas. A variation on a clay/water mix is to spray a film of water onto the area of interest until it is damp. The area is then dusted with clay that sticks in a thin film to the wet panel. More water is then sprayed over the clay and the car is driven. The process is repeated until the airflow pattern is revealed in the clay. I'll cover this technique, which I call 'eroding clay,' in more detail below.

Dust

If you live in an area that has dusty dirt roads, dust can be readily used in flow visualisation. It is particularly good for showing the size of the wake. A car driven along a dusty road, with the sun low in the sky behind it, will have a trailing dust

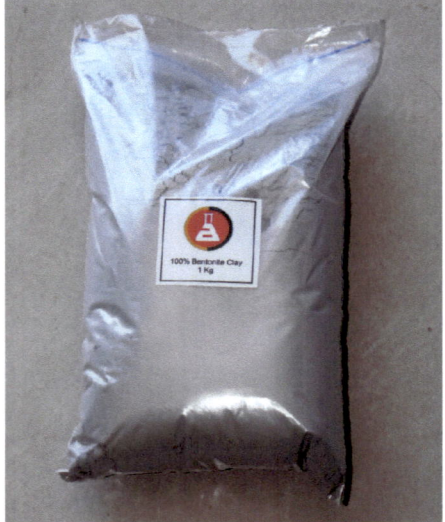

Figure 3-11: A fine clay like this bentonite (sold for cosmetic face masks) can be used to show areas of attached and separated flow.

Figure 3-12: What it looks like after the bentonite clay dust has been blown onto the damp surface – in this case, a front door window. Note its even spread. Wetting the clay further and then driving the car causes the clay to differentially erode, revealing the airflow pattern.

Figure 3-13: It may be a rally car but the shape of this Hyundai i30 is completely standard. Note how clearly the wake can be seen, and how the angle of the rear hatch extension does not decrease its size. The height increase in the wake about half a car length behind is interesting. Videoing and looking at the images frame by frame allows you to see instabilities in the wake.

Figure 3-14: Note the dust deposition pattern down the rear corner of the car – clean separation is not occurring. If it was, there would be a clear transition line between 'wake' and 'no wake.'

FLOW VISUALISATION

Figure 3-15: This Toyota Camry shows a very interesting dust pattern. Note the sharp edge on the tail-light (designed to promote clean separation) but how the separation that is actually occurring (evidenced by soft-edged dust line) doesn't actually match the tail-light's sharp edge!

Figure 3-16: In contrast with both the Hyundai and the Toyota, this Volvo XC90 shows a very well-designed separation edge, formed by the change of shape of the long light. You can see this abrupt transition in the dust signature, especially on the edge of the light. Note that the edge does not continue in the styling below the light – perhaps all this area was in disturbed flow from the wheel, small fender flare and mudflap, anyway.

cloud that shows the height and shape of the wake. In some cases, trailing vortices can also be seen. These are best photographed (or videoed) from a distance by use of a telephoto lens.

Another approach is this: the car is cleaned and then driven for some time on a dusty road. Dust will be deposited on the rear surfaces of the car that are located within the wake. (This process can be hastened by spraying the back of the car at intervals with a light water mist.) Separation patterns can sometimes be seen in the dust deposits. This is similar to using a clay/water mix (as described above) but in this case, the dust accretes rather than erodes, and the dust is provided free of charge and effort!

(I am told that snow depositions on the back of the car can sometimes also show similar patterns – but there's not much snow where I live, so I have never tried it.)

Figure 3-17: Dust revealing the wake of two motorcycles, travelling close together. Note how the frontal area of the wake of each motorcycle is about double the frontal area of the machine and rider. (Courtesy Ducati)

SPEEDPRO SERIES

Figure 3-18: This is not a motorsport book, but this photograph is perfectly at home in a chapter on flow visualisation. The rear wing creates an up-wash of air, so causing the development of downforce. (Courtesy Mercedes)

> **TUFTS ON LONG LENGTHS OF TAPE**
> If you want wool tufts that are stuck to a long length of tape, you can take this approach. Select a flat former (eg wood or plastic) about 150mm (6in) wide and a metre (yard) long. Wrap the wool around the length of the former in a spiral, keeping the gaps between the wool about 75mm (3in). When the spiral is complete, place tape along the length of the former, one side on the front and one side on the back, sticking to the former and the wool. Now if you cut down one side of the tape, you'll end up with lots of wool tufts placed on the tape an equal distance apart and of the same length.

WOOL-TUFTING

Let's say that you have decided to wool-tuft test your car for the first time. How, exactly, do you go about it?

First, decide on how you are going to record what you see. Unless you are performing a test where you can watch the behaviour of the tufts for an extended period (eg you've tufted the rear window and you are going to observe them in the rear vision mirror), it is best to photograph or video the behaviour of the tufts. It is possible to use a small camera positioned outside the bodywork (eg on an arm) but the influence of this equipment on the airflow patterns can be significant. Therefore, I prefer to use either a chase car, or photograph from the side of the road as the test car passes.

If you are going to photograph the test car from another car that is tracking it, almost any still or video camera will be adequate. You simply sit in the chase car, wind down the window, point the camera at the test car and shoot. Ensure that you fill the frame, so that you'll be able to see what is happening with the tufts without having to enlarge or zoom-in too much on the final images.

However, if you are going to photograph the test car from the side of the road as the car drives past, the camera requirements are more complex. For this application, the camera that you are using needs to have sufficient resolution that you can enlarge or crop – and so subsequently zoom in on interesting detail that you find when viewing the images. This is especially the case because you'll probably not be able to neatly fill the frame with the car as it goes past at 60km/h (35mph) or more.

In the case of a still camera, you also need to use a long focal length lens (eg the 35mm equivalent of at least a 200mm lens) and select a sufficiently high shutter speed to freeze the action – I typically use 1/1000th of a second. At high shutter speeds, you're likely to need a large lens aperture, so be careful that depth of field doesn't become too shallow. (And if none of that makes any sense to you, simply go out with whatever camera you have and try taking photos of cars driving past. The photos will need to be sharp and well-exposed.)

You will also need some yarn that contrasts with the colour of the car. I use wool, but most types of yarn will work fine. When considering colour, don't forget that if the tufts are to show up clearly, the glass areas of a car might require the use of a different colour tuft. Having available two balls of wool – one red and one white – will cover most situations.

To make the tufts, wind the wool around three fingers of your hand 30 or 40 times. Use a pair of scissors to cut off the ball of wool, then carefully cut through the wrapped wool. You should end up with lots of lengths of wool, all about 75-100mm (3-4in) long. Repeat the process to get the number of tufts you need. (To do a full car you will need about 300

FLOW VISUALISATION

Figure 3-19: The front three-quarters view of the medium-size 1979 W123 Mercedes. Flow is attached on the bonnet, after passing from the grille area and headlight. It is also attached on the front part of the guard (fender). However, directly behind the slight flare and mudflap, the flow separates and remains separated until part way along the lower front door panel.

Figure 3-20: Separation can also be seen around the A-pillar and side mirror. As a result, the flow across nearly the whole of the front door window is disturbed. Note also that there appears to be a separation bubble at the middle of the windscreen near its base. (This matches the pressure testing results we found on this car, covered in Chapter 4.) It was noteworthy that the driver said that here she could see that lower tuft whirling.

Figure 3-21: Flow separates at the end of the roof, although it appears to almost (but not quite!) reattach on the trailing edge of the boot lid.

Figure 3-22: Note that airflow is wrapping around the C pillar but it then separates, probably forming a large trailing vortex at this point. The claimed C_D of the car is 0.36.

SPEEDPRO SERIES

Figure 3-23: This Ford Escort shows a front-end flow pattern that is unusual in more recent cars. Note the separation that occurs around the sharp corner from the headlight onto the side panel and onto the bonnet.

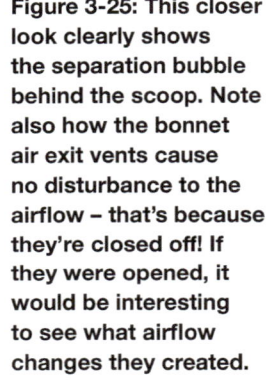

Figure 3-25: This closer look clearly shows the separation bubble behind the scoop. Note also how the bonnet air exit vents cause no disturbance to the airflow – that's because they're closed off! If they were opened, it would be interesting to see what airflow changes they created.

Figure 3-24: This 1999 Subaru Impreza WRX has been fitted with a front spoiler extension and an STi-spec rear wing, which is taller than the standard item. There is attached flow across the bonnet of the car – something common on most cars. (Note that if the flow were turbulent here, the amount of air passing into the intercooler scoop would be a lot less.) On the upper part of the guard (fender), the flow is also good (the single tuft able to be seen crinkled has been caught up on itself by a thread of wool). However, if you look closely, you can see something strange going on behind the intercooler scoop.

Figure 3-26: Certain types of wool fray very well when the tufts are exposed to turbulence. As can be seen here, even when the Impreza is stopped, this tuft tells the story of the behind-scoop turbulence.

Figure 3-27: The front corner of the Impreza. There is no obvious separated flow (remember that tuft on the guard is stuck in that position). It is interesting how the bumper and spoiler lip extension directs flow around the corner, rather than onto the upper surfaces of the car. Note also the three tufts showing flow through the radiator grilles.

tufts, or about 200 if you do only one side and the horizontal surfaces.)

Next, you will need some tape to stick the tufts to the car. Paper-based masking tape is good. This tape is designed for sticking to the paintwork of cars without then pulling the paint off when it is removed. It can also be easily ripped into short lengths, something that takes much less time than cutting off each piece of tape with a pair of scissors. If the car is heavily waxed and the masking tape will not stick, try good quality electrical tape.

FLOW VISUALISATION

Figure 3-28: This view of the back of the Impreza shows that, while the flow is attached across the roof, it separates only a short distance down the rear glass. On each side of the car this separation is further downward, and along the car's centreline it is closer to the roof. So does the airflow reattach on the boot lid? It's a little hard to see, but it looks as if the air stays separated right across the boot lid – some tufts can be seen whirling around in about the middle of the panel. However, what can also be seen is that there is attached airflow both sides of the tall wing – it's working fine. (Or, if it isn't, it's only because its shape is wrong, not its position.) So my guess is that a smoke stream along the car's centreline would leave the rear glass a little way down from the top and then pass beneath the wing, perhaps just reattaching itself at the very trailing edge of the boot.

Figure 3-29: This rear three-quarters view by artist Dave Heinrich best shows what's going on with the Impreza. This drawing was based on all the wool-tuft photos taken, including many not included here. Note the attached flows across most of the leading panels, with turbulence confined to three distinct areas: the (aftermarket) wheels, behind the rear vision mirror, and just behind the front wheel. The flow across the C-pillars and trailing edge of the roof is attached, but it separates only a very short way down the rear window, creating a major area of turbulence across the rear window and boot lid. However, the tall wing is sufficiently out of this turbulence to be in clean air, as can be seen by the attached flow across the aerofoil. Excluding the wing, the wake behind the WRX can therefore be estimated as being the full width of the car, with its height a little greater than the upper edge of the boot lid. Not shown here are the trailing vortices likely to be being shed by the C pillars. An independent Japanese wind tunnel measurement of the MY99 Impreza WRX indicated a C_D of 0.36. (Courtesy Dave Heinrich)

Stick the tufts to the area of the car where you want to be able to see the flow pattern. Keep the tufts far enough apart that they cannot whirl around and adhere to each other. (As you will find out after the first test, this is especially important in areas of separated flow, where the tufts can dance in all sorts of directions.) Remember that it's easy to remove and re-apply tufts, so in your initial test use a smaller number of widely-separated tufts. Once you've seen areas you'd like to explore in more detail, add (or shift) more tufts to those spots. Some people like to put the tufts in neat lines, but in terms of testing, it doesn't really matter.

To save time, it's normal to work on only one side of the car, on the basis that the car airflows are symmetrical around the longitudinal centreline. In some cases, however, this is not the case, so if you are exploring in detail an aspect like rear-end separation and flow reattachment, tuft the whole area. To wool-tuft half a car and take some photographs or video from the side of the road is quite a quick procedure. Doing so will take a few hours – that time includes applying the tufts, photography, and then removing the tufts. Don't leave the tufts in place (especially if the car is in the sun) as the tape will become hard to remove and may leave marks.

When you are interpreting your video or photographs, there are some important aspects to keep in mind. As already described, in areas where the flow is attached, the tufts will line up in the direction of the airflow. Thus, attached flow, whatever its direction, will be able to be seen. For example, air that is wrapping around A or C pillars can be identified. Flow that is separated will be indicated by surface turbulence, with the tufts whirling around and pointing in random directions.

If you are using high shutter-speed photography, you need to

SPEEDPRO SERIES

Figure 3-30: The first (early 1990s) Lexus was an impressive car from an aerodynamics perspective, with its conservative lines concealing a quoted C_D of 0.28. With its large frontal area, a top speed of around 245km/h (152mph) from only 190kW (255hp) showed that the C_D must indeed have been quite low. This view shows attached flow from the roof down onto the rear window, and flow wrapping-around the C pillar. However, it also shows a separation bubble on the central lower part of the rear window and boot lid. In this view, it is hard to see if the flow reattaches on the trailing edge of the boot lid.

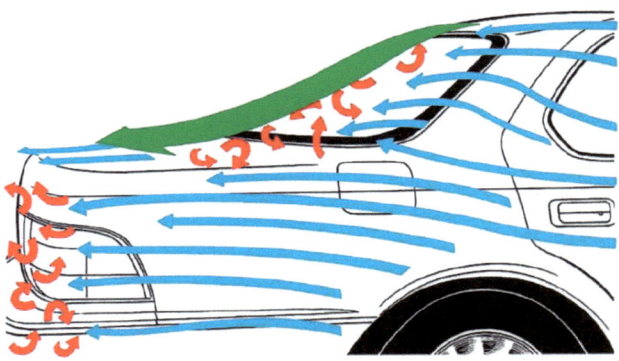

Figure 3-32: In this Dave Heinrich view, the flow paths have been drawn, based on the information gained from the tuft testing. Not shown here are the trailing vortices likely to be being shed from the C pillars. The rear window angle of the Lexus (27 degrees to the horizontal) and angle of the line drawn from the trailing edge of the roof to the trailing edge of the boot (18 degrees) shows the designers paid attention to rear-end flow reattachment.
(Courtesy Dave Heinrich)

Figure 3-31: In this photo, flow reattachment on the rear half of the boot lid can be clearly seen. Note how closest to the camera, the flow is wrapping around from the side panel onto the rear panel (ie there's not clean separation line on a vertical plane) and the created low pressure is dragging a tuft down from the boot lid surface.

Figure 3-33: Flow separation is occurring immediately behind the front wheelarch and the flow is also being disturbed behind the mirror. Note the tuft placed on the centre of the alloy wheel – all tuft photos of the Lexus show these tufts behaving well.

be careful to *look at a sequence of photographs*, not just one. A single shot might show a tuft pointing in a particular direction, but the adjoining photos in the sequence might tell a different story. Close-up video shows these sorts of subtleties well.

As described in Chapter 1, flow can separate and then reattach, forming a separation bubble in-between. In wool tufting, this is indicated by a tuft progression that, as you move downstream and along the car body, looks like the following: (1) parallel tufts, fluttering at their ends only a little, (2) random motion tufts, (3) parallel tufts, often with a little more tail movement than before. Remember that attached airflow wrapping around curves will create a low pressure on that surface; this can be resolved into lift and drag forces by looking at the resultant vectors (see Chapter 1).

FLOW VISUALISATION

Figure 3-34: The front of this 1997 Porsche 993 Turbo is covered in attached flow. Of course, given that the classic 911 shape has probably spent more years in a wind tunnel than any other generic car shape, perhaps that's not surprising! But it is especially interesting to see how the flow remains attached on the top of the guard (fender), even directly after what appears to be the sharp transitional change from the headlight to the upper surface.

Figure 3-35: A close-up shows more clearly the attached flows around the nose. Especially impressive is the attached flow on the side of the headlight closest to the camera – often in older car shapes, that tight transition around the corner will cause a separation bubble (see Figure 3-23). It is hard to see, but the wool tufts show a flow of air straight past the guard opening – in many cars, the wheel-well is a low-pressure area and you'll see the tufts being sucked into the opening.

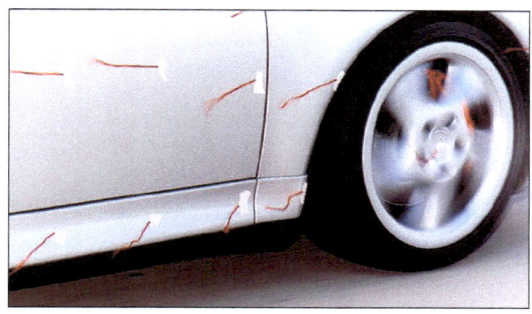

Figure 3-36: There is some separation directly behind the flared front wheelarch. It's interesting to see the airflow passing from the side of the car to its underneath area, despite the wide side sill panel. This flow may be promoted by the Porsche's flat undertray creating a low-pressure area.

Figure 3-37: This picture shows a number of interesting aspects of the 993's airflow. Despite its sophisticated shape, behind the rear vision mirror there is quite major turbulence, which can be seen by the tuft behaviour on the door glass. Also, the airflow

wrapping around from the door window, past the gutter and towards the roof does not stay attached – look at the two tufts standing almost straight up! However, the flow from the windscreen onto the roof remains attached, while quite oddly the airflow on the glass behind the exposed windscreen wiper doesn't look turbulent in any of the photos (it's wise to make a judgement based on several shots as the camera can capture a tuft in an atypical position) – instead it always indicates an airflow parallel with the wiper.

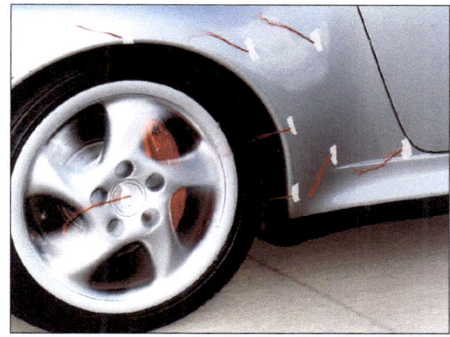

Figure 3-38: Further back along the side of the Porsche, the smoothly curved bulge of the huge rear guards (fenders) keeps flow attachment across their transitional shape. Note also that in all the photos (including many not reproduced here), the behaviour of the wool tufts positioned on the centre of the front and rear wheels show smooth flow across their front faces.

SPEEDPRO SERIES

Figure 3-39: The flow onto the huge rear guards and across the wheels may be good, but once those guards start curving inward, towards the centreline of the car, the airflow immediately becomes separated. In fact, the amount of turbulence that can be seen on the side portion of the rear bumper and just forward of the tail-lights is a little surprising – and it was consistent and strong. But what about the upper surfaces? As can be seen in this photo, the flow across the upper portion of the guard looks very good.

Figure 3-40: The flow on the roof is attached, but the transition to the rear window is poor. The added high-level brakelight causes a whole line of separation across the top of the back glass, before the flow reattaches itself. Look at the tuft in the upper middle that is actually being blown forward!

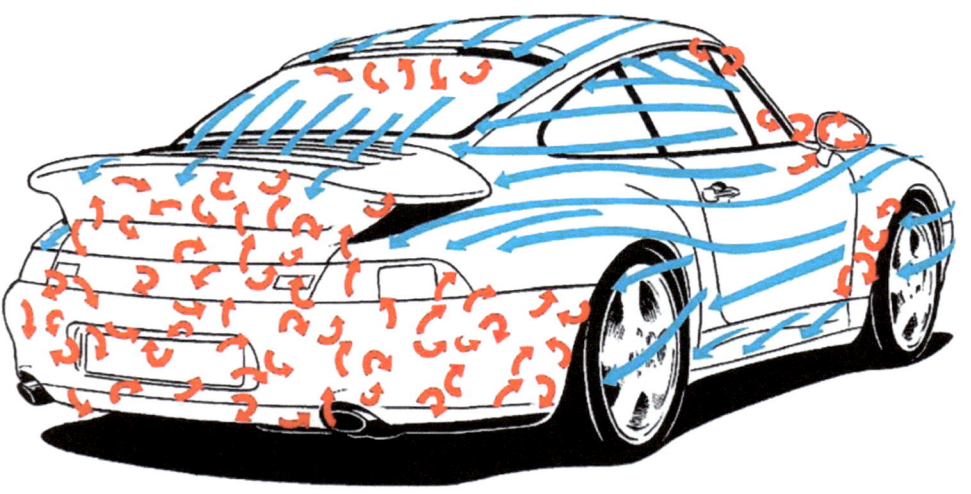

Figure 3-41: In this Dave Heinrich view, the flow pattern over the rear three quarters of the car is shown. The turbulence around the A-pillar (still equipped with an exposed gutter) and behind the front wheel are two blemishes on the flow over the front panels. The added high-level rear brakelight moulding apparently never made its way into the wind tunnel, while the flow into the louvres on the engine cover is massive (the engine cooling fan specification is 1210 litres/second at 6100rpm!). As can be seen, the wake extends the full lower width of the car – the narrowing rear bumper doesn't keep the flow attached at all behind the rear wheels. Note also the unusual flow pattern across the wheel openings, where the flow is not drawn into the wheel-wells. The 993 twin turbo has a factory quoted C_D of 0.34. The drag coefficient probably benefits substantially from the car's full-length flat undertray. (Courtesy Dave Heinrich)

FLOW VISUALISATION

Figure 3-42: With a factory quoted drag figure 0.39, the 1997 Beetle had an appalling C_D for the time. But couldn't they have done better than that? Well, in this case, having to hold true to a certain historical shape caused major constraints on the aerodynamic design.

Figure 3-43: As we've got used to seeing in these wool-tuft photos, the flow across the bonnet is attached (that wriggled tuft is just stuck in that position). The flow up the windscreen and onto the roof is also fine, as it is down the side of the car (directly after the bulging guard excepted). Even the flow from the windscreen around onto the side glass past that sharp-edged A-pillar isn't bad at all. So what makes the C_D figure so poor?

Figure 3-44: Well, it's not the side flow causing the problems – it's generally nicely attached. Even the flow directly behind the greatly flared front guard (fender) reattaches quite quickly.

Figure 3-45: So, the problem must be at the back? And it is. But it's a different problem to that we've seen before. Here the flow remains attached right down to the line of the guard/boot opening. And this results in a very small wake for the size of the car. But it also means that the airflow wraps in one long curve from the base of the windscreen right around over the top of the car to nearly the rear bumper. Not only will this shape have major lift, but much of the force will be upward and rearward – creating that high drag. In this view, the size of the small wake can be clearly seen. But also visible is that attached flow from the roof all the way down to the bumper level.

Figure 3-46: This diagram shows a classic airflow pattern for high lift – and so a great amount of lift-induced drag and probably very strong trailing vortices. This car, therefore, has both a small wake and high drag. (Courtesy Dave Heinrich)

Figure 3-47: This V6 Beetle RSi uses a huge rear spoiler and flow-directing upper window attachment to give 77kg (170lb) of downforce at a massive 250km/h (155mph). The C_D of this version of the car was 0.38, a little lower than the standard vehicle. (Courtesy Volkswagen)

SPEEDPRO SERIES

Figure 3-48: The Mazda RX7 has attached flow over the nose of the car, which as the photos in this chapter show, is quite common. Note the flows into the radiator and brake cooling ducts, and the smooth way in which the air passes from the front to the side of the car (below the side indicator). You can see also that the flow is being drawn into the wheelarch, with the slightly frayed tuft showing some turbulence at this location. (In some other photos this isn't happening – it may depend on airspeed.)

Figure 3-49: This view shows some of the same areas – but it also shows a major patch of turbulence behind the front wheelarch, especially over the vent. The turbulence surrounding the vent extends over a relatively large area. However, note the excellent flow patterns around the rear vision mirror – unlike most other cars, the Mazda has well-controlled airflow in this region.

Figure 3-50: And the back of the car? In this photo it can be seen that there is attached flow from the roof down onto the rear glass, with flow wrapping around the pillar. Other photos show that this attached flow extends across the upper surface of the hatch as far as its trailing edge. If the (aftermarket) wing were to be removed, the wake would consist only of the area below the trailing edge of the hatch, to the full width of the car. Surprisingly, there is some flow separation on the lower part of the rear bumper at the side of the car – perhaps the much larger than standard aftermarket wheels are contributing to this turbulence – but note that as with all cars of this era, there's also not a sharp edge to promote separation.

Figure 3-51: This Dave Heinrich view summarises the flow patterns that we found on this particular car – elevated aftermarket rear spoiler and all. (Courtesy Dave Heinrich)

FLOW VISUALISATION

Figure 3-52: This photo of the RX7 in Mazda's wind tunnel is very exciting – even though it's in B&W and a pretty poor photo – because of what it shows. Look very carefully at the wool tufts and you'll see that their behaviour duplicates what was found in the road testing. Specifically, note the attached flow over the bonnet and windscreen, the clearly demarcated area of turbulence over and around the vent behind the front wheel, and (as much as can be seen from this angle) the attached flow wrapping onto the rear glass from both the roof and side of the car. (Courtesy Mazda)

Figure 3-53: Flow pattern over a Mitsubishi Magna. Note particularly the excellent flow past the door-mounted mirror. (Courtesy David Bryant)

Figure 3-54: Pop-up headlights can cause major flow problems in their raised position. Here is the airflow pattern with the headlights retracted …

Figure 3-55: … and with the headlights raised, when top speed dropped by 10km/h (6mph).

Figure 3-56: This dust deposit was achieved quite quickly by spraying a mist of water on the rear of this W211 Mercedes. Note the clear separation achieved by the sharp edge on top of the boot lid, and how this edge extends on the panel above the tail-lights. It is surprising that Mercedes didn't extend the edge vertically down the tail-light – and possibly even then down onto the bumper. Note also the lighter layer of dust deposited behind the wheel, especially on the bumper cover behind the slight wheelarch flare.

SPEEDPRO SERIES

SMOKE STREAMS

I have tested with smoke outside of a wind tunnel. I used a commercial smoke machine, a long plastic pipe and a mains-powered generator – all placed in the back of a utility-type vehicle that was tracking the test car. (This was done on a racetrack hired for the purpose.) It worked especially well in identifying the area of the wake.

However, testing on a public road using smoke is difficult in terms of practicalities and preventing the vehicle carrying the smoke apparatus from influencing the airflow over the test car. It is also hard to see how it can be carried out with negatively impacting other road users and so being illegal.

However, sometimes luck can be on your side. I once talked to a man who was one of the university team that built a solar racing car. On a trial run, they were on a freeway well away from cities, and met the thick cloud of smoke from a bushfire. He told that me that from the chase car, they could see the smoke smoothly flowing around the ultra-low drag body of the solar racer, the streams coming together at the top and bottom of the long tail …

ERODING CLAY

I am now going to describe the process where a clay film is applied that is then differentially eroded by the airflow. This technique can show very fine details in the airflow over your car.

To use this technique, you will need to have some fine clay powder. The clay powder that I use is bentonite, sold for use as cosmetic face-masks. In small quantities it can be very expensive, but if you buy it in bulk (for example, 2kg (4.4lb) bags) it doesn't cost much. In addition, you will need a spray bottle of water. Quite a lot of water is used

COMPARING TUFTS AND ERODING CLAY

I did some testing on my 1979 W123 Mercedes, allowing the results of tuft testing to be compared with eroding clay. The testing was of flow separation on the front door windows. The Mercedes is a good car to use because there is clear flow separation behind the full height of the A pillar, not just aft of the door mirrors.

The photo above shows the mess that is the airflow over the Mercedes front door window. Not even one tuft on the window is pointing towards the rear of the car! One tuft is pointing forward, a number of tufts are pointing upward, and some tufts show the contortions indicative of turbulence. Clearly, there is lots of separated flow over the window, but how can we get to see things in more detail – like, where on the window, exactly, does the flow reattach – if at all?

Here is the pattern revealed by the eroding clay technique. At least three different zones of airflow can be seen. The first zone is directly behind the mirror, where a small area of clay has been removed. The second zone, a large one, is where the clay remains. This zone, that shows flow separation, extends all the way up behind the A pillar (so indeed it's not just the mirror that's causing the disturbance) but is most intense towards the bottom-left. A section of this zone extends along the base of the window. Zone 3, where there is attached flow, exists at its strongest along the top of the window, but there are also other areas of this zone lower down and towards the right.

FLOW VISUALISATION

I then re-tufted the window, using a greater number of shorter tufts (and swapping to white tufts to make them more visible). This photo shows that the left-hand third of the window (behind the A pillar) is in separated flow, and that the separation is strongest at bottom-left – as we found with the eroding clay. However, the tufts appear to show reattachment at about the middle of the window, rather than further to the right as the clay showed. Note the behaviour of the top tufts on the glass, which match the result of the clay testing.

This photo is taken from inside the car (the image has been flipped to allow easy comparison with the external shots). This photo uses flash mixed with a longer shutter speed, a technique that allows you to see how much a tuft is moving. The three zones mentioned in the analysis of the eroding clay can all be seen – Zone 1 (in the wake of the mirror) where there's little airflow, but what there is, is turbulent; Zone 2 (separated flow) where the movement of the tufts is great; and Zone 3 (attached flow), where the movement of the tufts is much less. Note in this shot, the top tufts are wrapping around onto the roof; looking at all the photos, these tufts either did this or pointed towards the rear of the car – but they never showed evidence of separated flow.

From this comparison you can see that the tufts give you a good sense of general airflow, but the eroding clay technique allows you to home-in on specific areas where separation and attachment are occurring. For example, if I were to make a moulding that smoothed air past the A-pillar (this picture shows the depth of the A-pillar rain gutter – it's huge!) then the clay technique would be very effective in showing the reduction in area of disturbed air. Flow changes caused through mounting a new mirror on a short support (so allowing air to flow between the mirror and door) could also be tested using the clay technique.

during the testing, so bring along a container from which you can refill the spray bottle. You will also need at least an hour and a location where you can drive reasonably quickly – 80km/h (50mph), minimum, and faster is better – but also safely pull off the road at frequent intervals. The bentonite clay powder I use dries to a light (almost white) colour, so if you wish to see airflow patterns on a white car, it won't work – you'll need a different colour clay. On glass areas, or on dark coloured cars, bentonite works well.

The first step is to spray a fine mist of water over the area that interests you. Don't spray so much water that running drops are formed on the bodywork. You may need to test on another panel first until you get the spray adjusted correctly and have a good technique. When the panel is wet, dust it with the clay powder.

To dust the panel with the clay powder, I use this approach, that requires two people and a compressed air hose with a nozzle. One person uses a kitchen sieve positioned above a tray, with some clay powder in the sieve. The person with the sieve shakes it so that a cloud of clay powder descends beneath the sieve. The cloud of clay dust is directed onto the panel by the action of the air nozzle, wielded by the second person. The dust that doesn't get pushed onto the panel by the air blast lands in the tray and can be re-used. In this way, an even layer of clay dust is deposited on the panel. A bit of a mess is made, so do this outside. (If you don't have a compressor, a strong fan could be used instead.)

After the clay has been applied to the car, spray it with water until it is wet, but not so wet that it runs. Immediately drive the car until the water has all dried. Stop the car, and

69

spray again, before again driving. The technique needs patience – it's easy to decide that nothing is happening, or that you should try to speed-up the process by spraying *lots* of water. However, don't do that! After about 15 minutes, you will see a pattern forming that differentiates separated and attached flows. The clay particles are more likely to remain in areas of separation and be removed in the areas of attached flow. Continue the test until you can clearly see the airflow pattern. When the test is complete, photograph the result and then wash the clay off with normal car washing equipment.

DUST

There are two ways in which to use naturally-occurring dust on dirt and gravel roads. To use either approach, you will need to either live in an area that has such roads, or visit one. In addition, the dust will need to be dry and, in most testing, also be a colour that contrasts with the paintwork of the car you're interested in.

The first approach is to simply look with curiosity whenever you see a dirty car. I am lucky enough to live in an area with plenty of dirt roads, and I often find myself intrigued with car dust patterns. Looking at the side of a car that has been driven on dusty roads (and started out clean) will show you where there is separated flow – the dust will be deposited in those areas. Especially where the wheelarches are slightly flared, or mudflaps are fitted, you will see these characteristic deposition patterns in the dust along the car sides.

However, it is at the back of the car where the most exciting views can be found. Dust shows with quite great precision the base area of the car (ie the area in contact with the wake), especially on cars where there is attached flow to the rear of the car. For example, square-back cars (hatches and wagons) will have clearly defined base areas, where the vertical side and roof separation points can be examined in detail. Many cars have quite poor separation, where you can see the airflow is wrapping around curves before separating. Often, the dust pattern shows that the separation line has been moving back and forth over a distance of perhaps 50-75mm (2-3in).

The second approach is to do specific testing using this dust. Wakes can be photographed quite effectively on dusty roads, using two different techniques.

The first approach is to locate yourself parallel to a dusty road, but at least 20-30 metres (yards) away from it. The sun will also need to be low in the sky and be positioned behind the road that you're facing. When a car drives along the road, you will get an excellent view of the dusty wake, back-lit by the sun. By doing this you will gain a good idea of the top and bottom extremities of the wake. The top extremity will show very clearly and, if there is plenty of airflow under the (smooth bottom) car, you should also be able to see the lower extremity of the wake. (The latter is made a bit more difficult because the rear wheels will also be kicking-up dust, and this will tend to obscure this bottom view.) Furthermore, you will also be able to see the patterns in the wake well behind the car. For example, these have the potential to show trailing vortices. (However, I should not overstate this; a dusty wake certainly doesn't show them like a textbook diagram!)

The second approach is to photograph a car travelling directly towards you on a dusty road. Again, it will help if the car is backlit. The width and height that the wake adopts will be visible: typically, it will be wider and higher than the car developing it. The width is especially important to note, as modern car aerodynamic optimisation involves quite a lot of effort in reducing the width of the wake.

Another way you can use the dust for testing the cross-sectional area of the car panels in contact with the wake is to add water to hold the dust in place on the panels. Clean the car, then spray the rear of the car with a fine mist of water. Drive the car on the dusty road, spray with some more water, drive again – and so on. Unlike the eroding clay technique covered above, this time material is building up rather than being removed – but otherwise the technique is similar. Continue the test until you have a clear indication of the rear separation zones. Sometimes, if the colour of the dust isn't sufficiently in contrast with the paint colour, using a photo manipulation program to increase the photo contrast will bring out the pattern.

Chapter 4
Pressure measurement

- **Pressure measurement aims**
- **Measuring instruments**
- **Magnehelic gauges**
- **Pressure probes**
- **Measuring pressures with a Magnehelic gauge**
- **Step-by-step pressure measurement**
- **Using pressure measurement data**
- **Using pitot tubes to measure airspeed**

Figure 4-1: A fully-instrumented wind tunnel can be used to measure the distribution of positive and negative pressures across the surface of the car. However, such an approach is out of reach for normal people. Instead, on-road testing with cheap instruments can give excellent results. (Courtesy Mercedes)

When we are discussing concepts like lift and drag, it's easy to forget that both of these forces are developed on the car by pressures acting on the car's external surfaces. If the pressures pushing backward on the car add up to a greater amount than the pressures pushing forward on the car (and of course that is the case) then drag will be developed. If the pressures pushing upward on the car are greater than pressures pushing downward on the car, aerodynamic lift will be developed.

Therefore, measuring pressures

SPEEDPRO SERIES

acting on the surface of the car gives you an incredible window into what is actually occurring in terms of the different aerodynamic forces that are in action. In addition, flows through heat exchangers like radiators, oil coolers and intercoolers are governed by the pressure differential across them. Again, being able to directly measure these pressures is extremely useful in assessing the likely flow of air through the heat exchanger.

PRESSURE MEASUREMENT AIMS

In wind tunnels, and where cost is no object, body pressures are measured by surface taps. Surface taps comprise small (eg 1mm) holes drilled through the panel, with each hole connected by small diameter hose to an electronic pressure sensor. Depending on how the car (or a scale model of the car) is instrumented, there might be hundreds of such holes. The pressure logging can therefore use hundreds of channels, all electronically sampled at high speed.

But for someone testing their own car, there are a few problems with taking this approach. The first – and most major – is that I don't know anyone who is prepared to drill small holes all over the bodywork of their car!

(An aside. About 20 years ago I was working as a freelance journalist for a UK car magazine. The editor of the magazine had just bought a car that had been the actual vehicle extensively tested in a UK wind tunnel. Unfortunately, it had a broken windscreen and since it was an Australian car, he couldn't get a replacement windscreen locally. But he knew I lived in Australia, and so arranged for me to buy a windscreen and freight it to him. The car was a Holden VN Group A Commodore, a vehicle with an awesome bodykit that had been developed for racing homologation purposes – the reason for the wind tunnel testing. After I'd sent him the windscreen, I suggested that he'd soon be proudly driving the streets – but he pulled me up short. "Only when the car comes out of the bodywork shop," he said. "I have to get all of those holes filled first!" As a wind tunnel mule, the car was covered in small holes …)

The second issue for most people looking at emulating the approach taken in professional wind tunnels is the high-speed electronic logging aspect. With cheap electronic pressure sensors and cheap logging modules, the problem is not so much cost but complexity. At minimum you will need to be confident with electronics and have a laptop configured and running in the car. In addition, you'll need to have a time-averaged display that you can view live, so you can quickly get a feel for whether the changes are heading in the right direction or not. And while all that sounds OK, I much prefer to take a simpler approach.

What is needed is a straightforward, cheap and effective way of showing pressures that are occurring on the bodywork of the car, while the car is being driven.

MEASURING INSTRUMENTS

The pressures acting on the surface of the car are typically very small. That's especially the case if you live in a country with speed limits where testing might typically occur at 80-100km/h (about 50-60mph). Many pressures will be less than 250 Pascals (Pa) – or just 1 inch of water. Let's take a look at what that means, because this is very important.

A pressure of 1 inch of water means that the measured pressure is the same as you would find at the base of a column of water 1in (25mm) high. *That is tiny!* If you are used to expressing measurements in pounds per square inch (psi), a pressure of 1 inch of water is just 0.036psi. So you can immediately see that a normal pressure gauge – like a car engine vacuum/boost gauge – is nowhere near sensitive enough to measure these aerodynamic pressures.

Low pressures of the sort that I am describing are often measured with manometers, instruments filled with liquid (and yes, you can use water – that's where 'inches of water' originally came from). It's easy to make a manometer, especially for measuring pressures below atmospheric. Here's how.

Get a large plastic soft drink bottle, and attach a thin vertical piece of timber to it with adhesive tape. Run a clear plastic PVC tube down the timber and place the bottom end of the tube in the soft drink bottle. Half-fill the bottle with water, and place markings on the plastic tube – 1 inch above the water level, 2 inches above, and so on. Run the free end of the tube to the area where you want to measure the pressure. If the pressure at the end of the tube is lower than atmospheric, the water will be drawn up the tube. A pressure of 1 inch of water below atmospheric will show as the liquid rising 1 inch up the tube. (2 inches of water below atmospheric, the water will rise 2 inches, etc.) You can add food colouring to the water to make it easier to see.

However, such a manometer won't be very effective in a car, because unless you drive extremely quickly, you'll be trying to read movements in the water level that are tiny.

PRESSURE MEASUREMENT

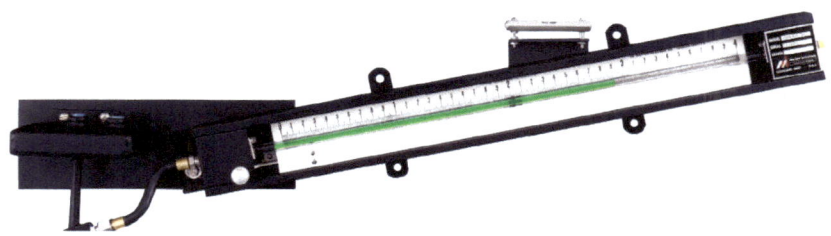

Figure 4-2: A high quality inclined manometer. An inclined manometer is more sensitive than a vertical manometer, but its use in a moving car is problematic. (Courtesy Meriam)

One way of overcoming this is to use an *inclined* manometer – one where the water is drawn up a slope, rather than vertically. The flatter the incline, the further the water will move for a given pressure change, so an inclined manometer is more sensitive than a vertical manometer. You can make your own inclined manometer (I once used an inclined manometer to measure the restriction of different secondhand mufflers, with the airflow provided by a vacuum cleaner) but normally you would buy a professionally made inclined manometer.

An inclined fluid manometer will have sufficient sensitivity (accuracy and repeatability are also typically very good) for aerodynamic car testing, but there are some negatives. Fluid-filled manometers can leak, especially when they are laid flat. For accurate measurement, they need to be held in a vertical plane when in use. Also, the fluid evaporates over time, and so needs to be periodically refilled.

So what is my suggested solution? For a do-it-yourselfer, I think that the Magnehelic gauges from Dwyer Instruments are the best way of measuring car aero pressures. They're accurate, cheap, robust and easily used in a car's cabin. They're also available in a wide range of pressures and include centre-zero models. Let's have a look at these gauges in more detail.

MAGNEHELIC GAUGES

The Magnehelic gauge was invented by Mr Dwyer himself in 1953. The gauges are large (100mm (4in) diameter), easy-to-read instruments that use diecast aluminium cases and plastic faces. Internally, the movement of a silicone rubber diaphragm is transmitted to a pointer without the use of gears or other direct mechanical linkages. This approach has some significant advantages over other gauge designs.

First, the use of a large diaphragm means that the pressure gauge can be much more sensitive than one using a traditional Bourden tube. Another reason that the sensitivity of the gauge can be so high is that the diaphragm movement is transmitted to the gauge pointer magnetically, avoiding physical contact that can also cause hysteresis (backlash) and jerkiness.

Second, the gauge is a differential design with 'high' and 'low' pressure ports. If you are measuring a high aerodynamic pressure (for example at the stagnation zone across the front of the car), you connect the pressure sensing hose to the 'high' port. If you are measuring a low pressure (eg at the transition from the windscreen to the roof), you connect to the 'low' pressure port.

Because the gauge is a differential design, you can also use it to measure the difference in pressures at two locations. For example, you can compare the pressures at the beginning and end of an alternator cooling duct to see if air will flow through the duct when the car is in motion. I am getting a bit ahead of myself, but in this situation, you would want to read a higher pressure at the mouth of the duct than at the alternator end of the duct. Using the differential function of the meter allows you to do this in one test. Run a hose from the 'high' gauge port to the mouth of the duct, and a hose from the 'low' gauge port to the area around the alternator. Drive the car – and the higher the needle moves on the gauge, the better!

Legend has it that the inventor of the Magnehelic gauge threw one down a flight of stairs to check the gauge for durability. I do not suggest that you do that, but I have dropped one of my Magnehelic gauges and it suffered no damage.

Magnehelic gauges are fairly

Figure 4-3: A 0-1 inch Magnehelic gauge. Magnehelic gauges from Dwyer Instruments are an excellent way of measuring car aero pressures. They're accurate, cheap, robust and easily used in a car's cabin.

SPEEDPRO SERIES

Figure 4-4: Magnehelic gauges are differential gauges, that is, they measure the pressure difference between their two ports. Two ports are located on the back of the gauge (as shown here) and two duplicate ports are located on the periphery of the gauge. Block the duplicate ports you are not using.

expensive if bought new. However, partly because they have been available for so long, and partly because many sellers don't seem to know what they have, these gauges are available quite cheaply on eBay. There are a few points to keep in mind when buying a gauge.

First, for aerodynamic pressure measurement, a gauge with a full-scale deflection (FSD) of 250Pa (1 inch of water) is a good initial choice. If you subsequently decide to buy a second gauge, a 750 Pascal (0-3 inches of water) FSD gauge will let you test at higher speeds.

Gauges are also available with a centre zero, and this can be very useful in two situations:

1. You are not sure whether the measured pressure will be above or below atmospheric. A centre zero gauge will show either positive or negative pressures, without the need to swap a hose.
2. The pressure that you are measuring changes from negative to positive (or vice versa) with changes in speed.

Note that most Magnehelic gauges have a legend on the face that says 'max pressure 15psig'. This means that the maximum pressure that the gauge can be subjected to as a whole is 15psi above atmospheric (this is not relevant in testing car aerodynamics). I mention this because many sellers will quote this as the gauge FSD, ignoring the much lower maximum reading on the scale. So if buying, it's best to be able to sight a photo of the scale so you can be sure of what you are getting.

Many Magnehelic gauges for sale secondhand will not have the needle pointing to zero. In the case of a normal Bourden gauge, this is a sign the gauge is defective and should not be purchased, but this is not necessarily the case with the Magnehelic gauge. The most sensitive of Magnehelic gauges are calibrated for a specific gauge orientation – normally vertical. When photographed lying on their backs, the needle might be showing more than half FSD. In this situation, check if there is writing on the faceplate that says 'calibrated for vertical position'. Even if this text is not present, minor offsets from zero

Figure 4-5: Plastic tubing can be used to connect the Magnehelic gauge to the probe and/or reference reservoir.

DIGITAL MANOMETERS

As indicated in the main text, I like using Magnehelic gauges – but is there a more modern alternative? There is!

Digital low-pressure electronic manometers are available, and while they're more expensive than a secondhand Magnehelic gauge, they also cover a wider range of pressures. For example, the meter shown here has a pressure range of ±10,000Pa (±40 inches of water) and a resolution of 10Pa (0.04 inches of water). However, note that this resolution is dependent on the units in which you are viewing the display – set to inches of water, for example, the resolution drops to 0.1 inches of water.

I compared the measurements achieved on the pictured electronic manometer versus a sensitive Magnehelic gauge, and the two instruments matched well in their readings.

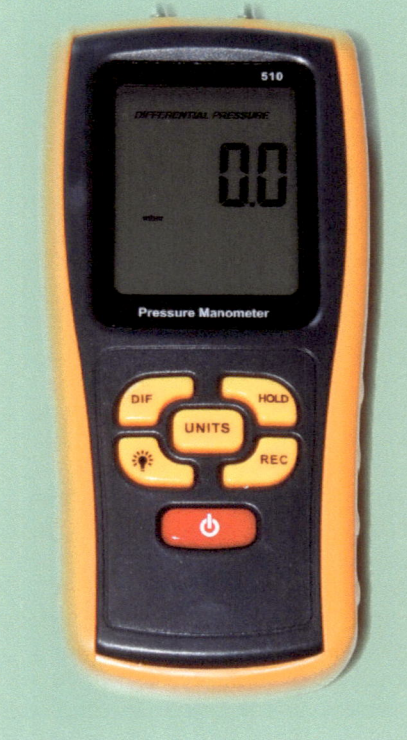

PRESSURE MEASUREMENT

Figure 4-6: To make measurements of high aerodynamic pressures (for example, across the front of the car) all you need is the gauge and a length of hose (but one that's longer than shown here!).

can be corrected with the zero set-screw located on the lower part of the faceplate.

PRESSURE PROBES

The simplest pressure probe is just the open end of the tube that runs to your pressure measurement device (eg a manometer, electronic pressure sensor or Magnehelic gauge). Position the open mouth of the tube in your area of interest, orient it at right-angles to the airflow, and hold the tube in place using tape. (Incidentally, I find that for quick testing, normal electrical tape works best. For longer term testing, use cloth-backed tape.) Such a sensing tube is typically about 6mm (¼in) in internal diameter and is made of rubber or PVC plastic.

I used this approach for years and achieved good results. For example, I developed an undertray to improve the flow of air through an intercooler located in the engine bay (see Chapter 8). The undertray worked – the lower measured intake air temperature reflected the much greater ambient airflow I achieved through the intercooler. I have also used this very simple probe to locate the best spot for engine air intakes (see Chapter 9). If you are just exploring this topic and want to get a feel for aerodynamic pressure measurement (especially across the front half of the car) then I suggest you start by using the bare end of the tube as the probe.

The 'next step up' over the bare tube end is to find a washer that is a tight fit over the end of the hose. The washer should be around 12mm (½in) in diameter. Grind or file a little less than half the width of the washer away, ensuring you leave enough material for the hole to still be properly formed. Push the hose through the hole and then glue it in place from the rear. Then, using a sharp blade, cut the hose off flush with the front surface of the washer. In use, the flat surface you've filed on the washer rests against the car bodywork. The fitment of the washer will in some cases settle readings that are jumping around a lot.

However, if you look at professional testing done in situations where drilling holes in the bodywork cannot be undertaken, a different probe design again is used. It comprises a small round disc just thick enough to take a side entrance tube. This tube connects to a centred hole drilled downward from the upper surface of the disc. Making a probe like this is fairly easy – although a little fiddly.

You will need some 25-35mm (about 1in-1.5in) diameter aluminium or plastic bar, some 2-3mm diameter brass tube (available from model suppliers) and some glue. Using a drill press, drill a sideways (tangential) hole toward the centre of the bar, starting from a flat you file on the side of the bar. Drill as

Figure 4-7: A pressure measuring disc probe like this can be made in a home workshop from small diameter brass tube and some 25mm (1in) aluminium bar. It's fiddly to make but the resulting probe is cheap and highly effective.

close as you can to the end of the bar without breaking through the end face. Ensure the hole reaches the centre of the bar (or goes a little past it). The brass tube will fit into this hole, so size its diameter appropriately to match the outside diameter of the brass tube you have available.

Now place the bar vertically in a vice, and mark the centre. Use a very small drill bit, eg 1mm (³⁄₆₄in), to make a hole axially into the bar (ie vertically downward into the end of the bar) until you reach the tangential hole you drilled earlier. Check the side and top holes are now connected by blowing through them.

Place the bar horizontally in the vice and carefully cut the end off, making the resulting disc as thin as possible without breaking through into the holes you've drilled. File or sand the disc faces square, and round the top edge. Finally, glue the brass tube in the side hole, ensuring you don't insert it fully. (A thread fixing glue works well, eg Loctite.) Depending on how good you are with your hands, and the size of the brass tube you can buy, you should end up with a disc that is only about 4mm (around ⁵⁄₃₂in) thick.

In use, the disc is temporarily

SPEEDPRO SERIES

Figure 4-8: The surface pressure measuring disc located above the windscreen in an area of low pressure. Note the small diameter hose that has been used close to the sensor, and how the tube has been orientated to reduce disruptions to the flow.

stuck in position with the hole pointing upward. A flexible plastic or rubber hose is run from the brass port to the sensor that you are using. Try to minimise the diameter of this hose near the pressure probe. While this probe design is not quite measuring the pressure of the actual body surface, it's very close.

MEASURING PRESSURES WITH A MAGNEHELIC GAUGE

As described earlier, when using a Magnehelic gauge, you run a hose to the area of interest and then connect it to the high or low port on the gauge. For example, let's say that you are measuring the pressure on the surface of the bonnet (hood), near to the front of the car. This is likely to be a low-pressure area (the airflow wraps around the leading curve), so you connect the tube to the low-pressure port on the gauge. You then drive the car at a fixed speed (eg 80km/h or 50mph), look at the gauge and note the reading.

However, remember that the Magnehelic gauge is a *differential* gauge – in this case it is comparing the pressure reading from the leading edge of the bonnet with … well, with what exactly? If the high-pressure port of the gauge is open, the *reference pressure* is the pressure in the cabin. And is the cabin pressure the same as outside atmospheric pressure? Let's say that you drop a few windows by a small distance. The airflow coming through the cabin vents will increase because you have decreased the pressure in the cabin – therefore, you will also have decreased your reference pressure. The result is that the reading of pressure on the surface of the bonnet will also change. In many cases, you can see this change in reference pressure by watching the gauge as you open and close the windows in the moving car – I told you these gauges are incredibly sensitive!

If you are making *comparisons* of pressures, rather than trying to read absolutes, this may not be an issue. Earlier I mentioned the simple 'tube end' probe that I used when optimising an undertray to improve intercooler airflow. In that testing, I just ensured that the cabin window openings were always the same amount, the cabin ventilation fan was always on the same setting, and that road speed was the same. That way, the reference pressure was constant, and so any changes I measured were in fact changes occurring on the surface of the body. (The presence of wind was taken into account by doing the runs in two directions and then averaging the results.)

However, the difficulty comes when you are not looking for changes in pressures; instead you want to ascertain whether the surface body pressure at a particular location is positive or negative. (You would do this if you wished to assess whether a particular panel was causing lift or downforce, for example.) In this case, you are after absolute values, and especially if the readings are small (eg on the rear half of the car), using the cabin as the reference pressure will not be sufficiently accurate. In this case, you need to provide another, fixed reference pressure.

A fixed reference pressure can be provided by connecting a sealed canister to the second port of the gauge. I chose to use a 5.2-litre (1.4-gal) aluminium tank that is an ex-Porsche Cayenne air suspension tank. (I just happened to have this left over after I fitted air suspension to one of my cars – see *Custom Air Suspension*, also published by Veloce). You could use a much smaller reservoir.

Place a T-piece in the hose to the reference reservoir line and add an on/off valve on the upright of the 'T.' A ball valve, as is commonly used in compressed air systems, is readily available, and is suitable for this use. The plumbing is organised so that when you open the valve, the reference reservoir is connected to the atmosphere.

The reason for the valve is this. Any variation in temperature of the air in the reference reservoir, and its plumbing to the gauge, will change the reference pressure. Therefore, just prior to making a test run, stop the car and open the valve, checking that the gauge shows zero. Close the valve, do the run, read the gauge and then come to a halt. The gauge should again read zero – if it doesn't, the reference pressure has changed during the run. In climates with extremes (eg you are substantially altering the

PRESSURE MEASUREMENT

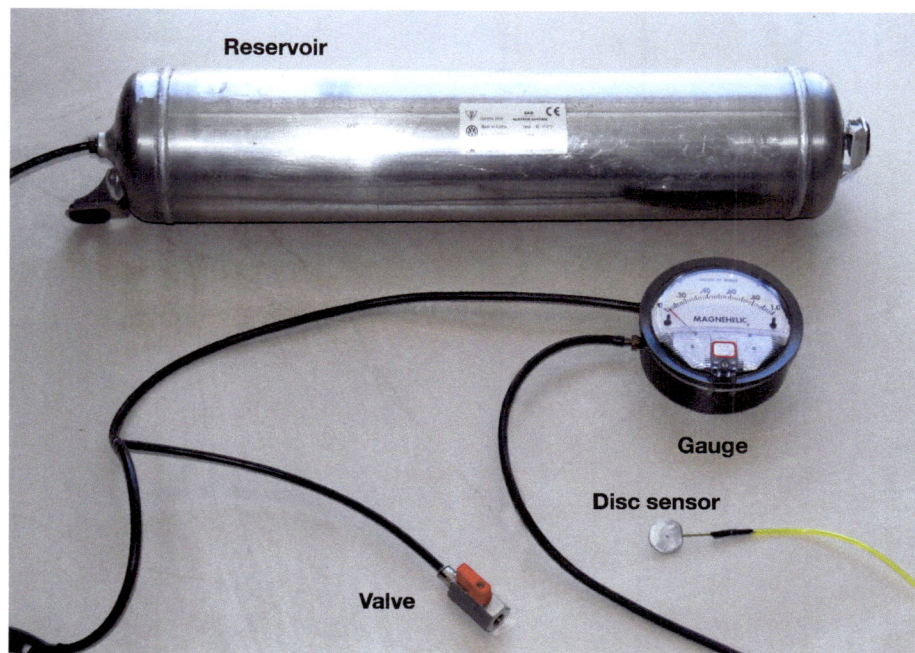

Figure 4-9: This equipment will allow you to measure even very small surface pressures. The reservoir provides a reference pressure that is set to atmospheric by momentarily opening the valve. The Magnehelic gauge, working in differential mode, is connected to the reference reservoir on one side, and the surface disc probe on the other.

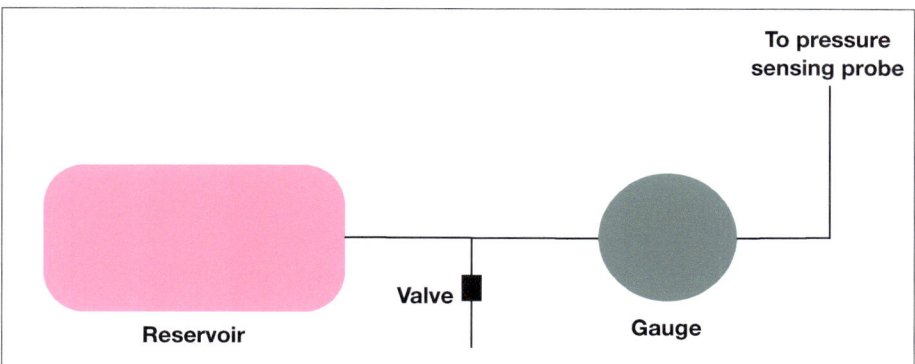

Figure 4-10: When measuring very small pressures with a Magnehelic gauge (or digital manometer), a reference pressure is provided by a sealed reservoir. The valve allows the reference pressure to be zeroed.

the system set up with a reference pressure reservoir.

Note that when using the Magnehelic gauge to simultaneously sense two bodywork pressures, no reference pressure is needed – the gauge is measuring the difference between the two aerodynamic pressures.

In what I have written above I have covered a wide range of complexities – from using just the end of a tube as a sensing probe, through to making a disc probe; from using just the cabin as the reference pressure, through to using a sealed reservoir that is vented prior to each test run. So what is needed in your particular case?

The approach you take depends very much on the accuracy and repeatability you need. Finding the best location across the front of a car for a high-pressure engine intake (covered in Chapter 9) can be done quickly and simply, using just the bare end of the tube as the sensor, and referencing against cabin pressure. As described above, just ensure that during the testing, you don't do anything that will alter cabin pressure. The pressures are relatively high (for car aero, anyway!) and when comparing locations, a higher pressure is better. However, what about measuring the pressure on the rear window of a fastback car, to see if it is negative or positive (ie is creating lift or downforce)? These pressures are much lower, and you want an absolute reading. In turn, this situation will require the use of a disc pressure probe and a reference pressure volume.

STEP-BY-STEP PRESSURE MEASUREMENT

This section describes tests that I carried out to measure the upper

cabin temperature with the heater or air conditioner), the reference reservoir may need to be insulated, or the cabin temperature allowed to stabilise for 15-20 minutes before doing the tests. If you find it hard to stabilise the temperature of the reservoir and its line, make the measurement as soon as you have reached the designated speed – this gives less time for pressure in the reference reservoir to vary with temperature.

Figures 4-9 and 4-10 show

SPEEDPRO SERIES

surface aerodynamic pressures on two different cars.

The first car is a 2003 model W211 E500 Mercedes with a quoted C_D of 0.27. The second is a 1979 W123 Mercedes 230, with a quoted C_D of 0.36 – some 33 per cent higher than the more recent car.

The plan was to measure the bodywork surface pressures along the centrelines of the two cars at a road speed of 80km/h (50mph). Averages of the readings were to be taken from runs in both directions over the test road, with the measurements taken at the following locations:

1. Front number plate (likely to be within the stagnation zone)
2. Leading edge of bonnet
3. Base of windscreen
4. Top of windscreen
5. Top of rear window
6. Base of rear window
7. Top of boot (trunk) lid
8. Rear numberplate (within the wake).

A Magnehelic 0-125Pa (0-0.5 inches of water) gauge was to be used, together with a disc surface pressure probe, as described above. A 5.2-litre (1.4-gal) reference volume was used for the gauge.

When performing tests, the key is to be prepared. I drew up sheets like the example shown in Figure 4-11, on which to record the results.

Don't be tempted to just scribble your results on a piece of paper; nothing is worse than spending an hour testing on the road only to find you cannot make sense of the notes that you've taken. (And yes, I speak from experience!)

So, how did the tests go? The more recent car – the W211 model – was tested first. At 80km/h (50mph) – the speed I'd selected for the test – the measured pressure on the front numberplate was exceeding the FSD of the meter. I therefore dropped the speed back to 70km/h (43mph) for all the tests. Taking about 45 minutes to complete, the rest of the W211 testing went smoothly. (More on the results in a minute.)

And the W123, the older car? On the second test, when I was measuring pressure at the top of the windscreen, I found the value exceeded the gauge's FSD – even at 70km/h. I was reluctant to go any slower in road speed, so I selected another Magnehelic gauge with a higher range (750Pa, 3 inches of water). Therefore, the two cars were measured with different gauges. However, I used the same gauge to perform two comparative test measurements on each car, and the results, while not identical, were fairly close. The W123 testing took a little longer to complete because I needed to swap gauges after the first two tests and then redo these tests before completing the rest of the schedule. However, that still meant well under two hours to complete the testing on both cars.

Let's look at the results.

The table below shows the measured data for both cars.

Figure 4-12 shows these measurements in graph form. Figure 4-13 (see page 80) shows the measurements on the cars, and Figure 4-14 shows the cars being smoke tested in the Mercedes wind tunnel. (Sorry about the poor quality

Surface pressure measurements, centreline

Car: Mercedes E500 W211
Date: 13/11/2017
Test road: North of Dalton
Weather: Fine, 10 km/h east wind, 23°C
Equipment: Magnehelic, Surface disc probe, 70 km/h

Location	Direction 1	Direction 2	Average
Front number plate	+120	+130	+125
Leading edge of bonnet	-130	-100	-115
Base of windscreen	+90	+70	+80
Top of windscreen	-140	-90	-115
Top of rear window	-70	-30	-50
Base of rear window	-10	~0	-5
Centre of boot lid	-20	~0	-10
Rear numberplate	-40	-10	-25

Reminder: reset reference volume pressure to atmospheric before each run

Figure 4-11: Test sheet from on-road testing of the W211 Mercedes. Always keep good records when testing.

PRESSURE MEASUREMENT

	Location	W211 pressures (Pa)	W123 pressures (Pa)
1	Front number plate	+125	+150
2	Leading edge of bonnet	-115	-125
3	Base of windscreen	+80	+60
4	Top of windscreen	-115	-300
5	Top of rear window	-50	-100
6	Base of rear window	-5	-100
7	Top of boot lid	-10	-50
8	Rear numberplate	-25	-50

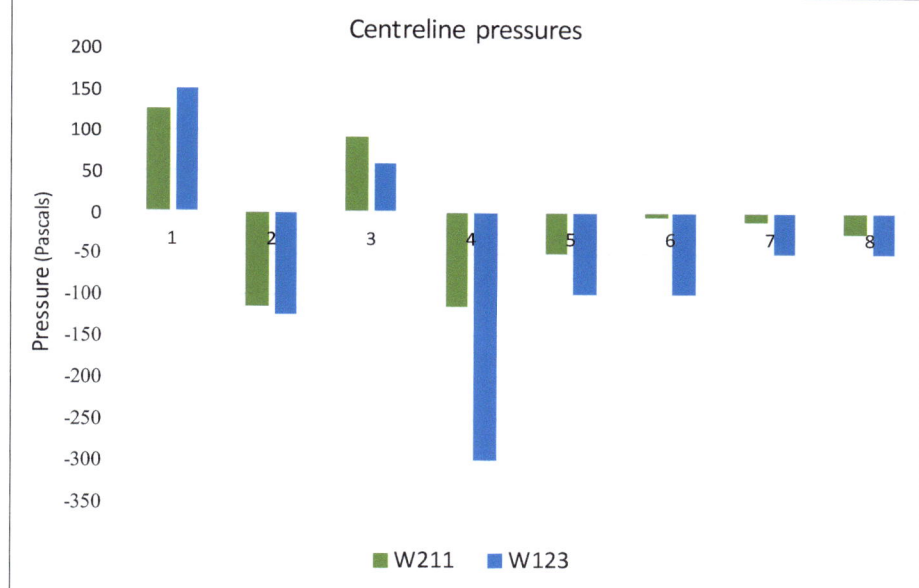

Figure 4-12: The centreline upper surface pressures measured on a W211 and W123 Mercedes, tested on the road with a surface measuring disc and a Magnehelic gauge.

of the W123 wind tunnel image; it is taken from a Mercedes video.)

Firstly, the measured pressures are *exactly what you would expect,* given the wind tunnel smoke streamlines. That's important, as it shows that the simple pressure measuring technique performed on the road actually works. Remember, you are finding out information that most people assume needs a wind tunnel and tens of thousands of dollars of instrumentation to gain.

Let's look at the pressure data for the W211 (the newer car) first. Note how the highest pressure is recorded at the front numberplate – often close, or at, the stagnation zone. Moving along the car's centreline, you can next see a substantial negative pressure at the point where the airflow wraps around the bonnet leading edge – a tight radius in an area where the boundary layer will be thin, so flow attachment will be high.

Moving towards the windscreen, you can see a high pressure – it's about 70 per cent of the pressure recorded at the front number plate. A low pressure is recorded at the windscreen header rail (of the same magnitude as found at the leading edge of the bonnet), and then smaller low pressures are recorded at the top of the rear window, at the base of the rear window and in the wake.

Now let's look at the measured data for the W123 – the older car. Note how the numberplate pressures (high) and the bonnet leading edge pressures (low) are very similar in magnitude for both cars. However, as can be seen in the wind tunnel photo for the W123, there is a separation bubble ahead of the base of the windscreen, and the lower recorded pressure shows this (this time, it was only 40 per cent of the pressure recorded at the front number plate).

At the top of the windscreen, the W123 bodywork makes a very sharp curve and the flow, that has reattached on the windscreen, has to get around this sharp corner. (With the boundary layer still relatively thin, it can do this.) The result is a very high suction peak – a low pressure of -300Pa, well over twice the value recorded on the W211.

Looking in more detail now at the wind tunnel photos, the smoke trails show that unlike the later W211, the W123 has flow separation from the rear end of the roof. This explains the low recorded pressures across the rear of the W123. To put this another way, because the flow remains substantially attached on the rear of the W211, the recorded pressures across the rear half of this car are *much* higher than the older car. The lower base pressure in the W123 will also be influenced by the rough underside, with airflow under

SPEEDPRO SERIES

Figure 4-13: The patterns of measured pressures compared. Note the lower pressures on the rear surfaces of the older W123 car (C_D 0.36) versus the W211 (C_D 0.27). The base pressure of the newer car is twice that of the older car!

Figure 4-14: Mercedes wind tunnel testing of both cars matches well with the on-road pressure testing. Note the flow separation occurring at the end of the roof of the older W123 car – this is responsible for the low measured pressures on the rear of the car. The sharp curve that the airflow negotiates around the windscreen header rail on the W123 is reflected in the strong suction peak measured in this area.
(Courtesy Mercedes)

the car being inhibited. These higher pressures on the rear of the W211 help explain the lower C_D value of the more recent car.

USING PRESSURE MEASUREMENT DATA
Pressure measurements are most useful when you want to:

1. Determine the existing pressure distribution to allow better decisions to be made about appropriate modifications.
2. Assess the changes caused by modifications, initially of those modifications quickly mocked-up for testing purposes, and then later in more detail.

Let's look at examples of each.
Pressures are important when you are deciding where to place bonnet vents (covered in Chapter 8). In short, if you want air to be vented out through the bonnet from the engine bay, you need to place the vents where the pressure difference is greatest – that is, where the pressure beneath the bonnet is highest compared with the pressure on the surface of the bonnet. Direct measurement of the two pressures (eg by using a Magnehelic gauge in differential mode) will give you this information quite quickly.

If you are working out the best location for an engine air intake (covered in Chapter 9), measuring the pressures at various points across the front of the car will give you this information.

The amount of air likely to be flowing through heat exchangers (including radiators, oil coolers and intercoolers) can be quickly ascertained by measuring the pressures on their forward and rear faces. If the pressure on the two sides is the same, no

PRESSURE MEASUREMENT

> ### UNDERCAR PRESSURE MEASUREMENTS
> Pressure measurements taken under the car are performed in the same way as upper surface pressure measurements. However, the logistics of positioning the probe are different, as normally you will need a way of lifting the car to gain access to the underside. You can use ramps or a hoist to do this. If using a jack, you must support the car on jack stands when working underneath it. I use wide, cloth-backed tape to hold the pressure sensing disc and tube to the underside of the car. Ensure the disc and tube are mounted securely, and remember that some underside surfaces (eg engine and gearbox undertrays) will grow warm when the car is running.

airflow through the heat exchanger will occur. Surprisingly, this is not uncommon, with designers and modifiers paying far more attention to the way in which the heat exchanger is fed air, and not enough attention to the way that air flows away from the heat exchanger. This topic is covered in more detail in Chapter 8.

Another example: in conjunction with flow visualisation testing (eg by wool tufting), pressure measurement will give you a good indication of where flow is attached and separated. In the case of the older W123 Mercedes shown earlier, we can be confident that a small spoiler or wing positioned on the trailing edge of the boot lid will do nothing – it's well inside the area of separated flow.

Incidentally, be wary of amateur CFD modelling that purports to show pressure distributions. Professional CFD modelling is likely to be very good, and the CFD examples I have shown in this book have been produced by either mainstream car companies and/or reputable software providers. However, in researching this book, I also came across many CFD images that I thought had quite doubtful validity. For example, cars modelled without a moving road or even moving wheels, and with resulting modelled C_D values that were absurdly in error compared with the real car. Directly measuring pressures on an actual car on a real road and with real levels of turbulence gives a high degree of assurance that what you are seeing in the data is actually occurring!

Now, what about assessing modifications by measuring pressures? As with modifications assessed through flow visualisation, modifications being assessed by pressure testing should initially be of the cheap-and-easy mock-up variety. For example, and especially if the initial testing is being done at low speeds like 80km/h (50mph), thin plastic sheet (eg ABS), or plywood and strong duct or cloth-backed tape will allow you to quickly and cheaply produce undertrays, side skirts, front and rear spoilers and the like for testing. In some cases, these appendages will need to be further braced, but in many cases, they will hold their shape sufficiently to allow you to perform the pressure testing.

Often, you initially just want to see if you are heading in the right direction. For example, I was curious to see if I could alter the pressures over the rear fastback hatch on my Gen 1 Honda Insight. I measured the pressure sensor in the middle of the rear hatch and drove the car at various speeds, finding a pressure of about -75Pa (-0.3 inches of water) at 80km/h (50mph). I then added a small Nascar-style flat plate spoiler at the trailing edge of the hatch, angled up at about 45°. I could immediately see the pressure change: the pressure on the hatch was negative at speeds of less than 100km/h (60mph) and positive at speeds higher than this. For fun, I then used a huge plywood panel as a spoiler, and gained positive pressures at all speeds over 60km/h (35mph).

These were 'quick and dirty' tests that in total took about 30 minutes – including the time to make the mock-up spoilers! The next step would be to do much more thorough testing, including recording the pressures at different speeds across the whole area of the hatch, and with different spoiler designs. Then would follow a lot more testing that also looked at the change in drag (eg through fuel consumption changes) and lift/downforce changes (eg through measuring ride height).

When assessing car aerodynamics and the effect of modifications, on-road pressure measurements are vital.

USING PITOT TUBES TO MEASURE AIRSPEED

As we saw in Chapter 1, a pitot tube measures airflow velocity. It does this using two pressure-sensing ports – one that faces into the oncoming airflow, and one (or more) at right-angles to this. Some references describe the forward-facing port as an 'impact tube,' and that's an easy way of remembering it – the air 'impacts' this port. Therefore, this port measures both dynamic and static pressures. The other port is a static pressure measuring port, of the sort that we have been using when sensing pressures acting on the surface of the car. If we subtract the static pressure from the dynamic pressure, we can see the pressure created just by the air movement – and so measure air speed.

SPEEDPRO SERIES

Figure 4-15: This miniature pitot tube is sold for use on model aircraft, but is highly effective in making car aerodynamic measurements. Use it in conjunction with a pressure measuring device (eg a Magnehelic gauge) to measure local flow speeds.

The best small pitot tubes I have found for use in car aerodynamics are the ones sold for model airplanes. I get mine from Eagle Tree Systems in the US (and also check out their tiny data logging systems) but these pitot tubes are also available from other suppliers. They are quite cheap – much too cheap to bother trying to make your own.

Reading the output of the pitot tube is easy – just use the Magnehelic gauge that we have been using in our other pressure measurements. A gauge with a FSD of 250Pa (1 inch of water) works well for speeds up to about 80km/h (about 50mph); for higher speeds you will need a higher reading gauge. (As described earlier, you can also use a digital manometer.) In use, the two pitot tube ports connect to the two ports on the gauge – if the gauge needle moves backward, swap the hoses. Note that you will need to make up some adaptors to connect the small output ports of the pitot tube to the Magnehelic gauge. It's easiest to slip hoses one inside the other, and then wrap the connections with electrical tape.

So, how can you calculate actual air speed from the Magnehelic gauge readings? There are a few approaches. The most accurate approach is to calculate the density of the air on the day, and then apply this correction factor to an equation (or table of data) that allows you to calculate speed from the differential pressure reading. Alternatively, you can do averages of runs in different directions (to cancel the effect of wind), and at different road speeds, and by the use of the car speedometer, work out what pressures correspond to what air speeds. (Remember though that this will be valid only at the time of testing.)

Or you could do neither of those things, and instead do what I do!

So what approach do I take? The first step is to select a test road speed. Then, at that speed, measure the pressure generated by the airflow *away from the car*. For example, you can measure this speed 500mm (~20in) above the roof or hood (bonnet). The support can be placed anywhere convenient – the probe just needs to be well above the car. Do two runs in each direction and then average the result.

Let's say that you do that at 70km/h (about 45mph) and the Magnehelic gauge reads 225Pa (0.9 inches of water). This is the *freestream* air velocity. Now, irrespective of what number was actually shown on the gauge, let's call this reading '100'. (It's easy to think of this as 100 per cent.) If we then measure flow speed near the car, and we find that the reading is (say) 100Pa (0.4 inches of water), we can do the following calculation. 100/225 = 0.44, multiplied by 100 = 44. That is, the airflow speed that we measured was 44 per cent of the measured freestream velocity. If the airflow speed were higher than the freestream, the figures would be over 100. Doing it in this way is called 'normalisation'. Normalisation must always be done against the freestream velocity measured in that particular test session. Taking this approach is easy, accurate and gives data with significance that is easy to picture.

Rather like the flow visualisation techniques covered in the previous chapter, and the surface pressure measurements covered above, using a pitot tube brings alive the patterns of airflow over the car. In use, the pitot tube is mounted at the location where the airflow speed needs to be measured. It would be best if the pitot tube were mounted in perfect free space, without the need for connecting tubes or a support. However, that's rather hard to achieve. So for measuring *relative* changes, just ensure that the set-up is consistent. For example, in the measurements shown below, the pitot tube was held by a spring clip to a piece of narrow wood taped vertically to the back of the car. The only aspect that was changed during the measurements was the height of the pitot tube on the wooden support.

Pitot tubes can be used to measure local flow velocities. For example, if you place the pressure sensing ports of a pitot tube near to the A-pillar, you will measure higher airflow velocities than the freestream. This is a good example, because it allows us to tie together all the approaches that we have been taking. What do I mean? Well, wool tuft testing showed us how airflow wraps around A-pillars, and surface pressure measurement showed us how pressures at this location are low. Now, if we add

PRESSURE MEASUREMENT

pitot tube testing, we can see that those pressures are low *because* of the higher air speeds! So it's another string to our bow – when investigating a particular issue, you can choose to use any combination of wool-tufting, surface pressure measurement or airflow velocity measurement.

But pitot tubes can also be used to sense what is going on in situations where the two other techniques may not be viable or useful. For example, in Chapter 7 I suggest that a wing (not a spoiler, but a true wing) when mounted on a three-box sedan should typically be located about halfway between boot lid and roof heights. At this location it will get good airflow. But how can we assess that on a specific car? Wool tuft testing won't show it, and neither will surface pressure measurement. But a pitot tube is very effective in this situation. Let's take a look.

I did some testing on my W211 Mercedes sedan. Air velocities were measured at different heights above the trailing edge of the boot lid at a speed of 70km/h (~43mph). As described above, these measurements were normalised against the freestream velocity, measured 500mm (~20in) above the roof. (To measure this, I taped a tall wooden support to the 'shark fin' aerial located above the rear window, and mounted the pitot tube atop this.) The following table shows the measurements.

Height above boot lid (mm)	Percentage of freestream velocity
400	80
200	75
100	44
50	25

Note that a height of 400mm above the rear boot lid is actually about roof height. I am not fitting a wing to the back of the Mercedes, but if I were, based on these figures it would be located around 200mm above the boot lid height, where the airflow is at 75 per cent the velocity of the freestream. (And yes, in this car with flow reattachment occurring on the boot lid, that turns out to be about halfway between lid and roof heights!) Figure 4-16 shows these measurements graphed. Figure 4-14 shows the Mercedes being wind tunnel tested with smoke streamlines. You can see that the proposed wing location matches what you'd expect from the smoke testing – and of course, you can do the pitot tube testing without needing to have a wind tunnel!

The second interesting aspect that can be explored with the pitot tube is flow direction. To read flow velocity correctly, a pitot tube needs to be angled directly into the flow. If the pitot tube is not aligned with the flow, a lower reading will result. Therefore, if the flow direction is not accurately known, alter the angle of the pitot until the measured value reaches a maximum – then, the tube will be aligned directly into the flow.

So why might this be important? As described in more detail in Chapter 7, the angle of attack of a wing needs to be defined not in reference to horizontal, but to the direction of the approaching airflow. By testing with the pitot tube at a fixed location and varying in its angular positioning to the oncoming airflow, the direction of actual airflow can be better determined. So not only can airflow speed away from the car's body be measured, the direction of that airflow can also be plotted.

With wool tufts, a surface pressure measuring disc, pitot tube and a few secondhand Magnehelic gauges (or a digital manometer), you will have a complete aerodynamic testing kit that is accurate, cheap and effective.

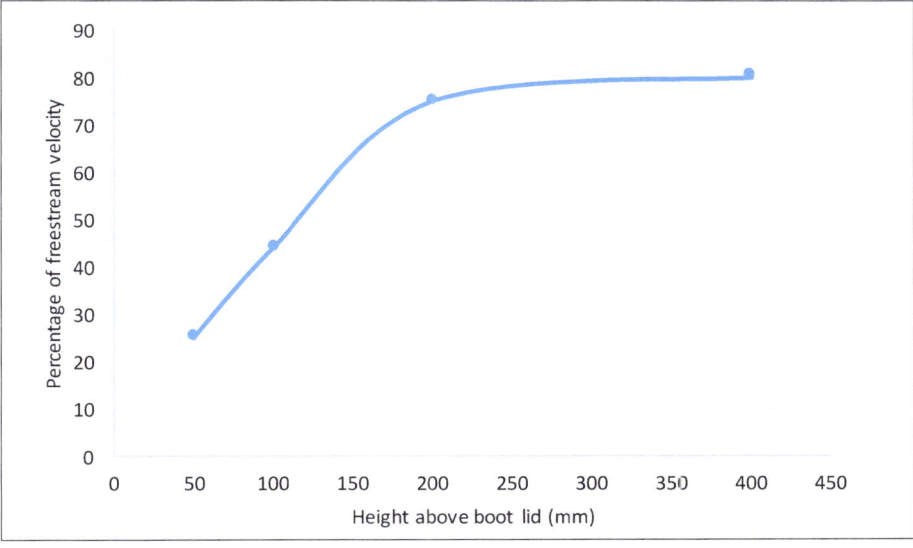

Figure 4-16: Measured airflow speeds at differing heights above the boot lid on a sedan. This graph indicates that a wing placed 200mm above the boot lid would work well. The airflow speeds were measured with a miniature pitot tube.

Testing the Jaguar XE

Unlike some current cars that are – at least to my eyes – rather ugly, the Jaguar is a stunning-looking car. Svelte, tightly-drawn, balanced, Jaguar-esque … it's all of those. The car I tested was fitted with a few optional additions – a tiny rear spoiler and side skirts. The C_D of the car is variably 0.26, 0.27 or 0.28 – the actual number depending on different wheel, powertrain and trim options. No lift figures have been released. (Courtesy Jaguar)

CASE STUDY: TESTING THE JAGUAR XE

Let's now put together the testing processes that we've so far described, and test a car to see how it compares with the 'official' data. Adrian Gaylard, chief aerodynamicist of Jaguar Land Rover, made available to me a presentation on the XE Jaguar's aerodynamic development, complete with detailed CFD images. Jaguar Land Rover Australia then lent me a Jaguar to do on-road testing of the car. (Thanks to both entities for being so helpful.)

Using the techniques described in the previous chapter, pressure testing was carried out on the car. The test speed was 70km/h (44mph). Nine test points were used on the upper surface and four points on the car's underside.

Peak measured pressure was 145 Pascals (Pa) at the front number plate. The relatively gentle curve at the front of the bonnet kept the front suction peak to a low -30Pa but the pressure at the base of the windscreen was much as you'd expect at this speed, at 75Pa. However, note the very gentle radius at the top of the windscreen, resulting in a higher than expected pressure here (on many cars it's even lower than this). Also, note the camber of the roof, where in the middle I measured -95Pa – that's a large upper area to be at lower than atmospheric pressure. The top of the rear window was -50Pa, and this dropped to -10 at the base of the rear window – implying this area was in a separation bubble. But look at what happens on the boot lid – a positive pressure of 30Pa! Interestingly, the base pressure in the wake is much higher than on many cars, at -5Pa. So, over the top surfaces of the car, we can see only few suction peaks of high magnitude, but lots of areas with lower-than-atmospheric pressure. We can also see an unusually high rear base pressure.

And beneath the car? Under the front apron and within the area formed by the rear diffuser, I measured atmospheric pressure. But on the undertrays positioned each side of the central exhaust, there was clearly lower than atmospheric pressure – in fact, -40Pa.

Overall? Looking at the profile from the windscreen over the roof, the shape is generating low pressures, and so lift. However, there is positive pressure on parts of the boot lid that would help offset some of this, and a large area beneath the car (albeit, not at low enough pressure to completely offset the upper low pressures) that would also reduce lift.

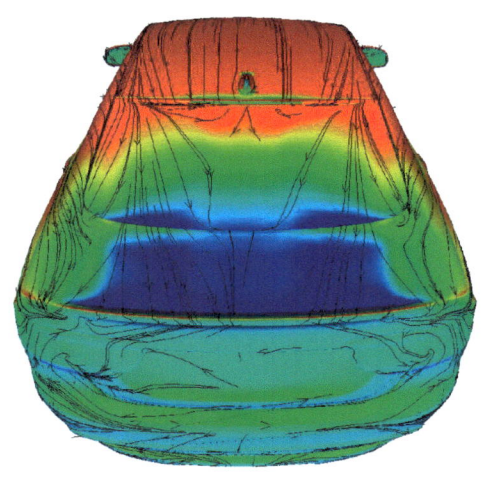

The CFD image provided by Jaguar, that covers the rear half of the car, is in strong agreement with the on-road surface pressure measurements. The CFD image shows the upper surface of the boot lid being subjected to positive pressure (blue), however it also shows the centre base of the rear window being slightly positive – which is not what I found. (The discrepancy is likely due to two factors: measuring on-road pressure at only a single point, and the challenges in CFD of precisely predicting vortex strength and location.) The CFD shows the roof and upper part of the rear window being subjected to negative pressures (red), which is just what I measured. (Courtesy Jaguar)

85

SPEEDPRO SERIES

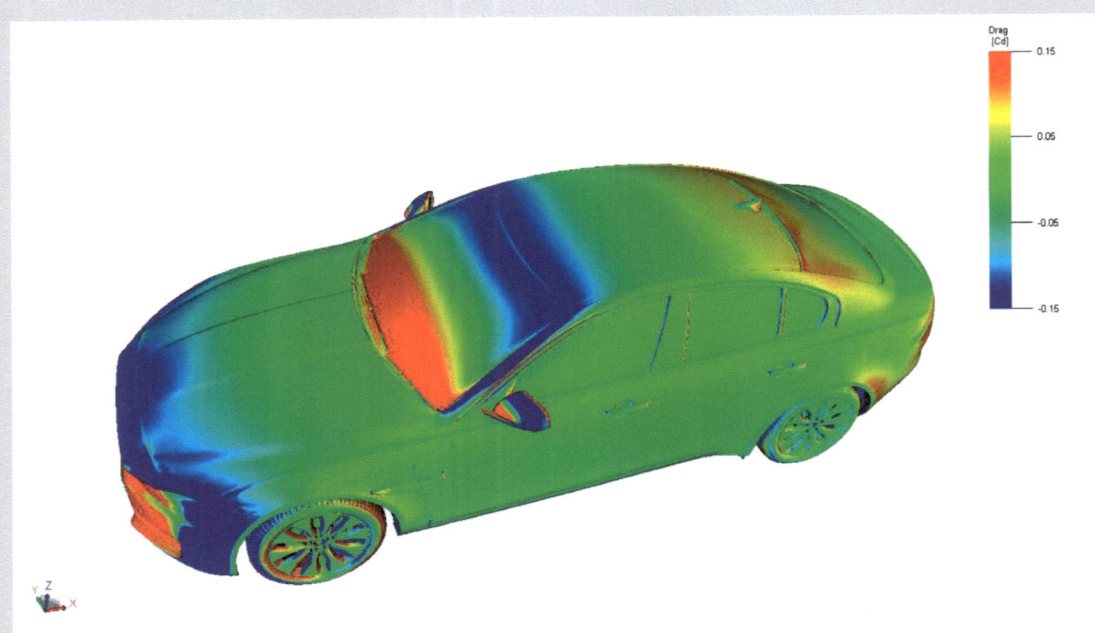

This CFD image shows not pressures but drag and thrust. The 'hotter' the colour, the more drag that area of the car is creating, and the 'colder' the colour, the more thrust. Note the pattern on the rear vision mirrors, and the small areas behind the rear wheels on the car's flanks – more on this latter area in a moment. (Courtesy Jaguar)

So, what does the underside that is generating these lower-than-atmospheric pressures look like? The car uses a front undertray that extends from the lower side of the bumper lip to roughly the firewall. Just visible in this photo is a downward bulge in the undertray: my measurements indicate that this creates a lower-than-atmospheric pressure. (Courtesy Wayne Medway)

The underside also features four deflectors: two on each side. One deflects air from impacting the front of the tyre, and the other – mounted on the lower suspension arm – appears to direct air to the brakes for cooling purposes. (Courtesy Wayne Medway)

CASE STUDY: TESTING THE JAGUAR XE

At the rear of the car, a diffuser is formed by the lower suspension arms, mufflers and the spare wheel well. (Remember the suspension is shown here at full droop; normally the lower arms would be near to horizontal.) The rear diffuser surfaces are angled upward at about 12°. This view also shows the forward and side undertrays. Note how the central muffler's lowest surface is also flat. Fitting undertray panelling to a car that did not originally have it can typically use the approach seen here. (Courtesy Wayne Medway

Next it was time for wool tuft testing. The first surprise was how attached the flow was immediately after the fairly bluff front. The low pressure already measured indicated that the flow was attached directly above the grille on the bonnet, and the wool tufts certainly backed that up. I'd measured the highest pressure on the number plate, and the wool tufts show the plate to be within the stagnation zone, with one tuft in this area going upward into the top grille, and one tuft going downward into the lower grille. (Unlike some XE models, this car didn't have automatic grille shutters.) The flow over the bonnet and up onto the roof is superb – testament to the smooth transitions I mentioned earlier. Notice also the flow in two other areas, both of which I will come back to. The flow around the A-pillar is excellent, and the flow down the side of the car behind the front wheel, while not perfect, is very good.

SPEEDPRO SERIES

CFD imagery (performed in Exa PowerFLOW) shows the same airflow pattern as observable with the tufts. Note how here, even with the grille shutters open, a high-pressure area forms in front of the radiator that helps direct oncoming air smoothly upward and onto the bonnet. (Courtesy Exa)

The XE Jaguar uses front air curtains, where air is ducted inside the bumper cover to slots positioned ahead and outboard of the front wheels. The photo on the left shows the intake to one of the air curtains …

… and this photo on the right, taken with one of the front wheels removed, shows the air curtain outlet. But do the air curtains work? (Courtesy Wayne Medway)

Here you can see the flow down the side of the car in standard form. Note that for this, and the next photo, more than 20 shots were taken before an indicative shot was selected. For most of the time, the airflow down the Jaguar's flanks looks like this – not perfect (note the tuft just aft of the lower wheelarch), but very good.

CASE STUDY: TESTING THE JAGUAR XE

I then plugged the intake to the air curtain with some foam rubber and re-photographed the car. The differences behind the wheelarch are clear – there is far more turbulence and the flow rarely reattaches directly behind the wheelarch. Also note the crinkled tuft ahead of the wheel in this photo – this was much more common than when the air curtain was working.

Another part of the car that was very impressive was the flow around the rear vision mirrors. In addition to reducing drag, Jaguar did a lot of work here to reduce noise. Unlike most cars, the flow around the A-pillars and onto the side glass was barely disturbed by the rear vision mirrors. Look at the two tufts directly downstream of the mirror – they're acting as if the mirror isn't there! Incidentally, the angle of the tufts near the pillar in both this and previous shots indicate the formation of a vortex off each A-pillar.

The Exa PowerFLOW CFD imagery shows just how good a match is achievable by zero-cost on-road testing with wool tufts, compared to the professional software. Note how little separation there is around the A-pillar, and how important the gap is between the mirror and body. (Courtesy Exa)

SPEEDPRO SERIES

The view of the mirror from inside the car. Note how the two tufts to the right (ie between the mirror and the main body) show little turbulence. You can also see in this, and the previous wool-tuft shot, how the mirror's trailing edges are shaped inward to reduce the size of the (unavoidable) wake. I don't think that this photo does the flow around the mirror justice – it was really quite superb.

So why does the mirror housing work so well with the A pillar? This photo shows some of the reasons for that. First, the frontal area of the pedestal on which the mirror sits is very small. Second, there is that quite large gap between the mirror and the bodywork, reducing the speed of the air that is passing through this space. Finally, note how the mirror has a forward-facing blunt point, and then grows upward and outward in full curves. As shown on page 86, the Jaguar CFD modelling shows the 'blunt point' causes drag, but most of the rest of the housing creates thrust! By the time you factor-in the wake area, the mirror as a whole wouldn't be generating thrust, but of the front-facing surfaces, there is more area of thrust than drag!

This side view shows a lot of the good aspects I've already described – excellent flow on the bonnet, windscreen and roof; good flow down the side, and excellent patterns around the side mirror and A pillars. But what is happening behind the rear wheel? Apart from a tiny separation edge on the rear lights, the side panels have nothing to promote clean separation. Further, the narrowing of the body (looking down from above) behind the rear wheels means this part of the bumper does not achieve flow reattachment. The result is clear turbulence. Note how the Jaguar CFD imagery (page 86) also shows this is an area of drag creation.

CASE STUDY: TESTING THE JAGUAR XE

And the rear three-quarter view? (Note that this was one of those shots where the flow reattachment behind the front wheel was not so good – but this is not typical.) However, the turbulence behind the rear wheel is typical, and we can also see a separation bubble at the middle bottom of the rear window, something that my pressure testing also showed. To an extent this might be nearly unavoidable, with the two rotating vortices that are coming off the C-pillars (started at the A-pillars? – you can trace the pattern of tuft swirl back to them) likely to leave this area of the car in separated flow. However, flow reattachment clearly occurs on the boot lid, as we saw in the pressure testing. As expected, within the wake, the tufts show turbulence, but the lowest tuft (that I stuck on the spare wheel-well diffuser) shows good undercar flow.

This Exa PowerFLOW CFD imagery show the same airflow pattern around the C pillar, and on the boot, that was revealed by wool-tuft testing. Of course, this simulation could be done before the car was built – something not possible with tufts! (Courtesy Exa)

The Jaguar is not perfect in its aerodynamics – but it's very good indeed. From its effective air curtains to the superb rear vision mirrors, from the undertrays that work well to its smooth surface transitions, from developing positive pressures on the rear upper surfaces to its relatively high base pressure, it's a good lesson in how to make a car both beautiful and aero-efficient. And that turbulence on its flanks behind the rear wheels, and lack of clear rear separation edges? I'll be curious to see what happens with the next model …

Testing the Tesla Model S

The most fascinating view of the Tesla is the rear three-quarters. The airflow down the rear glass is superbly attached, and with the slight kick-up of the rear deck, increased pressure will occur over this area, reducing drag and lift. (The SAE paper states that a small separation bubble exists at the base of the rear glass, however this is not visible in the tuft testing.) Also interesting is the behaviour of the air over the C-pillar. The tech paper states that much effort was put into reducing trailing vortices from these pillars, and while the paper does not say this, it looks to my eyes as if this has been achieved by adding a sharp edge to this pillar. This potentially alters the wrap-around of air from the side of the car onto the hatch, reducing lift. Significantly, this sharp line does not appear to be present on the original Model S concept car.

CASE STUDY: TESTING THE TESLA MODEL S

The Tesla Model S is one of the slipperiest cars on the road. The SAE paper that describes the aerodynamic development of the car (SAE 2012-01-0177) has the following listed specifications for a 2012 Model S with 21-inch wheels:

C_D	0.26
A	2.40m³
C_{Lf}	0.10
C_{Lr}	0.10

When independently tested (see Chapter 2), the car had an even lower C_D than the SAE specs indicate – truly one car worth aero testing! Owner John Lindsay kindly made available his Model S for wool tufting.

see on this black car is the vertical separation edge that starts on the tail-light and continues downward – but only as far as the top third of the bumper. Below this, separation is poor.

This view better shows the behaviour of the air around the C-pillar. Note how the air appears channelled in the 'corridor' formed by the upper part of the guard (fender) and the distinctly-edged hatch. Without the airflow wrapping around the pillar's curve, lift is weakened. Hard to

The poor separation occurring behind the rear wheels is clearly shown here. In this case, the airflow is worsened by the turbulence caused by the proximity of the chase car, but even in photos taken from the roadside, this area is a little problematic. Note the tufts being drawn into the wheel-well, perhaps indicative of airflow from the sides of the car joining the rapidly moving undercar airflow.

Here the rear shape of the Tesla is more easily able to be seen. Note the vertical separation lines and how they stop on the upper part of the bumper. You can also see the styling's inward pull of the bumper behind the wheels, causing the unclean flow separation shown by the tufts. Note also the sharp line on the C-pillars formed by the rear hatch. The large gap between the rear vision mirrors and the main bodywork can also be seen here – this helps reduce separation that would otherwise be caused on the window adjacent to the mirror. (Courtesy Tesla)

SPEEDPRO SERIES

The flow on the side glass and mirror housing is excellent. Not even one photo from the complete photo session shows separated flow in this area. However, the tufts nearest to the pillar on the side glass are angled upward in all photos, perhaps indicative of strong vortices being formed off these pillars.

Flow behind the front wheels shows separation. (Note that in this – and every other – shot, two tufts have been caught in the leading edge of the door, and so the behaviour of these two tufts should be ignored.) The Tesla does not use air curtains, and their absence can be seen in the flow separation that is occurring. Especially when the Tesla was in proximity to the chase car (but also present even in photos taken from the roadside), there is clearly separated flow immediately behind the wheel, with separation strongest behind the wheelarch lip.

This photo shows the beautifully attached flow from the stagnation zone up the bumper, across the bonnet, up the windscreen and onto the roof. The strong flow under the car can also be seen by the behaviour of the lower wool tufts being drawn under the car. Note the tufts to the left and right of the oval 'grille' that show the excellent flow around the front corners of the car – an area specifically mentioned in the tech paper as having had a lot of development. The flow separation behind the front wheels can again be seen in this view.

94

CASE STUDY: TESTING THE TESLA MODEL S

Airflow over a car is three dimensional, and few photos show that better than this one. Here you can clearly see that the airflow meets the front of the car and then changes direction: some going under the car, some over the car, and a good proportion flowing around the side of the car. This view makes it understandable how smoother wheels (the Tesla 'Aero' wheels) were able to reduce drag by 9 per cent (see Chapter 6).

The Tesla uses a flat underfloor with this rear diffuser, complete with longitudinal strakes. Many of the tuft photos show strong airflow exiting this diffuser. Of course, no front air dam is used on the Model S – flow under the car is encouraged rather than hindered!

Overall, the external bodywork of the Tesla Model S is conventionally optimised for low drag and lift, with attached flow over the upper surfaces of the car and largely attached flow along the sides. (I'd not be surprised to see front and rear air curtains used in the next model to improve flow reattachment after the wheels.) The A and C pillar treatments were aimed at reducing lift (and so vortex formation) and – especially in the case of the C pillars that are quite innovative – appear to succeed in doing so. A large amount of the responsibility for low drag and lift must come from the underside's undertray and diffuser – the underside is the best of any road car that I have seen. (Courtesy Tesla)

Testing the Mercedes-AMG GT

This fascinating view shows the mix of attached and separated flow that characterises the AMG GT. Across the bonnet and up the windscreen, the flow is attached (one tuft near the leading edge of the bonnet is stuck in that position). However, down the side of the car the flow patterns are not so good – to put it mildly. There is flow separation at the front corner of the car, and then major separation immediately behind the front wheel.

CASE STUDY: TESTING THE MERCEDES-AMG GT

The Mercedes-AMG GT is an interesting car. It has an official C_D of 0.365, among the poorest of any current car on the road. (In fact, you need to go back to the 1970s to find Mercedes models with similar drag coefficients.) But is that high drag a consequence of generating a lot of downforce? It's hard to tell, because AMG Mercedes has not released any downforce figures for the car.

Owner Paul Lawford generously made his car available for wool tuft testing, and was ably assisted by Paul Perez and Matt Grosvenor. All photos are by Paul Lawford, who commented that he was very surprised by what he could see through the camera's viewfinder!

Interestingly, some variants of the AMG GT use front air curtains – but not this model. It appears that the shape of the bumper was not greatly changed, irrespective of whether air curtains were fitted or not. The result is flow separation ahead of the front wheel – something that is uncommon.

With flow separation starting in front of the wheel, and with the presence of the large diameter, multi-spoked rim, airflow on the front fender is much disturbed. In this photo, taken late in the testing session, the frayed wool tufts show the flow separation is constant and strong.

SPEEDPRO SERIES

By the midpoint of the car, flow has reattached on the side. Flow around the side mirror is largely good, with the flow on the side window undisturbed by the mirror's presence. However, close inspection shows some separation occurring at the base of the A-pillar, apparently influenced by the mirror's mounting foot.

At the rear, separation occurs on the lower side of the car immediately behind the rear wheels. Separation also occurs outboard of the rear lights – without a clear edge, the separation point is probably also moving around. However, as with the whole car, the flow pattern on the upper surfaces is quite different to that on the sides, with attached flow down the rear hatch and under the optional fixed rear spoiler to the horizontal separation edge.

CASE STUDY: TESTING THE MERCEDES-AMG GT

This photo very clearly shows the two quite different flow patterns – separated flow in the wide wake, and attached flow down the rear glass (a few of the tufts on the hatch are stuck in that position – they're not showing separation). Note also the wrap-around of the airflow on the rear pillars, that would be creating drag and lift – I'd expect to see strong trailing vortices being shed from these pillars.

The cabin of the AMG GT is placed well rearward, and with its smoothly curved upper surface and attached flow, is likely to be developing substantial rear axle lift. Consequently, all AMG GT models have a rear spoiler of some type – either a pop-up design or (as here) a fixed spoiler. The blade of the spoiler has an attack angle of about 25 degrees to the approaching airflow, so it is much more likely to be acting as a spoiler than a wing. This is also supported by the separated flow that is visible on the underside of the spoiler in some of the photos that were taken.

Chapter 5
Measuring changes in drag and lift

- **Measuring drag**
- **Required power testing**
- **Coast-down testing**
- **Fuel economy testing**
- **Measuring lift and downforce on the road**

The two major aerodynamic aspects of their cars that people wish to change are the drag and lift coefficients – better known as C_D and C_L. Perhaps you're a hyper-miler – in that case, you will want to reduce aerodynamic drag to improve fuel economy (mileage). Or perhaps you are a performance driver. In that case, you'll want more downforce (or, more likely, less lift) so that you can corner faster and brake harder. Reducing lift will also increase high speed stability. However, while plenty of people say that their aerodynamic modifications have improved their car's drag coefficient or have increased downforce, how do you go about proving these things? That is, how do you make measurements of changes in these characteristics?

Figure 5-1: Manufacturers use wind tunnels equipped with sophisticated instruments to measure drag and lift. We don't have those tools available, so how can we go about measuring these forces on the road? (Courtesy Mercedes)

In a wind tunnel, which is where the manufacturer of your car developed their drag and lift

MEASURING CHANGES IN DRAG AND LIFT

figures, sensitive balances are used. These measure the force pushing in the same direction as the airflow (ie the drag) and the forces vertical to the road – either upwards (lift) or downwards (downforce). But an amateur working at home doesn't have access to either of those things – not the wind tunnel, nor the balances.

So how can we measure changes in drag and lift?

MEASURING DRAG

Drag can be indirectly measured in at least three different ways. I'll first cover them briefly and then explore the practical aspects of each.

Required power testing

At a constant speed, engine power is required to overcome rolling resistances (eg the resistance of the tyres and bearings) and aerodynamic drag. If you can measure how much power is needed to maintain a steady speed on a flat road, reductions in drag will result in a reduced power requirement.

Coast-down testing

If you place the car in neutral gear at (say) 100km/h (62mph), the time it takes for the car to come to a halt will be dependent on rolling resistance and aerodynamic drag. If you reduce drag, the car will roll for a longer time, as the resistance the air is posing will be decreased. Rolling down a constant-slope hill and measuring speeds is a variation on coast-down testing, and appears more accurate for amateur testing.

Fuel economy testing

As we've seen, the engine provides the power to overcome drag and rolling resistance. To develop its power, the engine must use fuel, and if the drag is reduced, less fuel will be used. Therefore, a change in aerodynamic drag can be shown by improved fuel economy (mileage).

REQUIRED POWER TESTING

The faster you are travelling, the higher the proportion of the total resistance that comprises aero drag. Barnard in *Road Vehicle Aerodynamic Design* suggests that for a modern car, the crossover point (that is where aerodynamic drag exceeds rolling resistance) is 60km/h (37mph). However, at any speed, if you improve aerodynamic drag, then the engine will be required to produce less power to maintain that speed. Note, though, that because the power to overcome aerodynamic drag increases very rapidly with speed, the faster you are going, the more influence lowering drag will have on the required power.

Unfortunately, directly measuring the amount of power being developed by the engine, while the car is driving down the road, is difficult. However, for a given air/fuel ratio, the amount of power being produced by the engine is also dependent on the instantaneous amount of fuel being used (eg instantaneous litres/100km or mpg), and also the mass flow of the air being consumed. Both of these are much easier to measure than engine power!

If you have a car with an instantaneous fuel consumption gauge, measuring changes in aerodynamic drag can be carried out by looking at changes in this figure. To do this, you will need to be driving on a flat road in zero wind conditions. The more resolution the gauge has, the better. (If you do not have such a gauge, you can instead measure injector duty cycle using a multimeter.) While both ideas sound great in theory, in practice, both are actually problematic. The main

Figure 5-2: Measuring the mass of air that is being consumed by the engine can be done by monitoring the voltage output of a mass airflow meter. If one is fitted as standard, this is easy to do and cheap. The greater the aerodynamic drag at a given speed, the more air the engine will consume.

problem is the requirement for a flat road and no wind. In the real world, almost no public road is flat, and very few places have dead-calm conditions! So you may, in this way, be able to measure changes caused by variations in aerodynamic drag, but the changes in drag will need to be very large.

Here is an example of when I did just that. In this case, rather than use instantaneous fuel economy or injector duty cycle, I measured airflow meter output voltage. A hotwire airflow meter measures air mass throughput, and, as an engine developing more power needs to breathe a greater mass of air, the voltage output of the airflow meter reflects the power being developed by the engine.

I carried out a test on a Volkswagen Golf cabriolet Mk3. Soft top cars have much higher drag when the top is down, due to the separation (and so turbulence)

SPEEDPRO SERIES

that then occurs. The testing was carried out at 100km/h (62mph) – as mentioned earlier, higher speeds would have been better.

In standard form, with the roof closed and the windows up, the output voltage of the airflow meter was 2.340V. I then wound down the windows, and the measured voltage rose to 2.385V; with the windows closed and the roof open it was 2.360V; and then, with the 'full drag' configuration of the windows open and roof down, it was 2.460V. Figure 5-3 shows these values.

Therefore, the measured airflow meter voltage with the roof and windows down was 5 per cent higher than the car in its normal configuration. However, airflow meters don't have an output voltage of zero at idle – it's more like 0.9 or 1.0V. If that were the case with the Golf, the airflow meter output would have risen by about 9 per cent in the highest drag configuration. Most airflow meters are at least roughly linear, so is this indicative of the drag

MEASURING REQUIRED POWER

I said above that directly measuring engine power in a moving car is difficult, and I talked about instead using the instantaneous fuel consumption figure or measuring engine mass airflow. There is, however, a way of converting either of these figures to actual power at the wheels of the car. This is how you do it.

You identify the instantaneous fuel consumption figure, or mass airflow figure, at a certain speed on the road. You then place the car on a chassis dyno, bring the car up to the same speed as you were going on the road, and then increase dyno load until the instantaneous fuel consumption (or mass airflow) figure is the same as you recorded on the road. You then read the dyno's scale to see the developed power at the wheels.

I did this for the aerodynamically standard EF Ford Falcon (fitted with a six-cylinder engine) shown here. This showed that the power developed at the wheels to maintain 105km/h (65mph) was 13.0kW (17.4hp). People often talk about the power to maintain a certain speed on a level road, and this is a good way of accurately ascertaining that figure. Note that it takes into account rolling resistance as well as aerodynamic drag.

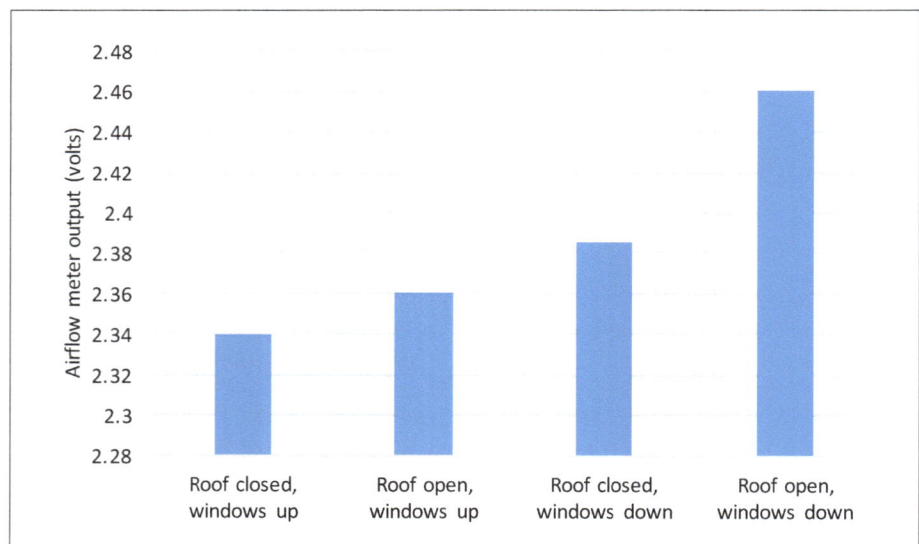

Figure 5-3: Changes in aerodynamic drag of a Volkswagen Golf Cabriolet, as measured by monitoring the airflow meter voltage output. You can see that the highest drag (and so greatest engine load) occurs with the roof open and the windows down.

of the Golf rising by about 9 per cent with the roof and windows down? That figure isn't unreasonable. But you can see how rubbery all the figures have become – and this for a variation in drag that is actually quite major!

When a home aero modification might reduce drag by just a few per cent, measuring it by monitoring engine power is difficult – at least at 100km/h (62mph)! Go 50 per cent faster in testing speed and the figures would be much more indicative of drag changes.

COAST-DOWN TESTING

My first experience of coast-down testing was not positive. At the age of about 23, I was testing the acceleration performance of my modified car. I'd found a back road

MEASURING CHANGES IN DRAG AND LIFT

to a power station, and was running 0-100km/h (~60mph) times. I'd hold the auto trans car on the brake with plenty of throttle, release the brake, then mash my foot to the floor. The time for a standard car was 8.4 seconds. I clicked the stopwatch at launch and then, as the speedo raced past 100km/h, clicked it again. The first run was 7.5 seconds. I refined the technique and got down to 7.0 seconds, with a best of 6.9 seconds. I did these times again and again. My new intercooler and exhaust, along with the increased turbo boost, were paying off – it was a good performance advance over stock. (But OK, pretty slow these days!)

Later that week, I was reading about aerodynamic testing, and, more specifically, coast-down testing. It sounded pretty good, so I went back to the same road on which I'd been conducting my acceleration testing (it was the only empty, near-flat road around), and did some coast-down testing. 0-100km/h had worked well for acceleration testing, so I did the same (but in reverse) for coast-down testing – 100-0. But the results were just ludicrous. I'd been doing the acceleration times with a time consistency of 1.5 per cent, but the coast-downs were just all over the place – as in, variations in recorded times ten times as great as I'd achieved with the acceleration runs. It was immediately obvious that no value at all could be placed on these times.

So what techniques can be used to make coast-down testing valid? The first point is that the higher the speed, the better. That's unlike most testing described in this book, where for example wool-tufting can be done at 80km/h (50mph). Secondly, the start and finish speeds should *both* be high.

Figure 5-4: Coast-down testing measures the time taken for the free-rolling car to slow from one speed to another. At high speed, most of the force resisting the forward movement of the car is aerodynamic drag, so a reduction in drag will result in a longer coast-down time. However, getting accurate results on real roads is difficult.

For example, if you live in a country that does not have speed limits, a coast-down test from 200km/h (120mph) to 160km/h (100mph) will potentially work well. Why? Because at these speeds, the aerodynamic drag is both high and makes up the greatest proportion of resistive forces the car is facing. That means that in turn, any drag changes you make to the car will be much more noticeable in the recorded times.

Secondly, and implicit in the point I made above, the coast-down time split should *not* include a return to standstill. At a walking pace, the contribution that aerodynamic drag has to the resistive force on the car is tiny, and so this sort of test is starting to spend a lot of time in irrelevant territory.

Finally, you need to do some baseline testing to ensure that you can, in fact, measure variations in drag. Earlier, I talked about measuring changes in required engine power by altering the drag complexion of a Volkswagen Golf cabriolet (roof up, roof down, etc)

and then measuring airflow meter voltage. A change in measured value could be seen that was consistent with what you'd expect. Now, what happens when you do a coast-down test with the windows open, and then shut? In any car, having the windows open will increase drag – so you should be able to see that change in your coast-down times.

Thoroughly warm-up the car before driving it, so that the tyres, bearings and oils are up to operating temperature. Select the highest speed that you can use on your test road. For example, that might be 120km/h. Do this speed, then select neutral gear so the car is rolling. When the speed has dropped to 110km/h, click the stopwatch. Click again when the speed has dropped to 80km/h. (In imperial terms, do something like this: get to 75mph, start the stopwatch at 70mph, and then stop it at 50mph.) Ensure that you do it over the same stretch of road, doing one test in one direction and a second test in the other direction. Average the results of the two tests. Now, open the windows and do it all again.

Here's the question: from a statistically significant view, are the averaged times of the two tests (windows up and windows down) clearly different? I am afraid that when I tried again, some 30 years after first attempting it, coast-down testing didn't work any better than I'd found previously.

I tested as described above, using a Skoda Roomster. Three runs were made in one direction, then three in the other. These were made with the car configured normally, and then another set of three runs in each direction were made with the front windows fully open. And the results? In one direction, the car slowed more quickly with the windows open (good!) but in the other direction, the car took *longer* to slow with the windows open (bad!). The consistency also wasn't very good, with for example a time variation of 6 per cent in the runs made in the same aero configuration and in the same direction.

But doesn't coast-down testing have a long history of being used to prove changes in drag? It does, but as two SAE papers describe (950626 – *ABCD –An Improved Coast Down Test and Analysis Method* and 940420 – *A Detailed Drag Study Using the Coastdown Method*), to gain accurate results requires the use of a sophisticated anemometer and highly complex interpretation of the data. The anemometer is carried on a probe in front of the car and measures both speed and direction (ie it takes into account the yaw characteristics of the airflow caused by cross-winds). The anemometer is also specifically calibrated for its near-car location.

Further, SAE 2016-01-1613, a very thorough wind tunnel study conducted by the Canadian government, makes the point that drag area (C_DA) can change with speed. The paper states:

"Experience has shown that for wind-tunnel tests on full-scale vehicles, the drag area drops as a function of speed, and reaches a plateau in the 100 to 120km/h [~60-75mph] range. However, for the SUVs and a minivan [that were tested], it was observed that the drag area continued to reduce as the speed was increased even up to 140km/h [~85mph]. This indicates that for these vehicles during coast-down testing, the drag area increases as the speed of the vehicle decreases and should not be considered constant."

I don't think that, in real-world conditions, coast-down testing is sufficiently accurate to be useful. However, if you live in an area that has at times dead-calm conditions, and you can test at very high speed, I think it would still be worthwhile pursuing.

A variation on coast-down testing, that appears to be more accurate, can be done if you live near a long, constant-gradient gentle descent. (These are not unusual on freeways.) You approach the descent at a certain speed, and then lift-off as you enter the descent, assessing your speed at one or more points on the slope.

I live near a freeway descent that, fortuitously, allows my Gen 1 Honda Insight to maintain a speed very close to the speed limit if I lift the accelerator, but leave the car in fifth gear. That is, the speed limit is 110km/h (68mph), and if I arrive at the crest of the hill at this speed and then lift off, the car will very nearly maintain 110km/h all the way down the hill, slowing only as the slope levels off. (Note that I run lots of Exhaust Gas Recirculation [EGR], and so engine braking is low. If your car has more engine braking, do it in neutral.) It was the fact that the car maintains such a constant speed while rolling down this hill that made me think it could be used for testing of drag.

I did some testing of the Insight on this slope, with the car in different aero configurations. The results are shown in the table below (and also in Figure 5-5). Speed 1 was the maintained speed on the constant gradient slope, and Speed 2 was measured at a specific mark where the slope was levelling off (both speeds rolling in fifth gear). Speeds were measured with a digital speedometer working from a 10Hz GPS input. The tests were done in quick succession and are the averages of multiple runs.

Configuration	Speed 1	Speed 2
With front winglets and rear wing	110	107
Front winglets removed	111	108
Front winglets and rear wing removed	110	106
Front winglets and rear wing removed, plus windows down	109	105

The results of multiple runs made in the same aero configuration were very consistent – much more so than the Roomster coast-down testing described above. This gives more confidence in the results! The higher the speeds, the lower the drag – so, do the numbers make sense?

First, the slowest speeds (ie highest drag) were recorded with the windows open – good! Second, the fastest speeds (lowest drag) were recorded with the car configured with the rear wing but without the

MEASURING CHANGES IN DRAG AND LIFT

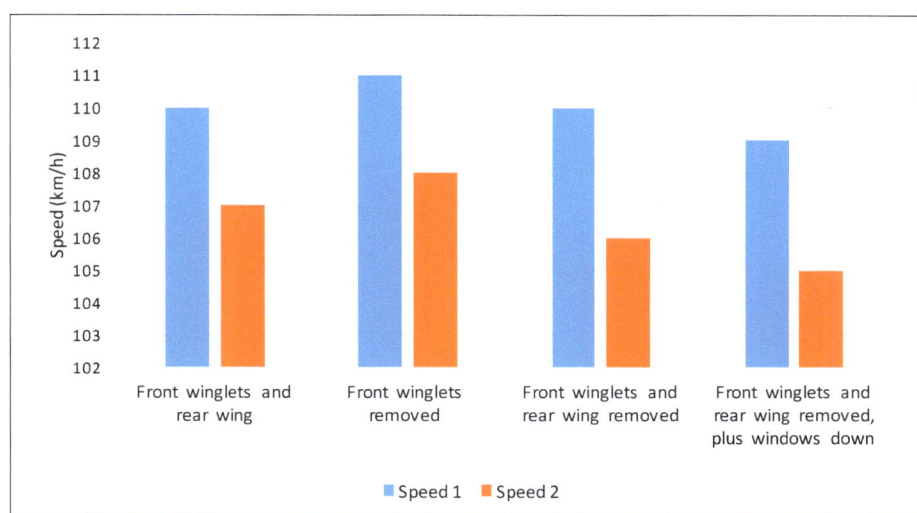

Figure 5-5: The speeds measured down a long descent in various aerodynamic configurations. The higher the speeds, the lower the drag. Speed 1 was the maintained speed on the constant gradient slope, and Speed 2 was measured at a mark where the slope was levelling off. Both speeds were while rolling in fifth gear. Speeds were measured with a digital speedometer working from a 10Hz GPS input. Taking this approach appears to be more accurate than coast-down testing.

front winglets. (The winglets were vertically-orientated, 300mm long wings positioned at the front corners and designed to generate some forward thrust. From this testing, they didn't appear to do that!) With both the front winglets and rear wing removed, drag appeared to rise again.

So, is it technically possible that the presence of the rear wing could reduce drag? Yes, it is – the car was fitted with a very effective undertray and diffuser, and a rear wing (a true aerodynamic profile with very low drag) may have caused the undercar aero to work better. But look – the change in speeds is very small, and I'd like to do a lot more testing before swearing that these results are indicative.

However, I do think that for comparing changes in drag, this type of hill descent testing is more valid than coast-down testing – at least runs made one after the other are very consistent.

FUEL ECONOMY TESTING

Measuring fuel economy is an excellent way of assessing changes in aerodynamic drag. That's the good news; the bad news is that it needs to be carried out over long distances and/or time to gain valid results. Let me give you two different examples.

Where I currently live, in Australia, I have a 'fuel economy loop' on which I perform testing. It's a rural, undulating road, normally empty of other traffic. The speed limit is 100km/h (62mph) but I normally do my test runs at 80km/h (50mph). I turn around at the same spot; I try to maintain the same speeds in the same areas; and I run the same tyre pressures, engine oil viscosity, etc. I don't test when it's raining or when it's windy, and I always test over the compete loop (ie not in only one direction).

I am mostly testing in my Honda Insight, that I have equipped with a MoTeC digital dash and MoTeC programmable engine management control unit. To calibrate the fuel economy of the dash, I have undertaken testing with a separate fuel cell plumbed to the engine, where I measured the mass of fuel used over a distance, with the distance determined by GPS. I then calibrated the dash fuel usage figure (calculated from injector duty cycle, and a GPS input) so that the dash figure was in agreement with the fuel economy figure calculated when I measured the GPS distance travelled and the mass fuel consumed. (When calculating fuel density, I even took into account the temperature of the fuel, a value needed when converting mass to volume.)

My fuel economy test loop is about 20km (about 12mi) long, and I use it when I am assessing the worth of aerodynamic modifications I make to the car. *But is this test valid?* The answer to that is, no, it's not – it's too short a test distance. If the sun is out, and the road is warm, the conditions will be different to if it's overcast and the road is cooler. If I turn to glance at the sheep shearing occurring at a farm, my speed will be slightly different at this spot. I have configured the fuel economy read-out of the MoTeC dash to be in litres/100km to one decimal place, and since on this loop I am running about 3.1 litres/100km (76 US mpg, 91 UK mpg) average, the smallest change I can see is 0.1 litres/100km, or 3 per cent.

But what if I make a car body change that results in an improvement in aerodynamic drag of 1 per cent? A 1 per cent improvement in aero drag does not convert to a 1 per cent improvement in fuel economy, especially at 80km/h (50mph), so I simply won't see this on my test loop fuel economy measurement. If I did something radical, like put a huge duck-tail spoiler on the back, I might see it immediately reflected in the fuel economy – but I could probably tell that would be the case before I even left the driveway! I could increase the resolution of the MoTeC display to two decimal places, but would this number tell me anything with certainty?

Now let's contrast that with another fuel economy test I used to prove or disprove the effectiveness of an aerodynamic modification.

In a previous time, and also a different location in Australia, each week for four weeks I travelled an identical morning route for a work contract. The car was again the Honda Insight, but back then it was largely standard (ie no MoTeC management, etc). The daily route comprised 86km (53mi) of mostly freeway travel, performed at the speed limits of 80 and 110km/h (50 and 68mph). The car had its standard trip fuel consumption display reset each day. The fuel consumption was incredibly consistent over this distance, every daily trip for weeks being 2.9 litres/100km (81 US mpg, 97 UK mpg) except for one trip that was 2.8 litres/100km (84 US mpg, 101 UK mpg) and another that was 3.0 litres/100km (78 US mpg, 94 UK mpg).

I then fitted some Airtab vortex generators (covered in the next chapter) to the trailing edge of the car's hatch. The measured trip fuel consumption immediately rose to 3.0-3.1 litres/100km, a 3 to 7 per cent increase over the 2.9 litres/100km average. (You can do the conversions for different units.) I am confident that this measurement on the 86km (56mi) daily trip was a valid indication of an increase in drag, and so I took the vortex generators off. (Note that I am *not* throwing the baby out with the bathwater; I think these vortex generators can work very well in other situations.)

You can see then that measuring improvements in aerodynamic drag through changes in fuel economy (mileage) can be done, but to be valid, these measurements need to be done over distances of something like 1000km (600mi) or more of repeatable trips. Not many people I know will do a full day of 10 or 12 hours driving, fit a spoiler – and then do another full day of driving the same route to make the comparison! Unless you are performing such driving for another purpose, it's too hard, too expensive, and it takes too long.

So why then do I even bother with my current 20km (about 12m) fuel economy test loop? Well, here's where it becomes more subjective. If you do a lot of car testing, I think you can start getting a feel for the results. For example, when testing the vortex generators on the long freeway drives, by looking at the interim figures appearing on the dash, I got a strong feeling – well before the end of the test – that the vortex generators were *not* making a distinct improvement to fuel economy. On my local test route (where, incidentally, I am often testing lean cruise mixtures in the engine management rather than aerodynamic improvements), I can usually tell in the first third of the trip whether there's going to be a pronounced change in final fuel consumption figure at the end of the test route.

So I would say that for getting a *feel* for where your aerodynamic modifications are going, a shortish local trip done consistently to the best of your ability in the same conditions is worthwhile. But would I go into print claiming a specific benefit for an aerodynamic modification tested in this way? No I wouldn't – I would want to do much longer back-to-back testing first.

In Chapter 2 of this book there's an interesting case study on the development of the Opel Calibra. In the quoted SAE paper on the car (written by the engineers doing the aerodynamic development), rear boat-tailing is stated to have caused a drag coefficient reduction of 0.011. With a final C_D of 0.26, that represents just 4 per cent of the total drag coefficient. *And that's a major change!* Could you measure on the road the improvement in fuel economy (mileage) from a 4 per cent improvement in aerodynamic drag? Yes, but to be certain, it would need to be a test carried out over a very long distance. (Incidentally, that's why fuel economy improvements recorded by long-distance trucking companies should always be paid close attention. If they're the result of aerodynamic changes, the sheer distances involved make these changes credible.)

I have tried to be honest and realistic in this section on measuring changes in whole-car drag. In short, I do not think that there is an easy solution for the home modifier that is accurate, simple, quick and easy! However, depending on your circumstances (eg local speed limits, available roads, the type of driving you regularly do), there are methods that will give you a least an indication of the drag changes that you are achieving as a result of your modifications.

COMPARING REAR VISION MIRROR DRAG

As described in Chapter 1, because of the effect of interference drag, the drag of various discrete components of a car alter when they're assembled into a complete

MEASURING CHANGES IN DRAG AND LIFT

Figure 5-6: Measuring the drag created by two different designs of rear vision mirror. If the drag of the mirrors is unequal, and the upright pole is placed equidistant between the mirrors, the pole will tend to rotate. By moving the pivot point closer to the mirror producing greater drag, the drag relationship between the two mirrors can be quantified. (Courtesy Georgina Edgar)

vehicle. However, it is still interesting and valuable to compare the drag of different components in isolation – for example, rear vision mirrors. Comparing the drag of two different rear vision mirrors can be done simply and accurately. Here is how to do it.

Mount the two different rear vision mirrors on a short cross-arm, eg one made of wood. Put one mirror at one end and the other at the other end, ensuring they're both facing in the same direction. The centreline of the mirrors should be about 500mm (~20in) apart. Drill multiple holes along the centreline of the piece of wood between the two mounted mirrors. Using a mounting screw through the hole, positioned mid-way between the mirror centrelines, screw the cross-arm to a vertical piece of timber to form a T-shaped assembly. The vertical pole should be about 2m (6ft) long.

To test the relative drag of the two mirrors, the assembly is exposed to airflow, with the mirrors orientated as they are when in normal use. I use a car with sunroof, with the pole projecting out through the sunroof and mirrors positioned about 1.2m (~4ft) above the roof of the car. Sit in the passenger seat and have someone else drive the car, preferably on a closed road. Start with slow speeds (eg 50km/h, 31mph). Hold the pole with one hand positioned low and the other high, eg at the height of the roof.

As the air speed increases, each mirror will pose increasing drag. However, if one mirror has more drag than the other, a torque will be developed on the vertical pole, tending to rotate it. If you use a square-section piece of wood for the pole, you can hold your fingertips on the flats of the timber to give pivot points that will allow the pole to turn easily. If the pole rotates, stop the car and then move the position of the upper mount so it is closer to the mirror giving more drag. Retest and again check the pole's rotation. Continue to move the mounting point until there is no discernible torque on the pole. You can do a final check that this is in fact the case by going a bit faster – then even a tiny torque can be clearly felt.

Now, how can the relationship in drag between the two mirrors be quantified? Simply work out the ratio between the two distances from the centreline of each mirror to the final pivot point location (the pivot location that gave no torque on the pole). For example, if the two distances are 205mm and 305mm, 305/205 = 1.49, so the mirror *closest* to the pivot point has, in this case, about 1.5 times the drag of the other mirror. Note that this is total drag – that is, $C_D A$.

And can we use this approach to measure the difference in C_D values? Yes we can. The above test was done using a Gen I Honda Insight mirror and a Yamaha R6 motorcycle mirror. Measured with a tape, the Honda mirror has a frontal projected area of about 260cm^2, and the Yamaha a projected area of about 140cm^2. Therefore, if the two mirrors had identical drag coefficients, the total drag of the Honda mirror should have been 260/140 = about 1.9 times

SPEEDPRO SERIES

the Yamaha mirror. But in fact, the Honda mirror has a total drag only 1.5 times the Yamaha mirror, showing that the Honda mirror has a C_D that is about 20 per cent lower than the Yamaha design.

In this case, the Honda mirror has a lower C_D than the Yamaha mirror, but because it is much larger in projected frontal area, total drag is still about 50 per cent higher. Remember, though, that's when tested in isolation – on the car the story will be a little different.

Note: this measuring technique can be made more sophisticated if you are looking for small differences. For example, you can place the pole on bearings (or use bearings at the top) so that the cross-arm can move more freely. You can also compare test items against a square flat plate (C_D = 1.17) of a known area, allowing you to calculate the C_D of your test object – and all at correct Reynolds numbers and degrees of turbulence! Incidentally, this measuring approach was first taken by the Wright brothers, when they were testing differently-shaped aerofoils for lift and drag. They used a bicycle as the vehicle.

MEASURING LIFT AND DOWNFORCE

Unlike changes in total car drag, lift and downforce can be measured accurately on the road over a short testing period, and without it costing a lot. You will still need to do some careful testing (ie keeping constant the atmospheric conditions, speed and car loads) and you will need to install some specific sensors and do a little bit of simple electronics work – but it is straightforward to do.

Before we look at how to do it, let's look at some fundamental concepts. As I've described earlier, the aerodynamic forces that are applied to the car are through the mechanism of pressures felt on the car body. They might be pressures pushing down, they might be pressures pushing up. In either case, they act largely on the suspended mass – that is, the bodywork that is suspended on the car's springs. As a result of that, if larger aerodynamic pressures push down on the body than up, the springs will be compressed to a greater extent than will occur if the car is stationary. If greater pressures push the car body upwards than downwards, the springs will be extended. To put this another way, if your car develops front lift, the front ride height will increase as you go faster. If your car develops downforce at the front, the front ride height will decrease as you go faster. Ditto for the rear.

The amount that the ride height will increase or decrease is dependent on two factors: the stiffness of the suspension and the magnitude of the lift or downforce. The softer your suspension, the greater the change in ride height will be at speed for a given aerodynamic force. This is easier to understand in real numbers. Open the rear load area of your car and place a weight over the rear axle line – say one of 10kg (22lb). How far does the rear suspension sink? That is the amount the rear ride height will decrease if you can develop 10kg (22lb) downforce at the back.

As I said, if your suspension is soft, it will be easier to measure lift or downforce. Harder suspension will extend or compress by a smaller amount for the same aero forces. However, the difficulty is in this: how do you separate normal suspension movements over bumps from the aerodynamic change in suspension height?

And before we even get to that, how do we measure ride height? The good news is that ride height sensors are readily available – they're used in cars with air suspension. The sensors are connected between the car's bodywork and the suspension arm or axle, and thus constantly measure the height of the suspension.

MEASURING LIFT AND DOWNFORCE ON THE ROAD

As mentioned above, lift and downforce can be directly measured by using sensors mounted on the

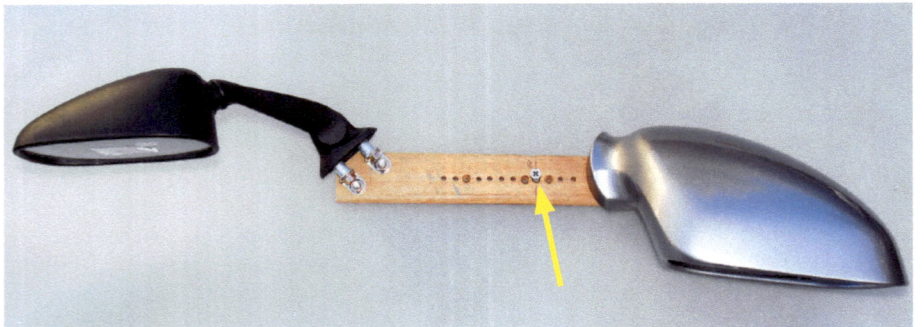

Figure 5-7: This shows the location of the pivot point (arrow) that was required to balance the drag of the two mirrors. The distances to the centrelines of the mirrors are 305mm (left) and 205mm (right), indicating that the mirror on the right has a total drag that is nearly 1.5 times the mirror on the left. However, the mirror on the right has about 1.9 times the projected area of the mirror on the left, indicating that in fact the larger mirror has a lower drag coefficient. (Courtesy Georgina Edgar)

MEASURING CHANGES IN DRAG AND LIFT

Figure 5-8: It looks good – but does it develop downforce? Unlike the measurement of total car drag, downforce or lift can be measured accurately and at low cost on cars being tested on the road. (Courtesy Honda)

suspension. Let's now look at how to do this.

The best used suspension height sensors that I have found are the Dunlop units that were fitted to P38 1994-2001 Range Rovers. These sensors comprise potentiometers (pots), where the arm connected to the suspension moves, causing the pot to rotate. Unlike more complex designs, these sensors use a simple three-wire design, allowing them to be fed a regulated voltage and then output a varying voltage signal. If buying these, ensure you pick the ones that come with a short section of loom and a specific plug. You can then cut off the plug and hard-wire the sensor into place, or add a new plug and socket. (More on wiring in a moment.)

As standard, these Range Rover sensors have a rod-shaped arm with a 90-degree bend at one end. I cut off the bend and then used a 6mm die to thread the rod. (A ¼-inch die can also be used.) Threading the rod allows miniature ball-joints to be screwed onto the arms. Brackets and links can then be made that allow the sensors to connect to the suspension arms. There must be no 'slop' in this linkage; you need to use quality parts and do a good job if you are to get accurate readings.

These sensors need to be installed so that their movement stays within the linear region – about 110° of rotation. The amount of movement that the sensor arms makes through the full suspension travel depends on the leverage ratios in the links. This is best determined through trial and error. Ensure that the ball-joints do not bind at any suspension position – it's best to remove the springs and use a jack to move the suspension when testing for this. If you are using the Range Rover sensors, note that they are not 'handed' – both left-hand and right-hand sensors are identical. This means that on one side the cable will exit at the bottom, while on the other the cable will exit at the top (important to note if clearances are tight).

You can fit two, three or four sensors. The minimum should be one sensor at the back and one sensor at the front. If you mount the sensor so that it is rotated by

Figure 5-9: Suspension height sensors can be used to accurately measure changes in ride height caused by aerodynamic lift or downforce. These are the pot-based sensors from a Range Rover that I use. Note the white dots on the sensor bodies – the sensors are linear only within this angular range.

SPEEDPRO SERIES

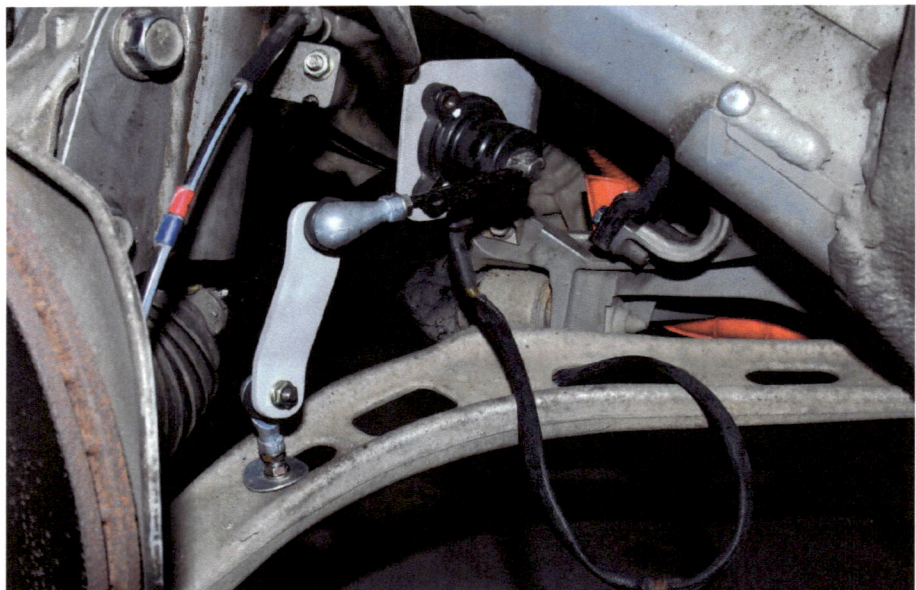

Figure 5-10: A Range Rover suspension height sensor being installed on the front suspension. Note the use of the two miniature ball-joints and the additional link. Installation is fiddly but once installed, the sensor can be used to provide you with a continuous readout of ride-height. There must be no slop in the linkage!

the central part of an anti-roll bar, an average of the left and right suspension movement at that end of the car will be gained. However, in some cases it is simpler to mount the sensor on just one wheel at the front and one wheel at the back – normally, the aerodynamic lift or downforce is symmetrical from side to side of the car.

Do not underestimate the task of fitting such sensors to the front and rear suspension. It's not difficult, but it is fiddly and tends to take much longer than you expect. I don't recommend using a conventional pot, and rigging up something temporarily – water and/or dust will wreck the open pot quickly, and if you are making aerodynamic modifications aimed at altering lift, you will want to be able to undertake testing over a period as you refine the modifications. The resistance track on a normal pot will also wear out quickly.

With the pot fitted and powered, we now have a sensor that outputs a voltage which varies with ride height. So, when the car is stationary, the voltage output of the sensor on the rear suspension might be 2.3V. But as soon as we start driving down the road, the voltage coming from the sensor is going to dance all over the place as the suspension springs deflect with bumps. What we need is a way of showing the *average* voltage output of the sensor, say with the voltage averaged over 5 seconds. This sounds complex to achieve but is actually quite easy – and cheap too.

You will need a multimeter (to measure the voltage coming from the averaging circuit), a 5V regulated power supply that will operate off the car's 12V system (to feed the circuit, including the sensor), a 220μF 16V electrolytic capacitor, and a 10 kilo-ohms potentiometer that is wired as a variable resistor. (Note: If you have not played with component level electronics before, none of this is hard or expensive.)

The circuit is shown in Figure 5-11. Note that I have not shown the 5V power supply in detail – simply buy a cheap module from eBay capable of outputting 5V (only a low current is needed) and operating off the car's 12V supply. Note also that the capacitor is polarised, with its negative terminal marked by a line of negative symbols down the body, and so it must be connected into the circuit the right way around.

In use, rotate the circuit's pot (variable resistor) until the averaging period is as you want it. This can be tested by pushing the car up and down, and then seeing how fast the multimeter display returns to a static reading when you stop pushing. I think about 5 seconds works well.

On the Range Rover sensors, the wiring colour codes are:

- brown – ground
- blue – 5V
- green – signal

With a 5V feed, rotation over the linear range gives a 0.8-4.5V output. Ensure that the voltage increases with increasing ride height, or you could be measuring lift and think that you are measuring downforce! If the voltage falls when you want it to rise, swap the brown and blue wires.

The system can be used in a number of different ways. Simplest is to leave the output in just the raw voltage. The change in voltage is indicative then of the change in ride height, reflecting lift or downforce. For example, if the voltage measurement of the front suspension is 3.70V at normal ride height, and 3.55V at 100km/h (62mph), the suspension has decreased in height by '0.15V.'

MEASURING CHANGES IN DRAG AND LIFT

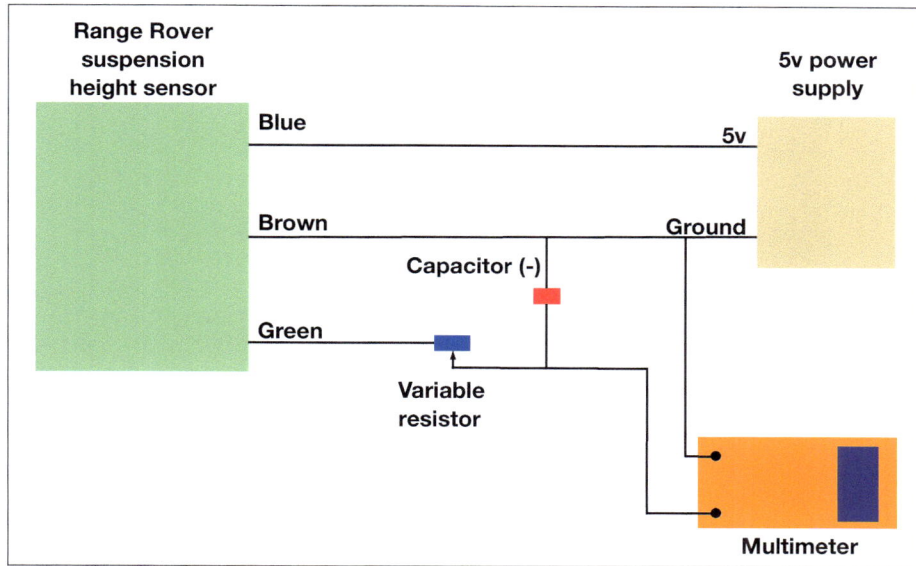

Figure 5-11: This simple circuit can be assembled very cheaply using parts from eBay. It averages the output of a ride height sensor, allowing bumps and road imperfections to be ignored, but still allowing you to see even small changes in ride height caused by aerodynamic lift or downforce. The variable resistor can be set to give the desired averaging time. A multimeter is used to monitor the sensor output.

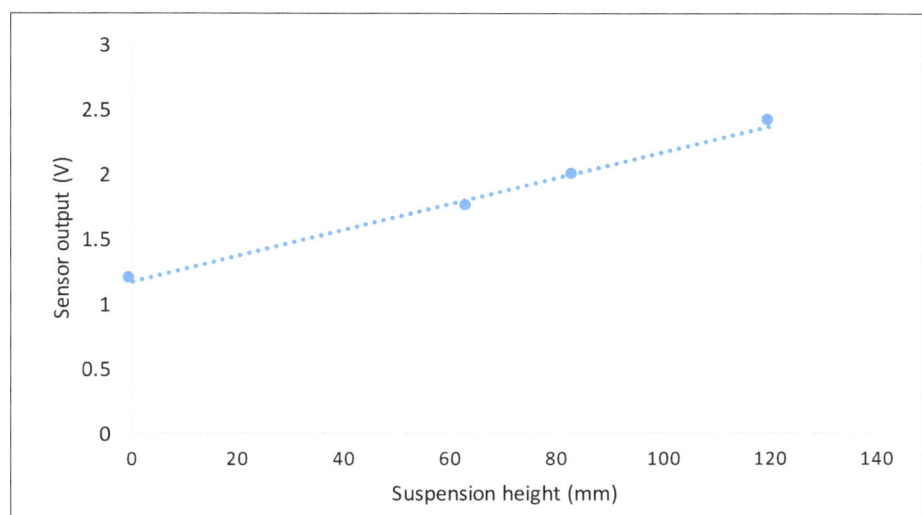

Figure 5-12: The calibration curve for a Range Rover height sensor mounted on the front suspension. As can be seen, the relationship isn't perfectly linear – but it is close.

in ride height and the wheel rate to calculate the lift or downforce. (Note that when doing this for one end of the car, double the wheel rate because in fact the suspension of both wheels is being acted upon at that end of the car.)

You can also choose to calibrate the sensor voltage against ride height. (If you are good at programming, this will allow you to use a small micro-controller module to display actual ride height, eg on a dashboard display.) To do this, you will need to remove the springs and then manually move the suspension through its full travel, from maximum compression (ie on bump stops) to maximum extension. You can then develop a table that correlates sensor output voltage with ride height. I choose to measure suspension height from the bump-stop position, that is, if my display reads '50' (I read the sensor outputs on a MoTeC programmable digital dash), it indicates that the suspension at that end of the car has 50mm (2in) of clearance to the bump stops. Calibrating in this way is a lot more work, but if you have the suspension in pieces to fit a ride height sensor, this is a good time to do such a calibration. Incidentally, note that the geometry of the linkages to the sensor will mean that the relationship between suspension height and voltage output probably won't be perfectly linear – but it will usually be close. Figure 5-12 shows the actual calibration curve for a sensor fitted to a car.

An example – Honda Insight
I have fitted my Honda Insight with custom air suspension. The control system uses three of the previously described Range Rover height sensors, feeding signals to a dedicated suspension controller.

With the car stationary again, you can add weights to the car at that axle line until this change in voltage is replicated – this will show you how much downforce was being developed. (Or you can use a spring balance – or digital crane scale – to provide the equivalent lift.) If you know the wheel rate of the car (ie the suspension stiffness recorded at the wheel), and the springing is linear, you can also use the change

111

SPEEDPRO SERIES

BUMPY ROADS, AND GETTING ON AND OFF THE THROTTLE

When using on-road ride height measurements to ascertain downforce and lift, there are two aspects to be careful of.

First, if you test on bumpy roads, you may find that ride height changes not because of lift or downforce, but because of the action of the dampers. Dampers are typically asymmetric in their behaviour, with more rebound than bump damping. The bumpier the road, the longer the average time the suspension stays compressed. If you are measuring ride height with a short-term average, you can see this effect. Either test only on smooth roads, or make A/B comparisons where you change the aerodynamics and then see the change in ride height, on the same road and at the same speed.

Second, you may be find that the front and/or rear ride heights change, depending on whether the engine is powering the car or you are free-rolling at the same speed. The reasons for this are complex: they involve the height at which drag is being felt (ie the height of the longitudinal centre of pressure) versus the effective height at which the tractive effort force is reacted. The locations of the centre of gravity and the pitch centre (the point around which the car rotates) are also relevant. It is therefore best to always make comparative measurements of aerodynamic changes when the car is under power. Having said that, it's also important to see the aerodynamic and vehicle dynamic effects of a throttle lift, so this is a set of ride height measurements you can additionally make.

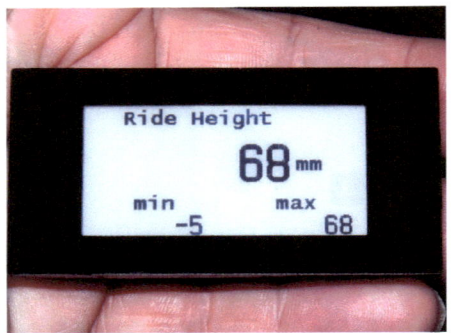

Figure 5-13: Rather than use a multimeter to display ride height in volts, you can use this Lascar SGD 21-B PanelPilot panel meter. The meter can be powered from the same 5V supply being used for the ride height sensor. The display is easily programmable, allowing a seamless conversion from volts to mm (or inches) ride height. The meter can also show maximum and minimum values, and have high and low ride height alarms. Note that while clear even in low light, the e-paper display is not backlit.

This varies suspension height via air solenoids. I also have 'live' readouts of front-right, front-left and combined-rear suspension heights (reading in millimetres height above the bump stops) on my MoTeC dash. I use a running averaging time of 5 seconds for these displays.

To measure lift/downforce, I temporarily disconnect the air suspension control solenoids, preventing correction of suspension height occurring real time. I then drive the car on a level road at a low, constant speed and read the suspension heights. This approach is needed for two reasons:
- to get rid of any stiction that might be holding the suspension in a fixed position when parked
- to assess heights when the car is level and when aerodynamic forces are so small that they will not influenced measured ride height

I then increase speed to, in this particular case, 115km/h (just over 70mph) and read the heights of the suspension again *when travelling at a constant speed in a straight line on a level road*.

Note that if your suspension is sufficiently soft, and your instrumentation has sufficient resolution, you will easily see squat and dive from acceleration and braking (including the dive caused just by lifting-off and rolling at high speed). In some cases, you will also see the change in suspension height caused by climbing or descending inclines, and if you are reading the left and right heights separately, you will also see the influence of road camber.

With the ride heights noted at speed, I then again drive slowly on a flat road to ensure the initial suspension heights re-occur. (This is important with air suspension, as a change of air temperature within the springs can also alter ride height.)

With the Insight aerodynamically standard but for a large, extended undertray at the front, the following results were gained. (More about the undertray configurations of the Insight in Chapter 7.) The rear ride height did not change with speed. On the other hand, the front ride height clearly decreased, indicative of front downforce being developed. At 110km/h (68mph), I consistently saw a decrease in front ride height of 5mm (just over 3/16in). To statically achieve this suspension deflection required the placement of 15kg (33lb) on the front axle line, so at this speed, 15kg (33lb) of downforce was being developed – a lot for what is a small and light car (850kg, or 1870lb). Note: you should be careful if adding downforce at only one end of the car – it could make the car unstable

MEASURING CHANGES IN DRAG AND LIFT

at high speed. This example is taken half way through a process of developing a full undercar aero package for the Insight – it was not the final configuration.

To add verisimilitude to this overview, here's another thing to consider. I have designed a function that allows a 10mm (just over ⅜in) lower ride height than normal to be selected by turning a dashboard knob. I also have a 'low suspension' warning set in the MoTeC dash, that flashes a warning if my suspension height is below a minimum height for more than a few seconds. (This is to warn me if there is a problem with the air suspension system.) In normal speeds at the low ride height, no warning is displayed on the dashboard. But with the front undertray fitted, if I drove fast (even on a smooth road) the 'low suspension' warning came up frequently before the air suspension system compensated and put more air into the springs.

An example – Lexus LS400

When assessing how effective a trial front spoiler was in reducing lift, I measured front ride height in a Lexus LS400 I then had. This car already had front and rear ride height sensors fitted; these were used to control the levelling of the HID headlights. Rather than use a filter on the output voltage as I did above, I simply unplugged the pot-type sensor and then measured the changing resistance of the sensor that occurred with varying ride heights. To gain an average reading in the different aerodynamic configurations, I used the 'averaging' function on a Fluke 123 Scopemeter. In addition, this meter displayed a real-time graph of the signal.

Aerodynamic testing at speed was done in the following way.

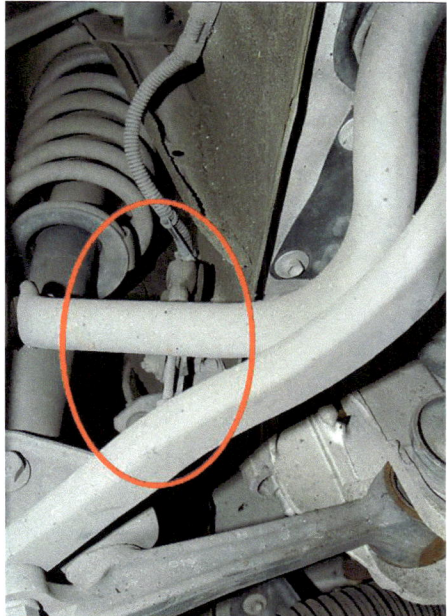

Figure 5-14: This Lexus LS400 has ride sensors front and back as standard – they are part of the headlight levelling system. As a way of monitoring ride height, I unplugged the front sensor and directly measured its resistance.

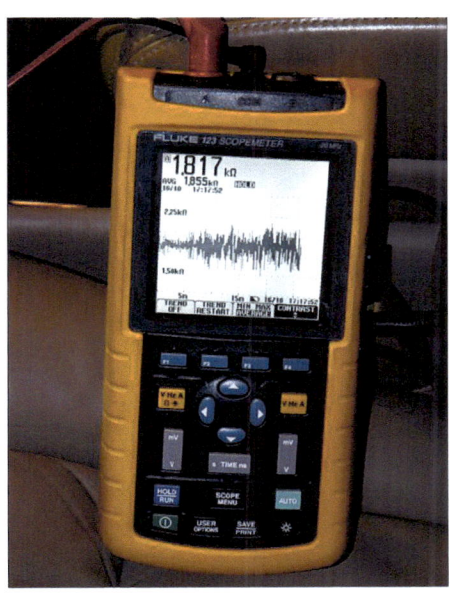

Figure 5-15: This Fluke 123 Scopemeter was used to log and average the front Lexus ride height, as indicated by the varying resistance of the sensor.

Firstly, the car was accelerated to the required speed and then held at that speed. Once the speed had stabilised, the logging was activated. The car was then held at the speed for several kilometres, and then, prior to slowing at the end of the run, the logging was halted. The testing was carried out on a smooth, straight and nearly flat freeway.

Testing with the car in standard trim gave these results:

Speed (km/h)	70	100	130	140
Average front ride height (sensor resistance in ohms)	1839	1839	1838	1832

The average ride height is expressed as sensor resistance in ohms: the lower the resistance reading, the lower the ride height. The trend is downwards as speed rises – since (approximately) 4 ohms = 1mm of ride height, you can see that the average ride height has dropped by 1-2mm as the speed has gone from 70-140km/h (about ¹⁄₁₆in from 43-87mph).

The next step was to make a simple front spoiler. This comprised a flat piece of aluminium sheet 800 x 150mm (31 x 6in). It was placed a little forward of the front axle line, so well back from the leading edge of the car. I made it remotely adjustable in angle, primarily so that I could lift it on bumpy roads and deploy it on smooth roads. In the following test, it was oriented vertically. Testing was carried out at 100 and 140km/h.

Speed (km/h)	100	140
Standard ride height	1839	1832
Spoiler ride height	1870	1883

SPEEDPRO SERIES

As can be seen, the ride height of the front of the car increased with the front spoiler fitted. At 100km/h (62mph) the front of the car was being lifted by about 8mm (5/16in), and at 140km/h (87mph) by about 13mm (1/2in). So the spoiler provided lift, not downforce! (The spoiler was mounted too far rearwards, and so air pressure was being developed upwards on the undertray in front of the spoiler, causing the lift.)

Unlike measuring changes in drag, I think that there are practicable, cheap, accurate and effective ways of measuring changes in lift (or downforce). The faster that you can go, the greater the change you will see, but at even 100km/h (62mph) you should be able to measure changes in ride height resulting from vertical aerodynamic forces.

Figure 5-16: The test spoiler on the Lexus LS400 – it wasn't always as battered as this, that's what low testing can do! The on-road measured results showed that this spoiler created front lift, caused by its rearwards placement.

Figure 5-17: As we've seen, lift and downforce is caused by varying speeds of airflow above and below the car causing different pressures. However, it's often easiest to find out what the action of these pressures are by directly measuring front and rear axle lift or downforce. (Courtesy Mercedes)

Chapter 6
Reducing drag

- **Reducing frontal area**
- **Reducing rear vision mirror drag**
- **Reducing cooling drag**
- **Ride height and rake**
- **Reducing the area of the wake**
- **Reducing the strength of the wake**
- **Achieving clean separation**
- **Reducing separation bubbles**
- **Undertrays**
- **Wheels, tyres and wheelarches**
- **Trailing vortices**
- **Add-on modifications**
- **Testing multiple drag-reducing modifications**
- **Reducing drag – development of a new car**
- **Reducing drag – modifying an existing car**

In this chapter I want to look at techniques that are specifically aimed at reducing drag. If you are a hyper-miler, this is for you! But if you're also a high-performance driver, there is important information in this chapter that is relevant to what you want to achieve. I think that for a fun car on the road, the best approach is one that has low aero drag but also good stability and grip. In this chapter I'll look at reducing drag, and in the next chapter I'll look at reducing lift and increasing stability.

REDUCING FRONTAL AREA

As described in Chapter 1, the drag area of a car is found by multiplying the C_D by the projected frontal area. You would therefore assume that if you wished to reduce drag, an easy way of doing so would be to reduce frontal area. However, this assumes that the C_D does not increase at the same time!

Let's explore that idea in more detail. Prevalent in older modified cars was the use of a low front spoiler (sometimes called an air dam or chin spoiler). This prevented air passing under the car – with very few exceptions, the underside of the car was rough and so caused great drag. Decrease the amount of air hitting these obstructions, and the overall drag was reduced. But hold on – adding a low front spoiler also increases the frontal area of the car! So for the front spoiler to be effective in reducing overall drag, the gains in C_D made by the spoiler have to be *greater than the losses* caused by increasing the frontal area. And how will you know if this is going to be the case? You can make a guess based on the undercar roughness (the rougher the underside, the more likely a low front spoiler will reduce overall drag), but to be certain, you would need to test a low front spoiler to see what changes to drag actually resulted.

SPEEDPRO SERIES

Figure 6-1: Using a laser in the Mercedes wind tunnel to calculate projected frontal area of a 116 Mercedes S-class. Looking at the image behind the car, you can see the proportions of the area made up by the external rear vision mirrors and the width of the tyres. (Courtesy Mercedes)

I use this example not to make things difficult, but to show that changing one aspect of aerodynamics can influence other areas. So what reductions in frontal area are almost certain to reduce overall drag – that is, there will *not* be a proportionate increase in C_D with the decrease in area?

One approach where you can be fairly confident is the removal of external rear vision mirrors. For example, if in your country you are required to have only one external mirror, you could remove the other. Or, if in your country it is legal to use cameras and interior display screens, you can reduce the frontal area created by the rear vision mirrors by an even greater amount. One car modifier – Pascal Dunning – estimated a reduction in frontal area of 2.5 per cent with the removal of both the mirrors on a small car. A head-on photograph and an overlay of graph paper will let you calculate the change (just count the squares!).

And the change in C_D with the removal of the mirrors? While it's an intriguing truth that many mirrors develop aerodynamic thrust (that is, they have low pressure areas on forward faces), overall, mirrors create drag. Audi, when testing its A2, found a reduction in C_D by about 4 per cent with the removal of the mirrors. No matter what the shape of their housings, the requirement that mirrors have for a large, flat rear surface means that they generate considerable wake – something able to be seen by wool-tufting the adjacent front door glass in most cars. In addition, mirrors disrupt the flow around the A pillar. It seems inevitable that cameras and screens will replace external rear vision mirrors in the low-drag cars of tomorrow. (That said, there are some challenges. The first is to ensure that the 'on all the time' current draw of a camera and display system doesn't use more energy than that saved by the lower drag. The second is that mirror housings often provide the home for other systems, eg puddle lamps, indicator repeaters and lane departure sensors/cameras – even, in some SUVs, depth sensors for water wading.)

Another area where you can be fairly confident that a reduction in frontal area will not be accompanied by an increase in C_D is in reducing tyre widths. Tyres make up a proportion of the frontal area, and so a reduction in tyre width decreases frontal area. SAE paper 950627 deals with a 1990 Rover 820Si, a car with a standard C_D of 0.34. It had its tyre width increased from 195 to 205 section; C_D increased by 0.007 – just 2 per cent. (Conversely, going from 195 to 175 section width dropped C_D by 0.013 – nearly 4 per cent.

Just a few words on using narrower tyres. First, when considering a narrower tyre, it's worth making a physical measurement of the tyre mounted on the rim and appropriately inflated. Not all tyres with the same section width are actually the same width! Second, to maintain the same rolling diameter with the narrower tyre, you

Figure 6-2: In this car, the external rear vision mirrors have been replaced with cameras and internal screens. Taking this approach reduces frontal area. (Courtesy Pascal Dunning)

REDUCING DRAG

Figure 6-3: A low front spoiler on a car like this 1970s BMW will reduce drag, even though the frontal area is increased. (Courtesy Shannons)

will need to go up in aspect ratio (tyre profile). Higher profile tyres tend to handle more poorly, are often made of harder rubber, and with the narrowing of width, you've also changed the shape of the contact patch. (In a well-balanced, smaller car I don't have a problem with relatively narrow tyres. In fact, I always prefer a car that handles well on relatively slippery tyres, rather than one that grips and grips and then when it does let go, is very hard to control.) But – and here's the key point – if you change to narrower tyres, *you will be changing the handling characteristics of the car*, and you should be aware of that. This is even more important if you are using very high tyre pressures to reduce rolling drag – be careful in these areas!

With older cars, it used to be often implied that lowering the car's ride height substantially decreased the frontal area. This was premised on the idea that the underside airflow was so bad that the frontal area might as well have been the height multiplied by the width of the car – and in many cases, that was accurate! With smooth underfloor cars, that idea doesn't apply. However, it should be noted that lowering *any* car does decrease frontal area by a small amount, as there is less tyre area exposed. (I'll talk more about lowering the ride height in a minute.)

You might think that areas of the car that allow air to pass through them should not be categorised as frontal area. After all, if there's a big opening at the front of the car (the radiator grille) and the air can pass through that grille, through the radiator and the exhaust to the atmosphere, surely that isn't the same as having solid bodywork? The answer is that no, it's not the same – it's actually *worse*. This is because the frictional drag created by the heat exchanger, and the turbulence generated by its exhaust airflow, create a lot of drag More also on the drag of cooling systems shortly.

In summary, if you can decrease frontal area without then adversely affecting C_D, you will reduce overall drag. However, in some cases – eg a front spoiler on a car with a rough underbody – increasing frontal area may in fact reduce overall drag.

REDUCING REAR VISION MIRROR DRAG

Earlier, I talked about removing the external rear vision mirrors as a means of reducing frontal area. However, a lot of legal jurisdictions require the use of external mirrors – cameras are illegal. So what can be done to reduce drag by altering

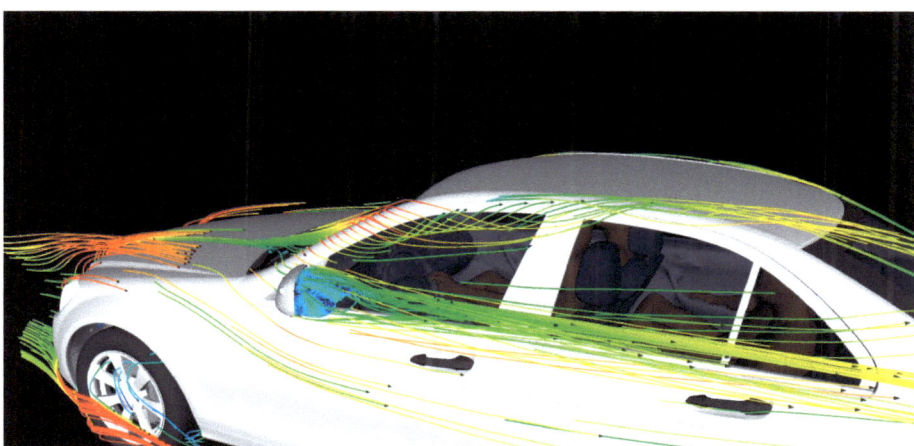

Figure 6-4: This CFD graphic of airflow speeds shows the wake of the mirror (blue), and also how airflow is influenced well downstream of the mirror wake itself. The flow behind the mirror does not return to 'full speed' until well onto the rear passenger door window. (Courtesy Mercedes)

SPEEDPRO SERIES

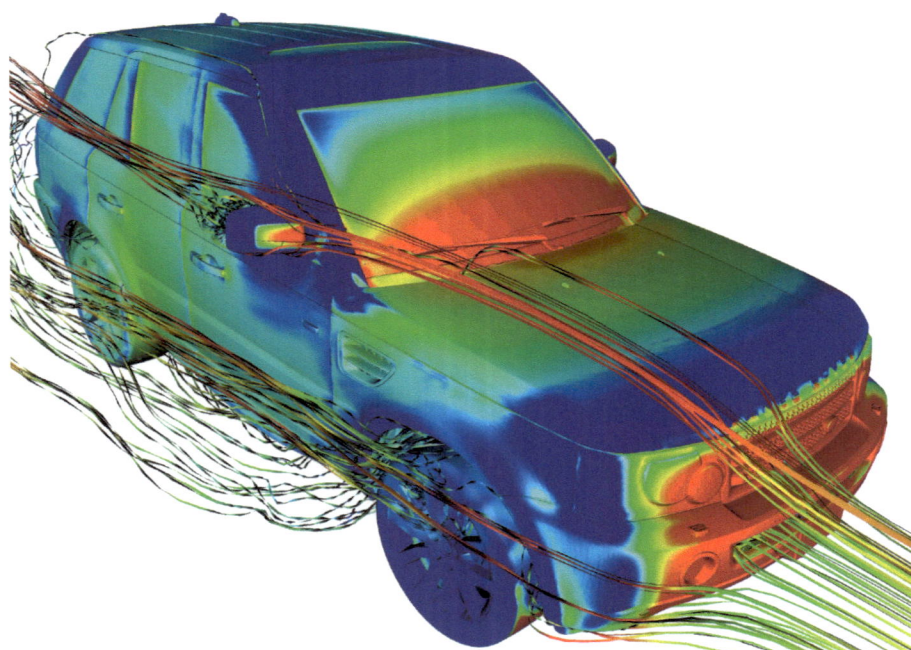

Figure 6-5: In this CFD graphic, red shows highest pressures and blue, lowest pressures. Note the area of high pressure on the rear vision mirror (red), but also note that as airflow wraps around the curves, low pressures are developed. If the resultant force has any forward component, the mirror is creating thrust on at least part of its surface. Unfortunately, though, the low pressure wake behind the mirror will much more than cancel any forward thrust developed on the front face. (Courtesy EXA / Land Rover)

the shape of the mirror housing? In addition to the size of the reflecting glass itself, three inter-relating aspects need to be considered. These are the influence of the housing shape on:

- drag
- noise
- water and dirt deposition on the side glass

Here I'll concentrate on drag reduction (noise is covered in Chapter 10) but be aware that making changes to mirror housing shape may also influence water and dirt deposition.

So what are the potential drag benefits of optimising the shape of the mirror housing? They exist, but unfortunately they are very small.

Volvo produced a very interesting SAE paper on this subject (2000-01-0486 *Drag and Dirt Deposition Mechanisms of External Rear View Mirrors and Techniques used for Optimisation*) that showed that the lowest increase in C_D they could expect to achieve from a single mirror was 0.0027 (referenced to a frontal area of 2.23m²), and that each then-current mirror caused a change in C_D of about 0.007.

What does this mean? Well, if the car as a whole had a C_D of 0.28 (an arbitrary figure), a typical single mirror was responsible for about 2.6 per cent of this, while the best-shaped mirror would reduce this to 1 per cent. Clearly, it is much better to remove the mirrors altogether than to tweak their shape!

To optimise the shape of the mirror housing for drag, three aspects need to be considered.

The first is that, just like the car as a whole, early flow separation should be avoided. That is, if the flow separates prior to the trailing edge around the mirror glass, the mirror wake will be larger, and so drag is more likely to be higher. This separation can occur if, in plan view, there is a contraction in shape just prior to the trailing edge. Figure 6-6 shows an example of this type of mirror, where the shape creates a negative pressure gradient and so promotes early separation. Note that while I haven't tried it, an onboard smoke generator and wand should be able to inject smoke into the mirror wake of a moving car and clearly show its size.

The second aspect to consider is the creation of thrust (negative pressure) on the housing surface, where this negative pressure has a forward-facing component. As touched on above, some mirror housings develop areas of forward thrust where, again when seen in plan view, the airflow wraps around the outside curve. Typically, you want to have a 'full' curve positioned as far outward as possible. Figure 6-7 shows a mirror housing that does *not* have this feature, while

Figure 6-6: The edge (arrowed) is likely to create early separation, so increasing the size of the mirror's wake.

REDUCING DRAG

Figure 6-7: This mirror housing has a relatively flat outer housing, and so is likely to develop little forward thrust.

Figure 6-8: This mirror housing has a 'full' outer curve and so is likely to develop some forward thrust. However, the upper and lower radii need to be sufficiently gentle to keep attached flow on the top and bottom of the housing.

Figure 6-8 shows the full outside curve that helps develop a forward thrust. (Note that, of course, there are still much larger areas of positive pressure on the front-face of the mirror housing than there are negative pressures.)

And can these pressures (both high and low) be measured in the same way I described in Chapter 4? Probably not. The areas we're talking about are very small and the surface pressure measurement techniques covered in that chapter (eg using the circular puck) would disrupt the very flow you're trying to measure.

However, mirror housings are also much cheaper than car body panels, and so if you can get hold of a spare mirror housing, you could drill the housing to take surface pressure measuring taps, with the sensing hoses fed though the mirror mount to your measuring instrument (eg a Magnehelic gauge).

The third aspect to consider is the size of the 'foot' – the mirror housing mount. In short, the smaller that this can be made, the better.

Not covered so far, but still quite important, is the location of the mirror mount. Two different approaches can be taken – either mount the mirror housing on the 'sail' (the triangular area at the forward edge of the door) or mount the mirror on a pedestal that attaches to the door skin. The latter approach allows airflow to wrap around the full extent of the A pillar – the flow can then pass largely unimpeded between the mirror and pillar. The wool tufting photos in Chapter 3 show good lower A-pillar flow on the Porsche 993, Mazda RX7 and Mitsubishi Magna – all cars that have the mirror mounted on the door and not the sail panel. When considering the mirror mount, remember that the speed of flow around the A-pillar is high (that's why there is a low pressure) and so if the mirror disrupts this flow, the effects are much greater than if the mirror were simply tested for drag by itself.

Trial changes to mirror shape, especially where fuller curves are needed, are easily made using plasticine. If the mirror housing is

Figure 6-9 (Left): One of the Volkswagen XL1 rear vision cameras, located near the forward edge of the doors. (Right): The interior screen for the rear vision camera in the Volkswagen XL1 is located on the door. (Both photos courtesy Volkswagen)

SPEEDPRO SERIES

Figure 6-10: Here smoke has been injected into the wake of a rear vision mirror. Where possible, the rear extremities of the mirror should be slightly curved inward to reduce the wake size, much as is achieved with boat-tailing of the car body. (Courtesy Mercedes)

too large, start with a smaller mirror. Note that if you use mirror housings from another car, you may need to do some tricky modifications to retain remote control of mirror angle and operate the indicator, mirror heating, lane departure warning equipment, etc. If you don't wish to retain any electrical functions, and if legal requirements of the original mirrors don't need to be met, the mirrors from motorcycles are smaller and can more easily be spaced away from the bodywork. Finally, note that as described in the previous chapter, it's easy to make comparisons of total drag (ie C_DA) of different mirror designs – at least when they are measured off the car.

REDUCING COOLING DRAG

How much drag does the cooling system in a car cause, and how can that drag be reduced?

Hucho (*Aerodynamics of Road Vehicles*), quoting 1981 research, states an increase in C_D of 0.025 from fitting a normal front-mounted radiator, with exhaust air dispersing out past the engine. This is reduced to 0.02 if a ducted radiator is fitted that exhausts into the front wheel wells, while a ducted radiator picking up from low down at the front of the car, and exhausting air through the bonnet, causes the lowest change in C_D of 0.01.

Hucho doesn't state what the overall C_D is of the car to which he is referring, but it was probably around 0.40. This would then indicate the cooling system causing about 6 per cent of total C_D. Hucho also states that an optimal cooling system can increase drag by only 2 per cent, but that a poor design can increase drag by as much as 10 per cent.

Incidentally, although it is not often referred to, some cars do exhaust a lot of radiator air out through the front wheel housings. My Gen 1 Honda Insight (C_D = 0.25) has openings that connect the engine bay to the wheel housings. The openings are much larger than are needed for the driveshafts and steering arms. SAE paper 2017-01-1529 (*Complete*

> ### LENSES AND MIRRORS?
> SAE paper 2017-01-1358 (*A Rear-View Side Mirror with Exterior Lens to Improve Field of View and Aerodynamics of Automobiles*) describes an approach that maintains most of the simplicity and low cost of traditional external rear vision mirrors, but substantially reduces drag. The novel rear vision mirror system comprises an external, specially-shaped lens that is used in conjunction with a mirror placed largely inside the vehicle. The lens, which is concave on both faces, protrudes from the side of the car by only 70mm (2¾ in). The system can be configured to provide a wide field of view (40°), or alternatively can use a flat (and so, in many jurisdictions, legal) mirror on the driver's side and provide a 25° field of view.
>
> If you are adept with lenses and mirrors, it may be worth exploring this option, especially if legal constraints mean the use of a camera and screen are prohibited.

Body Aerodynamic Study of three Vehicles) models cooling airflow, and shows the following exit proportions that are probably indicative of a recent car's behaviour: wheelarches – 45 per cent; transmission tunnel – 55 per cent. This paper also has the cooling drag figures for two more recent cars: the Jaguar XJ has a cooling drag change C_D of 0.025 and the Land Rover Range Rover a cooling drag change of 0.035. (Both cars are presumably 2016 models; if so, the XJ with a CD of 0.29 would have a cooling drag of 9 per cent.) So in summary, most references suggest a cooling airflow drag of 5-10 per cent. Note, though, that this proportion is creeping up as total vehicle drag drops.

Incidentally, while the drag

REDUCING DRAG

caused by the cooling system can be measured by simply blocking the radiator, comparing this method with using a strain gauge to measure changes in the actual forces acting on the car does not necessarily correlate well. That is because changes in cooling system airflow may affect the airflow around the vehicle body as a whole. As described above, in the case of the lower front spoiler, it's not always wise to try to isolate the effects of different aerodynamic changes. To put that another way, you cannot say that by rule of thumb you gained a C_D reduction of 0.01 *here*, and another 0.015 *here*, and 0.02 *here* – and thus you've dropped C_D by 0.045!

Optimal versus actual

It's important to differentiate cooling systems that are optimal for low drag from those fitted to the vast majority of normal road cars. An optimal system is likely to use a fully ducted radiator. The intake duct will be positioned at an area of high pressure and will be relatively small in frontal area, for example, 25-30 per cent of radiator area. The duct that connects the opening to the radiator will progressively expand (a 7° expansion angle is good) and with this expansion rate, the airflow will remain attached to the duct walls. The radiator, which might be angled from the vertical to allow the final duct cross-sectional area to be smaller than that of the radiator, will have equally high pressure developed across its face. On the other side of the radiator, the air will again pass through a second duct, this time one that exhausts into an area of low pressure – for example through a vent in the bonnet that directs the exhaust air along the panel, so not causing localised flow separation.

There is nothing at all wrong

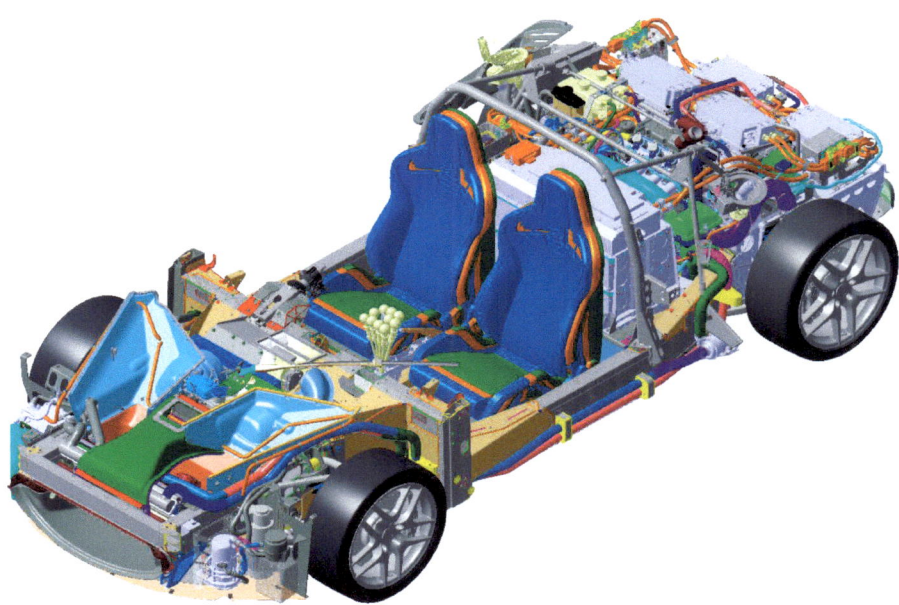

Figure 6-11: A mid-engine car like this Lotus has much more room at the front to use good ducting to and from the radiator. (Courtesy Lotus)

with those ideas – except for the fact that implementing them in a road car you're not building from scratch is likely to be exceedingly difficult! For example, space requirements normally dictate that the radiator (or whatever the front-most heat exchanger actually is) is positioned quite close to the front of the car, preventing the use of the optimally expanding intake duct. Furthermore, the vast majority of cars do not use ducting on the rear face of the heat exchangers – instead there are fans and perhaps just a few finger-widths of space to engine components! (The situation is a little different with a mid-engine car with a front-mounted radiator, where there is room in the front of the car to duct air to the radiator and then exhaust it more appropriately – see Figure 6-11.)

So what actually happens to the radiator airflow in a typical car? SAE paper 2002-01-0712 (*Use of a Pressure-Based Technique for Evaluating the Aerodynamics of Vehicle Cooling Systems*) contains some interesting information on the airflow prior to the radiator. The car was a typically-configured front engine, rear-wheel drive car of the 2000 model year (Ford AU Falcon). This car has cooling air openings above and below the bumper, and the radiator is vertical. Exhaust air passes into the engine bay. Ducts are not used to connect the front openings to the radiator.

To measure the distribution of pressure across the front and rear faces of the radiator (and from this infer the distribution of flow), 24 pairs of pressure sensing tubes were used. For each pair, one tube had its mouth placed flush with the rear of the core and the other flush with the front face. (That is, the tubes were inserted from each side of the radiator and pushed through until their mouths were flush with the other sides.) The higher the difference in pressure measured with

Figure 6-12: The dual radiator duct design of a 2017 Ford Focus. Sealing the front openings to the radiator reduces turbulence ahead of the radiator, but also reduces the area of the radiator that is actually working! (Courtesy Mountune)

these tube pairs, the greater the flow at that point. In addition to pressure measurement, the flow directions in the space between the radiator and the front of the car were ascertained with wool tufts.

If you want to accurately measure flow distribution across the radiator, both of these techniques are easily undertaken. If less accuracy is required, rather than using pairs of flush-mounted probes, you could use simple bare-end sensing hoses moved around to various places on the front and rear of the core (ie measure the differential pressures) or even more simply, measure just the pressure variation on the front face of the core. Wool tuft movements could be monitored by a small video camera (eg of the GoPro type).

And what did the paper find? Very uneven flows across the radiator! In this case, the top part of the radiator had the best flow and there was low to zero flow in the middle of the radiator, the area behind the bumper beam. The lower part of the radiator flowed more poorly than you would expect, given its position behind the relatively large lower bumper opening. Behind the bumper beam, a pair of rotating vortices in front of the heat exchangers (the air-conditioning condenser was also present) could be noted. At least in this case, there was no reverse flow occurring through the radiator. It appears that, as found in Chapter 8, the pressures on the rear face of the radiator were roughly similar across its area. The lesson for all this is that there are likely to be major pressure (and so flow) variations across the face of a radiator that has simply been placed at the front of the car and located behind some front-facing openings.

To make the best of a bad situation, some manufacturers choose to use two separate ducts connecting two front openings to two areas of the radiator – one duct for the above-bumper opening and one duct for the below-bumper opening (see Figure 6-12). This approach gets rid of the large 'free' volume in front of the radiator and so reduces the influence of the wake behind the bumper beam. In effect, the cooling capacity of the radiator is simply split into two, one area behind each duct.

Grille shutters, blockers and variable air outlets

Chapter 8 covers how to improve the flow of air through heat exchangers. As also described above, this is achieved by maximising the aerodynamic air pressure across the front face of the heat exchanger, and reducing the pressure behind the heat exchanger to as low a value as possible. Doing this creates the greatest airflow through the core – excellent for cooling. On the other hand, to reduce *drag* to the greatest extent, the heat exchanger should have no airflow through it at all!

The way that these contradictory requirements are achieved in an increasing number of cars is to use a radiator shutter. That is, a 'venetian blind' arrangement is installed in front of the heat exchangers, so that when little cooling is needed and low aerodynamic drag is preferred, the blinds are shut and airflow blocked. At times of highest cooling requirement, the blinds are fully opened, so creating little obstruction to the cooling airflow. Variations in requirements between these extremes is catered for by varying shutter openings.

And the potential reduction in drag from using a shutter? That clearly depends on what proportion of the total drag is made up by the cooling system airflow, but SAE paper 2011-01-0155 (*Application of Design of Experiments and Physics Based Approach in the Development of Aero Shutter Control Algorithm*) suggests a maximum change in C_D of 0.027, which seems reasonable. (Assuming a total C_D of 0.30, that would indicate a C_D reduction of 9 per cent.)

Note though, that a Canadian study (SAE paper 2016-01-1613 – *Evaluation of the Aerodynamics of Drag Reduction Technologies for Light-duty Vehicles: a Comprehensive Wind Tunnel Study*), that was carried out on vehicles in a full-size wind tunnel, found that while grille shutters could typically reduce drag by 1-5 per cent, the actual reduction depended on two factors:

- The obstructions present even when the grille shutter was open. If obstructions were already

REDUCING DRAG

Figure 6-13: This Mercedes uses an electronically-controlled shutter to control airflow through the cooling heat exchangers. (Courtesy Mercedes)

present, the gains in closing off the cooling entrance were less – *de facto* closed grille shutters already existed!

- The yaw angle of the oncoming airflow. The reduction in drag was not as great at yaw angles other than zero.

In addition to these factors, there's another complicating influence, which is most easily summarised by the question: *with the grille blocked, what happens to the air that would otherwise have flowed through it?* In a paper prepared by Jaguar Land Rover engineers (*An experimental investigation into the flow mechanisms around an SUV in open and closed cooling air conditions*), wind tunnel testing was carried out on a 13MY Supercharged Range Rover. The testing showed that base drag was generally greater for the closed cooling cases. The authors comment:

... *this follows intuition, as when the cooling apertures are closed, the air which would otherwise lose some momentum going through the engine bay all passes around the exterior, thus having a greater potential for creating drag elsewhere.*

Therefore, while I stated earlier that to reduce drag to the greatest extent, the heat exchanger should have no airflow through it at all, the actual situation is more complex than that, with both pluses and minuses occurring with this change in airflow. (That said, in all 21 configurations in which the Range Rover was tested, C_D decreased with taped-up cooling air inlets.)

Further influences on the drag reduction gained by the use of a shutter are the position of the shutter (further forward is better) and how well the shutter seals, with any leakage reducing its efficiency.

Fitting a radiator shutter system to a car that did not originally have one is possible, however it may be tricky and expensive until scrap parts are more readily available. (Note that the above 2011-01-0155 paper also covers some of the logic required to develop electronic shutter control maps.)

Instead of the complexities of variable radiator flow achieved through the use of active grille shutters, many people seek to decrease cooling drag by simply fitting a 'grille blocker.' As the name suggests, this prevents airflow through the grille. Especially in cars that have both upper and lower

Figure 6-14: A removable upper grille block in place. This one is made from shaped foam board, coated with epoxy and then painted. (Courtesy Darin Cosgrove, EcoModder.com)

openings for cooling system airflow, blocking only one of the openings is likely to be acceptable in all but hot climates. I have seen many people make elaborate blockers that cover the front face of the grille, but it's much simpler to use dense foam rubber cut to the appropriate shape, painted black (if required) with a spray can, and then inserted into the opening. The result is near-invisible, and can be easily removed in summer, if towing, etc.

If there are measured pressure variations across the radiator, and some areas of the radiator appear to be doing nothing, an improvement in drag might be possible by installing a duct between a front opening and the radiator, and then blocking the rest of the radiator core. Choose the front opening that has the highest measured pressure, and remember that the duct can be of the expanding type so that the area of the duct at the radiator is larger than the area of the intake. The duct will need to be sealed to the radiator, eg by the use of a foam rubber strip.

The approaches that I've so far covered involve regulating flow through the entrance to the cooling system – but what about instead varying *outlet* airflow? Barnard (*Road Vehicle Aerodynamic Design*) makes the point that controlling the outlet flow is actually more beneficial to drag than controlling inlet flow. This is because a grille blocker or shutter simply throttles the airflow, while a carefully-designed variable flap on the air exit can cause the airflow to accelerate. This faster airflow helps recover lost longitudinal momentum.

In a 2004 paper by Barnard et al (*Fixed and Variable Cooling Outlet Geometries for the Minimisation of Associated Drag*), the cooling exit from the engine bay under the car was varied by either (1) a sliding plate with a rounded rear edge, or (b) a flap that pivoted from a hinged front edge, dropping down at the rear. The paper found that at small openings, the flap was more efficient than the sliding plate, but at large openings, the sliding plate was more efficient. (The latter was the case because when the flap was opened a long way, it caused an increase in projected frontal area.)

And how effective was the flap in reducing cooling drag? Compared with the standard vehicle's modelled 0.047 cooling drag, the use of a flap reduced this to just 0.0012 at 90km/h (56mph), rising to 0.02 at 202km/h (125mph) (the car's top speed, where maximum cooling was required). To put this another way, when referenced against the car's C_D of 0.3, by using the variable air exhaust flap, cooling drag decreased from a standard (and high) 16 per cent to a maximum of 7 per cent and a minimum of just 0.4 per cent!

Controlling the air exit flow has other advantages for home modifiers. Typically, under the car there is more room available than in front of the radiator. Secondly, the area is concealed from view, so cosmetics are not so important. Finally, constructing a forward-hinged flap is easy, especially if it is mounted on the trailing edge of a

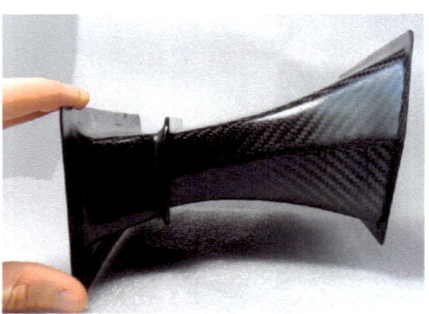

Figure 6-15: (Far left) This duct was created to channel all cooling air through just a small proportion of the radiator. If taking this approach, ensure that adequate cooling performance is still retained. (Left) The new carbon fibre duct was integrated with a large grille-block panel. (Both photos courtesy Pascal Dunning)

REDUCING DRAG

> **IMPLICATIONS OF REDUCING COOLING AIRFLOW**
>
> Before making any changes to cooling airflow you must have a dashboard gauge that accurately reflects coolant temperature. Many cars now electronically hold the dashboard coolant temperature gauge at a fixed value eg 90°C (194°F) until things are going catastrophically wrong. If you have reduced airflow through the radiator, you *must* be able to see if temperatures are higher than normal, or are rising. Either fit an independent coolant temperature gauge or use an OBD reader to monitor coolant temperature. In a high-performance application, it is a good idea to also monitor engine oil temperature.
>
> Note also that reduced cooling system airflow will reduce the efficiency of *all* the heat exchangers located at the front of the car – they might include not only the radiator, but also the air-conditioning condenser, transmission oil cooler, hybrid power converter cooler, and so on. Keep this in mind before making a complex, permanently-fitted grille blocker.
>
> Finally, Jaguar aerodynamicist Adrian Gaylard made an interesting point to me about altering cooling airflows. Closing cooling intakes also changes lift balance. For cars that have the cooling exit flow on the underside (as opposed to wheelarches or bonnet – ie most cars) front lift will reduce substantially, with a reduction in C_{Lf} of around 0.04-0.05. This will dial-in aerodynamic oversteer; more of a concern for fitting blockers as most grille shutter systems will (and should) be designed to open at high speed.

new engine bay undertray. The flap could be given no control (set its angle to cater for worst-case cooling requirements), manual control (eg by an engine-bay adjusting rod) or automatic control (eg by a geared DC motor).

RIDE HEIGHT AND RAKE

You might intuitively think that a lower ground clearance would result in decreased drag. However, the relationship isn't quite that simple. Scibor-Rylksi (*Road Vehicle Aerodynamics*) plots C_D versus ground clearance for smooth and rough underside vehicles. He shows that for smooth underside vehicles, C_D decreases as ground clearance *rises*. However, for vehicles with rough undersides, C_D decreases as ground clearance *falls*. This is the case for the rough underside vehicle because the drag is higher when more airflow must negotiate the rough surface.

Tamai (*The Leading Edge*) looks at ground clearance for minimum drag for solar race cars. For cars of this type, he suggests a ground clearance of 200-300mm (8-12in). However, he also cites research that suggests, for a more generic car model, an h:L (ground clearance to body length) ratio of about 0.02 is desirable. To apply this, simply divide the length of your car by 0.02. For example, that gives a desired ground clearance for my Gen 1 Honda Insight of 79mm (3.1in). The standard ground clearance of most vehicles is about 150mm (6in), so perhaps it *is* the case that the lower you can set the practical ride height for your road car, the better! However, Tamai also highlights research that shows streamlined bodies have a drag curve that drops quickly as ground clearance increases, but past a certain ground clearance, drag starts to slowly rise again – see Figure 6-16.

SAE paper 2016-01-1613 has already been mentioned in this chapter. In the paper, the Canadian researchers trialled lower ride heights for a range of 2013-2015 model year cars. This study, where the vehicles did not have full underbody covers showed that dropping ride height could give large reductions in overall drag. The greatest incremental reductions in drag occurred when the nose was dropped in ride height, eg by 20mm (just over ¾in). A 2-3 per cent reduction in drag was measured on most of the test vehicles with a front ride height drop of 20mm; this increased to 5-6 per cent with a 40mm (just over 1½in) front and rear ride height drop. Interestingly, the Tesla Model S, a car with the smoothest underbody I have ever seen on a road car, also achieves a drag reduction by lowering the

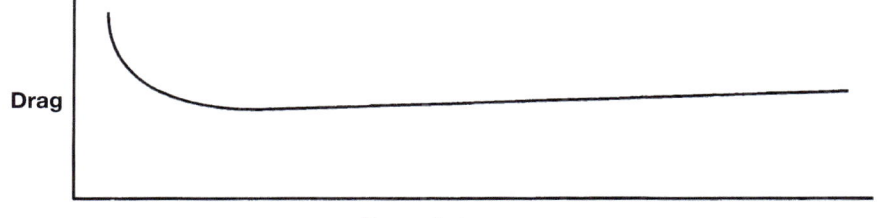

Figure 6-16: Streamlined bodies have a drag curve that drops quickly as ground clearance increases, but past a certain ground clearance, drag starts to slowly rise again. (Based on data from Tamai)

125

SPEEDPRO SERIES

 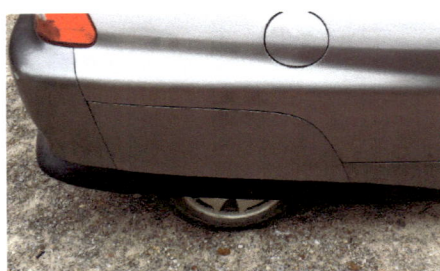

Figure 6-17: Using custom air suspension allows you to adjust ride height at the turn of a knob. The view on the left shows the rear of my Honda Insight at its highest position ... and, on the right, at its lowest position. The variation in ride height is 130mm (just over 5in). Air suspension, with its rising rate springing, is also advantageous when you want to run a low ride height most of the time.

car using its height-adjustable air suspension.

I did some testing on my Honda Insight, a car that has custom air suspension that is easily able to be adjusted for height. For this test, the car was equipped with the full undertrays and rear wing/fins described in the next chapter. Two tests were done – with the suspension at the lowest height that is useable on real roads, and at the greatest height, again able to be used on real roads. The minimum ground clearance at lowest height was 100mm (4 inches) and at the greatest ride height, 190mm (7½in). On a windy day, over a loop of about 15km (just over 9mi) at 90km/h (about 55mph), the fuel economy was a significant 11 per cent better at the lower ride height. (And the actual ride height I run? A minimum clearance of 130mm (just over 5in), primarily in order to gain sufficient bump capability on my rough roads.)

So taking all of this into account, it's impossible to state an optimal ground clearance for drag reduction – to find it, you will need to do some testing of your car at different ride heights. That's especially the case if you are running full underbody covers. That said, on normal road cars where at least 100mm (4in) of ground clearance is needed, lower usually equals better.

In most cars, altering ground clearance is not an easy, cheap or quick proposition. It requires the substitution of new springs that, if they are to give a lower ride height, will also need to be stiffer to avoid bottoming-out over bumps. Trialling two or three different ride heights to see which will give the lowest drag becomes difficult. However, there are two approaches that can be taken to give easily-adjustable ride heights. The first, in a car with struts all-around, is to use adjustable height spring seats. These used threaded struts and a threaded locking collar, the height of which can be adjusted with a suitable spanner. As mentioned, another approach is to use air suspension, where the height can be adjusted by driver control, or automatically altered based on speed, etc.

Another point on ride height. If you choose to lower the car for drag reduction, and you are also seeking downforce, you need to ensure that the downforce doesn't cause your suspension to compress to the point where bump clearance is poor. The amount the suspension compresses will depend on both the amount of downforce and the stiffness of the springs, but a lower ride height will give you less room to manoeuvre. More on this next chapter.

And vehicle rake? As found in the previously-cited Canadian study, most of the technical references suggest a decrease in drag and lift with an increase in rake, that is, with the front of the car lower than the back. For example, Hucho (*Aerodynamics of Road Vehicles*) shows that moving from 0° to -2.5° of rake can change C_D by about 0.03 – however, that was with rough underbody cars having drag coefficients in the 0.38-0.45 range. It can be said, though, with a reasonable degree of certainty, that positive rake (the front of the car higher than the back) should be *avoided* if you are chasing low drag and lift.

If you are easily able to adjust the height of the front and rear of your vehicle, I suggest that you trial different rakes. However, note that to achieve even small angles of rake (eg 2°) the change in ride heights is likely to be greater than you first think. For example, if the car has a wheel base of 2.8m (110in), and if the standard rake is zero, to achieve a -2° rake will require that the front suspension height be dropped by nearly 100mm (4in) (or the rear be lifted by the same amount), or the front be dropped by 50mm (2in) and the rear lifted by 50mm (2in). Note that smaller cars will require an even greater change in suspension height – 2° only sounds small! (The tangent of the desired angle change, multiplied by the wheelbase of the car, gives you the required height change at one end.) Rake is also important to consider if the attitude of your car changes significantly with load variations.

Finally, if you are assessing the drag changes of different ride height, and rakes through fuel consumption changes, you will need

REDUCING DRAG

to ensure that rolling resistance is not changing at the same time. For example, cars that change in front or rear toe with ride height variation will need a wheel alignment at each trialled ride height if rolling resistance is to remain constant.

REDUCING THE AREA OF THE WAKE

As described earlier, the wake is the turbulent air behind the car. It is normally thought of as being the area bounded by the vertical and horizontal separation lines, but the wake can in fact extend to the ground and be wider than the car. A smaller wake normally indicates a lower drag.

It is easy to see the significance of the wake to drag by considering a few points. Firstly, the pressure in the wake is lower than atmospheric. The vertical surfaces to which this lower pressure is applied are literally pulling backward on the car. Therefore, reducing the area of those surfaces exposed to the wake reduces drag. The other way of thinking of it is that we want to have as a high a pressure as possible on these rear surfaces. For example, if we can keep the flow attached down the inclined rear window in a sedan, the pressure on the window will be higher than if it were part of the base area in the wake. Some of this pressure has a forward component, resulting in a reduction in drag.

A teardrop shape has a minimal wake – the streamlines come together at the end of the long tail. Solar race cars are the best examples of vehicles with minimal wakes that you will ever see on a public road. On the other hand, a truck – that usually has no taper in the tail at all – leaves behind a large wake.

Car manufacturers reduce the size of the wake by:

Figure 6-18: Boat-tailing of the rear of the Lexus LS (right of image). This boat-tailing is about 10° each side. (Courtesy Lexus)

Figure 6-19: The reduction in the width of the wake caused by rear boat-tailing can be seen by these smoke streamlines. (Courtesy Mercedes)

- having attached flow to the smallest rear cross-section of the car, eg to the trailing edges of the boot on a sedan
- boat-tailing, where the car narrows toward the rear when seen in plan view
- using downward-angled roof extensions (eg on the roofs of hatches)
- using smooth undertrays so that there is airflow under the car and so the wake does not extend down to the road

Hyper-milers, and other people chasing major decreases in drag, fit long, narrowing tails that have attached flow to the smallest vertical cross-sectional area they

SPEEDPRO SERIES

Figure 6-20: This temporary tail extension was made by Darin Cosgrove, and fitted to his Gen 1 Honda Insight. It weighs 14kg (31lb) and uses the following angles: top: 13-16°, bottom: 6-13°, sides: 16-24°. A 9.7 per cent improvement in fuel economy was measured at 80km/h (50mph). (Courtesy Darin Cosgrove, EcoModder.com)

Figure 6-21: Earl Poon uses a single wheel trailer to support the aerodynamic extension of his 1998 Toyota T-100 Xtra cab SR5. The trailer uses airbag suspension, and the tow hitch has dual air shocks to stop the bucking associated with towing a small wheelbase trailer. The best fuel economy from the V6 petrol automatic has been 7.4 litres/100km (38.4 US mpg) at 105km/h (65mph). (Courtesy Earl Poon, EcoModder.com)

can manage. (Note that unlike decreasing the height of the roof, boat-tailing does not cause lift.)

The reductions in drag that can be made by altering the size of the wake depend on the magnitude of the changes made to the car. Historically, the largest change in wake size was achieved when sedan (three box) vehicles gained flow reattachment on the trailing edge of the boot. The wake area was therefore reduced in size from one that was nearly the area of the height multiplied by the width of the car, to one that was only the height of the boot lid multiplied by the width of the car. Notable first sedans to do this were the NSU Ro80 and the Audi 100. We don't have drag coefficients for these cars when they did *not* gain flow reattachment (as they probably didn't in the initial design process), but we do have figures for hatchback cars with varying hatch angles.

Hucho (*Aerodynamics of Road Vehicles*) describes a vehicle that had the angle of its hatchback varied. At angles where the rear window was greater than 30° (ie closer to vertical), flow separated at the trailing edge of the roof. The C_D in this configuration was 0.40. At angles of less than 30° (ie becoming more like a fastback, with flow attached to the end of the hatch, so giving a smaller wake), the C_D was much lower – in fact, at 10° hatch angle, it was just 0.34. This C_D is 15 per cent lower than with the larger wake. (And at 30° hatch angle? Drag rose to 0.44, with observance that separation line was jumping back and forth between the end of the roof and the end of the hatch.) Figure 6-24 shows the general relationship between rear hatch angle and drag. Note that the change in C_D associated with the change in hatch angle is not just the result of altering

REDUCING DRAG

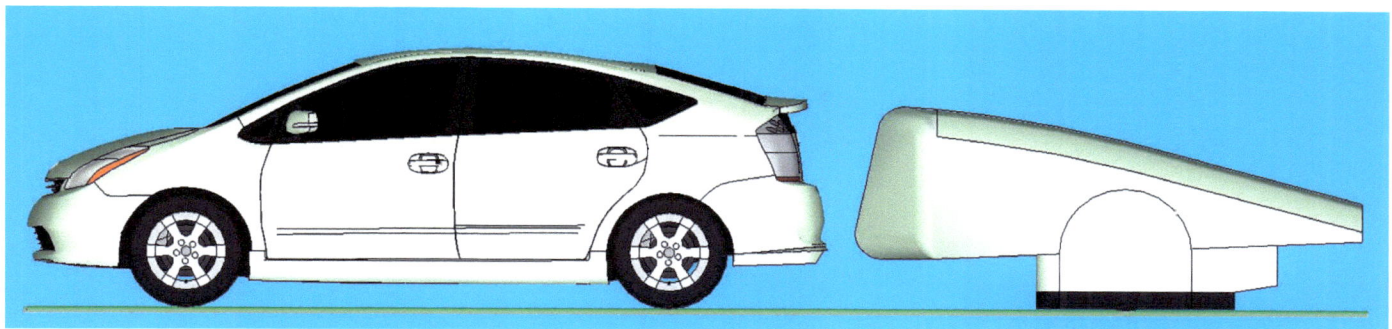

Figure 6-22: This trailer was originally designed to be towed behind a Toyota Prius, and add extra storage space for use on long trips. As can be seen, aerodynamics was a key driving factor. The trailer closely follows the taper of the Prius. The rounded front edges help air to attach easily. The front end is bellied forward at the bottom centre to help air go over the top of the trailer rather than being drawn down and under (a key finding from previous aero-trailer builds and real-world testing). The distance between the car and trailer is as close as possible, and allows for a 15° grade to be climbed, and for the car to be steered fully in either direction while going forward. The trailer has a smooth underside, and completely faired wheels, with rubber skirting near the ground. The final built trailer is a reasonable likeness, though much more boxy, but still gives open-road mileage that is only about 8 per cent worse than without trailer. (Courtesy Wyatt Fisher, EcoModder.com)

Figure 6-23: The angle of the roof extension on this Mitsubishi Mirage reduces the size of the wake. (Courtesy Mitsubishi)

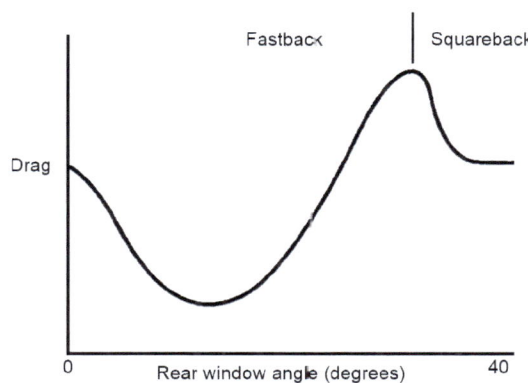

Figure 6-24: Drag is at its lowest with rear hatch angles of 10-15° (ie a fastback configuration), and reaches a maximum at about 30-35°, before dropping again at hatch angles of about 40° and higher (ie a squareback). (Based on data points from Hucho and others.)

wake area – it also influences the behaviour of the trailing vortices.

Incidentally, while Figure 6-24 refers to fastback and squareback vehicle shapes, a similar idea can be applied to notchback (3-box) sedans. In the case of those vehicles, it is the Effective Back-Light Angle (EBLA) that becomes critical. The EBLA can be found by drawing a line that connects the trailing edge of the roof to the trailing edge of the boot, and then comparing the angle that this line forms with the horizontal. Jaguar Land Rover aerodynamicist Adrian Gaylard comments:

"*Typically, for lowest drag, the EBLA is around 12 degrees. As this increases, so does drag, until around 30 degrees where the rear-pillar vortices burst and the rear flow fully separates. With an effective backlight angle approaching 30 degrees, it's often better to separate it as the drag can be lower for a fully separated rear flow, compared to one where rear pillar vortices are keeping the rear screen flow attached on a high screen angle.*"

SPEEDPRO SERIES

Figure 6-25: The arrow points to a long piece of wool that is showing the Effective Backlight Angle (EBLA) of this car.

Figure 6-25 shows a Mercedes W211 with a long length of wool attached to the rear centreline of the roof. You can see how the wool replicates a smoke streamline, and so the EBLA is able to be visualised through airflow, as well as measured.

The Opel Calibra, described more fully in Chapter 2, used a rear width 180mm (7in) narrower than the rest of the car body, giving a reduction in drag of 0.011 (about 4 per cent). The Gen 1 Honda Insight has a narrower rear wheel track to accommodate its boat-tailing. A rough measurement of the Honda indicates the base area (the area in the wake) is only about 60 per cent of the area that would have occurred if the airflow separated at the highest and widest points of the car.

Hucho, quoting previous research, suggests an optimal boat-tailing angle of 22° to the car centreline (ie a 44° included angle). This is a greater angle than seen on most production cars, where 10-15° to the centreline is used.

SAE paper 2016-01-1621 (*Effect of Side Taper on Aerodynamics Drag of a Simple Body Shape with Diffuser and without Diffuser*) discusses boat tailing on a simple squareback (wagon) shape, with the testing performed by CFD.

The paper found that on the squareback car:

- Best drag reduction came from the use of 10° of boat-tailing. Increasing the boat-tailing angle over 10° caused increased drag, despite a smaller wake. The increased drag occurred because of the creation of two strong counter-rotating vortices.
- As the boat-tailing taper length increased, so did drag reduction, up to the maximum of 15 per cent of the car's length.

As the photos on the previous pages show, wake reduction techniques can vary from the mild to the wild. (Incidentally, my favourite wake reduction technique – one not since apparently repeated – was a 1930s German bus project. The long-distance bus was equipped with an inflatable tail that extended as required at high speed. In cities, where it would be awkward, it was deflated and retracted – all automatically, apparently.)

I did some work to one of my cars to reduce the size of the wake, and I wish to recount that project here. It's at the 'mild' end of wake reduction approaches, and could be expected to be successful on any squareback car.

Wake area reduction on a squareback

My incentive for exploring ways of reducing drag on squareback cars came from an engineering paper written on drag reduction techniques being used on semi-trailers (sometimes called tractor-trailers). In the paper *Summary of Full-Scale Wind Tunnel tests of Aerodynamic Drag-Reducing Devices for Tractor-Trailers*, Jason Leuschen and Kevin Cooper of the National Research Council of Canada tested what happened

REDUCING DRAG

to the drag of a full-size Volvo semi-trailer in a 9 x 9m (about 30 x 30ft) wind tunnel. The results of no less than 28 drag reduction devices were shown by the tests that were conducted in 2006. Most interestingly, the aero drag was shown not only as a change in C_D, but also in the amount of diesel fuel that would be annually saved if the truck travelled 130,000 kilometres (about 80,000mi) per year at a cruising speed of 107km/h (66mph).

The greatest saving in fuel (and so therefore the greatest drag reduction) was caused by the fitment of a Transtex Composite folding rear trailer deflector – a saving in fact of just over 3000 litres (nearly 800 US gallons) of fuel per year! The change in C_D caused by fitting this deflector was 0.05. Now known as the TrailerTail, this commercially available product comprises folding extension pieces that mount on frames attached to the rear doors of trailers. The extension pieces are angled inward toward the centreline of the vehicle. If the airflow remains attached until the trailing edges of these flaps, the resulting wake will be smaller. (Furthermore, and while this is not stated by the company or the engineers who tested the device in the wind tunnel, it appears from images provided by the company that trailing vortices normally being shed by the vertical rear edges are altered in their characteristics.) The company claims that fitting the TrailerTail reduces fuel consumption at 105km/h (65mph) by 6.6 per cent.

So what about trialling some of these ideas on a squareback car – a Skoda Roomster that I owned? A simple roof extension looked like it had the best promise – it could likely be fitted without hindering the opening of the hatch.

However, rear extensions that are angled toward the centreline

Figure 6-26: This semi-trailer TrailerTail reduces the size of the wake by the inward sloping side and top pieces. After reading a research paper on its effectiveness, I decided a similar approach could be taken on a squareback car.
(Courtesy Georgina Edgar)

Figure 6-27: Despite its appearance, the Skoda Roomster has a relatively low C_D (for a brick!) of 0.34 and gets excellent fuel economy from its 1.9-litre diesel. But could its wake be reduced in size?

131

to reduce the size of the wake will only work if the flow remains attached to these guides. To put this another way, an abrupt change of angle of the sort that would be achieved with a roof extension angled sharply downward will likely cause flow separation at the change of angle. In addition, the flow will remain attached only if the airflow is already attached to the body prior to the extension. In other words, if the airflow has separated from the body ahead of the extension, the angled extension is unlikely to cause it to become reattached. Therefore, extensions must have attached flow – both by being able to work with air that is already attached, and also in keeping the air attached across their change in angle. Wool-tuft testing can show what is actually happening.

To make trial extensions, I bought a piece of Corflute plastic sheet from a sign maker. The piece was 1800 x 600mm (71 x 24in) and 5mm (about 3/16in) thick. This plastic sheet, that uses two thin layers sandwiching a corrugated core, is light, easily cut with a serrated knife, and sufficiently rigid to model extensions. I made a short roof extension with bracing side pieces and attached it to the car with cloth-backed adhesive tape. This was sufficiently rigid and weatherproof to allow a long drive, and in fact a 400km (250mi) freeway drive was needed for work purposes. (Due to aesthetics and design considerations, I didn't make side or bottom extension pieces.)

With the rear extension in place, I thought the aero noise level (from the back of the car, aero noise tends to be a dull roar) was lower than previously, and the fuel economy on this trip was very good – a two-way average of 5.0 litres/100km (47 US mpg). On

Figure 6-28: I made a rear roof extension using Corflute, a lightweight but stiff plastic sheet that can be cut with scissors or a sharp knife.

the basis of previous trips, I would have expected a best of about 5.2 litres/100km (45 US mpg). So this was encouraging – although not absolutely definite.

I then made the roof extension even longer (ie the maximum that was practical on the car) and wool-tufted the area. With the car being driven by at 60km/h (37mph), I was able to photograph from the side of the road the behaviour of the tufts. The tufts showed a mix of good and bad:

- the flow across the roof and onto the extension was attached
- the flow across the extension itself remained attached
- however, the flow along the sides of the car around onto the supporting pieces became separated

I then altered the design, making the extension a bit shorter (and more in keeping with what I wanted in terms of rear vision and external appearance) and altering the side pieces to better pick-up flow coming off the side of the car. (However, this meant the rear hatch could not be lifted, as the extensions stuck to the body either side of the hatch.) After these changes, the flow across both the top and the side pieces remained attached.

But was the design actually reducing the size of the wake? In a wind tunnel the size and shape of the wake can be determined by injecting smoke into it. Because the area is a lower pressure, the smoke tends to stay in the wake, so making it visible. Driving the car on a dirt road can also reveal the size and shape of the wake, with the dust generated by the tyres filling

Figure 6-29: The first roof extension was long and supported by side flaps that attached to the edges of the hatch.

REDUCING DRAG

Figure 6-30: However, wool tuft testing showed that while flow from the roof (1) was attached onto the extension (2), there was separated flow on the side panels – the change in angle across the light was too great.

Figure 6-31: I then moved the side pieces so they better lined up with the side of the car. This gave attached flow, but also meant the hatch couldn't be opened!

Figure 6-33: The roof extension that was used for a lot of highway testing. Note the middle support, needed to maintain the correct angle.

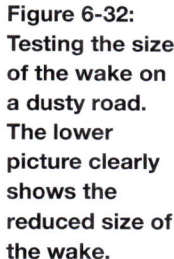

Figure 6-32: Testing the size of the wake on a dusty road. The lower picture clearly shows the reduced size of the wake.

the wake. Testing on a dirt road with and without the roof and side extensions showed very clearly that the wake was smaller in modified form.

I then did a lot more open-road road testing, measuring fuel economy. With one configuration of roof extension, I noticed that the fuel economy improvement seemed to have disappeared. However, careful observation showed that the unbraced rear Corflute extension was being lifted by the airstream, resulting in no downward angle to the extension when travelling at highway speeds. I made another version, braced in the middle and with a slightly steeper, shorter extension. Note that this version attached only to the hatch – not so good for side aero but much more practical for a longer Corflute trial.

However, testing with this design showed again that fuel economy was no longer improved – it was back to standard. So what was going on? As we've seen, when attached airflow wraps around a corner, low pressures are generated. If the airflow was remaining attached across the roof and rear extension, it would likely be generating a force with an upward component (lift) and also a backward component (drag). If the drag created by this low pressure more than

SPEEDPRO SERIES

Figure 6-34: The red arrow shows the airflow and the yellow arrow the force developed. Note that the force has both drag and lift components. The trick is to get a greater gain from the reduced wake size than increased drag.

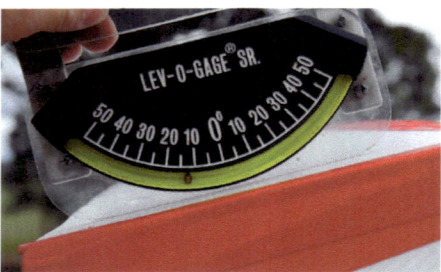

Figure 6-35: An angle for the roof extension of 12 degrees from the horizontal proved to work well.

Figure 6-36: The final mock-up. A 'production' version could be made from cut, heated and folded ABS sheet.

offset the reduced drag caused by a smaller wake, the result would be no improvement. I measured the angle of the rear extension and it was about 20° to the horizontal. I then made a revised version that decreased this angle to about 12°. (Interestingly, in SAE paper 2011-37-0012 – *An Investigation into the Wake Structure of SquareBack Vehicles and the Effect of Structure Modification on Resultant Vehicle Forces* – a 12° chamfer on the roof trailing edge of a squareback is used to good effect to achieve a reduction in wake size.)

Many weeks of highway testing were carried out with the Roomster extension in place. During this test I consistently gained fuel economy a few per cent better than I would have expected. However, perhaps more significantly, the best-ever fuel economy results were clearly better – in fact 8 per cent better than achieved without the rear extension.

The best outcome from practicality, appearance and fuel economy perspectives was a roof extension with these characteristics:

- angled downward by 12° from the horizontal
- 200mm (8in) wide
- curved in plan-elevation to match the profile of the roof
- mounted completely on the hatch

It's an approach that could be trialled on many similar-shaped cars.

WAKE AREA REDUCTION ON A PICK-UP

A pick-up truck tends to have a very large wake, as the air separates at the end of the cabin roof and upper sides. One way to reduce this is to make a sloping cover that goes over the cargo tub, keeping attached flow on the top and the sides until the tail-gate is reached. This substantially reduces the area of the wake.

Jim Michler of Phillips, Wisconsin, USA has built a high-quality cover for his GMC Canyon. Jim says that he wanted a cover (sometimes called a 'topper') that:

- one person could install or remove
- had provision for large loads, with easier access than a standard topper
- was lockable
- was lightweight
- reduced drag

The topper is made of four separate pieces that are bolted together. Each piece is plywood-on-timber-frame construction. The vertical front panel immediately behind the truck cab is bolted to the truck cargo tub and the two side pieces. The hinge for the lid is fastened to the front panel with wood screws. The front panel weighs about 2.3kg (5lb).

To keep the lid rectangular, the side pieces were produced with a twist – that is, the sides are closer to vertical at the front than at the rear. This required some tricky joinery in the frames, work that Jim has completed beautifully. The sides are bolted to the truck cargo tub and to the front panel. The side pieces, that have holes to receive the latching

REDUCING DRAG

Figure 6-37: All pick-up trucks have large wakes as airflow separates at the back and sides of the cabin roof. (Courtesy GMC)

pins for the lid, weigh about 6.8kg (15lb) each.

The lid is plywood-on-frame construction. Unlike the side pieces, the frames for the lid are also made from plywood. Note that the lid is curved in end-view. The major portion of the lid is made with two layers of 5mm (³⁄₁₆in) lauan plywood. This plywood was used because it is cheap, readily available, and easily formed to the required curvature. The smaller radius at the sides of the lid are created by narrow wood strips glued together, then planed and

Figure 6-38: Jim Michler has built a high-quality cover for his GMC Canyon. It substantially reduces the size of the wake, and, as shown in the table overleaf, results in an average fuel economy improvement of 15 per cent.

Figure 6-39: The side frames incorporate some very tricky joinery, with the angles changing along the length.

Figure 6-40: The lid uses curved plywood ribs and two layers of plywood.

Figure 6-41: When not chasing low drag, the pick-up can still carry a substantial load. (All photos courtesy Jim Michler, EcoModder.com)

135

SPEEDPRO SERIES

sanded to a smooth curve. All wood was sealed with epoxy inside and out, then painted white on the inside and with hardware store Flame Red enamel on the outside. The red is not an exact match for the truck – but is close enough.

The lid has in interior light, an external brakelight, and a lockable latch. The lid weighs about 16kg (35lb). It can be removed in 5 minutes: pull the electrical connector, pop off two gas struts, pull the hinge pin, and lift off. Jim comments that while he has removed the lid himself, it's much easier with a helper.

So does it work aerodynamically? Unlike many aerodynamic modifications aimed at reducing drag, the topper should be able to make a substantial reduction. The truck is driven in a climate that varies in temperature from -30°C to -7°C (-20°F to 20°F) in winter, and to 21 to 32°C (70° to 90°F) in summer. Fuel consumption is higher in winter (ie mileage declines). Testing consisted of recording the average gas mileage for the three coldest months in winter and the 3.5 warmest months in summer. Total distance for each winter ranged from 4000-4800km (2500-3000mi), and 5600-6400km (3500-4000mi) each summer. The driving was generally 15 per cent in the city and 85 per cent on the highway. Highway speed limits were almost all 89km/h (55mph), with only a short distance in 105km/h (65mph) speed limits.

Jim has made a number of changes to the truck that have improved mileage (GMC Canyons don't normally get these types of figures!), but what we're concerned with here is the before and after results of just the wake reduction device. These figures are shown in the table below.

This is an improvement of 18 per cent in winter and 12 per cent in summer, giving an average improvement of 15 per cent. Jim is a member of the www.ecomodder.com forum, and you can find out more about his truck there.

A LARGER WAKE

What difference does an *increased* wake make to fuel economy at 100 and 110km/h (60 and 68mph) freeway travel? One day I had an easy way of finding out. The car was the six-cylinder Ford EF Falcon I then owned, and I needed to travel around 100km (60mi), the whole distance on freeways or main highways. I was making the trip to pick up a power saw that I'd bought, which I thought would fit into the boot – although probably with the boot lid up a little.

I got onto the freeway and reset the fuel computer. I then selected fifth gear and stayed in that gear nearly all the way. There were some down-changes for a toll booth, and later in the trip, some traffic lights, but otherwise the trip was about as fuel-economical as it's possible to get. The Falcon achieved 7.5 litres/100km (31.4 US mpg) for the trip.

I reached the destination, loaded the scroll saw into the boot, and then closed the lid as far as possible. As expected, the lid wouldn't shut – the gap was about 200mm (8in). Because the Falcon is an aero-efficient design, you'd expect attached flow across the boot lid, so the raised boot lid would increase the size of the wake.

Given that the route is quite flat and the changes in speed minor, the extra mass of the scroll saw would have made very little difference to the fuel consumption (rolling drag goes up very slowly with weight). There was also little wind, similar overall grades, and the return route was completed driving in the same way as the initial journey.

So, was there a measurable change in fuel consumption? Yes! The average fuel consumption with the boot lid propped open a little was 8.6 litres/100km (27.4 US mpg), an increase of no less than 15 per cent!

Note that, while I did not measure it, the raised lid would also have been acting as a rear spoiler, increasing pressure on the rear window, and so slightly offsetting the larger wake. However, this was insufficient gain compared with the much larger loss caused by the increased wake area (and possibly changed trailing vortex behaviour).

Configuration	Winter	Summer
Without topper	9.5l/100km (24.7 US mpg)	7.3l/100km (32.1 US mpg)
With topper	8.1l/100km (29.2 US mpg)	6.5l/100km (36.0 US mpg)

REDUCING THE STRENGTH OF THE WAKE

We have seen that reducing the size of the wake can decrease drag, but are there methods that can increase

REDUCING DRAG

the pressures acting on the car within the wake? Any increase in such pressures will decrease drag.

SAE paper 2011-37-0015 (*An Investigation into the Wake Structure of SquareBack Vehicles and the Effect of Structure Modification on Resultant Vehicle Forces*) covers an interesting technique. The approach is premised on the fact that the pressures acting on the car in the wake are influenced by more than just wake size alone. In fact, vortices acting within the wake, especially those that cause the movement of high-speed air close to the vehicle base surface, can further reduce pressures. So, how can these vortices be disrupted or slowed?

Tests were carried out in wind tunnels on a full-size vehicle (a 2011 Ford S-Max). To disrupt the vortices, horizontal slats were positioned laterally across the rear of the vehicle. These slats were 30mm (a little under 1¼in) deep. The slats, extended across the full width of the rear of the car, were positioned about 125mm (5in) apart, and protruded at right-angles to the surface. On the mid-size vehicle being tested, a maximum of six slats was used.

Significantly, before testing with the slats, the vehicle was first modified by fitting a complete undertray. (This alone dropped the C_D by 0.032!) Notably, without the flat underfloor, the slats did nothing. When looking at both the drag and lift results, the best outcome was achieved with all six slats fitted. This resulted in a decrease in C_D of -0.004 and a decrease in C_L of -0.01.

It would appear that if you have fitted undertrays giving a flat floor (more on undertrays later in this chapter), then experimenting with horizontal slats within the base area may be worthwhile. Remember though that the test vehicle was a squareback with a simpler vortex pattern within the wake than occurs with fastback or sedan body shapes.

Figure 6-42: The ridge on the rear of this Subaru Levorg may be present to disrupt the flow of vortices against the rear panel, so increasing rear wake pressure. (Courtesy Subaru)

ACHIEVING CLEAN SEPARATION

As I've described earlier, the airflow separates from the car to form the wake. Let's explore separation in a bit more detail. Think of the airflow on a three-box sedan, where the flow is attached down the rear window and onto the boot lid. Now, at what *exact* point does the flow separate from the trailing edge of the car? Does it wrap a little around a curved surface onto the back vertical panel? If it does, it will generate a low-pressure zone (a suction peak) running across the car at this point. That low pressure will create a force with drag and lift components – not good. Now, rather than having a curve at this trailing edge, let's style a sharper edge. In this configuration, the airflow no longer wraps around the corner – instead it leaves the car cleanly at this point. The result is reduced drag and lift.

Let's look at an example of where one car model had this curved rear edge, and then the updated model replaced it with a squarer form, giving cleaner separation. The 2000 VX Holden Commodore was produced with a rounded trailing edge to the boot lid – poor for clean separation. The car had a C_D of 0.329. It was replaced with the VY model in 2002, a car with sharper rear edges. The C_D decreased to 0.319 (3 per cent less). In addition, the VY model had a lower rear lift coefficient, decreasing from 0.160 to 0.128 (a substantial 20 per cent drop). The table below shows these figures.

	C_D	C_{Lr}
VX Commodore	0.329	0.160
VY Commodore	0.319	0.128

Most manufacturers have for many years been aware of the need for clean separation on upper car surfaces. (In fact, it's one of the fascinating games you can play if stuck in traffic – look for the cars with and without the styling to give clean separation on trailing edges of boots or roofs.) However, the

SPEEDPRO SERIES

Figure 6-43: The very clean rear separation line down the side of this Toyota Prius C can be clearly seen. (Courtesy Toyota)

Figure 6-44: Intriguingly, Honda added a tiny side spoiler to one of its Civic models to promote clean separation. (Courtesy Honda)

Figure 6-45: The Lexus LS+ Concept car's approach to gaining clean rear separation. (Courtesy Lexus)

equivalent sharp change in shape to give clean separation on the *sides* of the car has only much more recently started being adopted. Initially, it was confined to tail-lights, with an edge moulded into their shape. If the tail-light (eg on a wagon) went the full height of the rear, even better. The use of the separation sharp edge then spread to the rear panels, and then finally to the bumper.

This means that if you own a car that does not have such clean separation edges on the sides and roof (or boot in a sedan), then you can add them. To find the required location, use the visualisation techniques of wool tufting and/or dust signatures. Dust is especially effective at showing a vacillating separation point, where perhaps because of the effect of crosswinds giving a yaw component to the airflow, the separation point moves back and forth over a short distance. Wool tufting will show you how well-attached the flow is prior to the separation point. While you might assume that the flow will be attached prior to the separation point, the negative pressure gradient that the airflow is pushing into means that cars with excellent flow attachment down both sides, right to the trailing edge, are rare. For example, directly behind the rear wheel on the bumper is an area where on many cars, the flow has already separated. As a result of this, the flow separation edge that you add may need to follow the actual separation points, and so not be a straight, vertical line.

You can also use pressure measurement to ascertain the magnitude of the suction peak (if any). The pressures are very low, so a circular pressure puck will need to be used, rather than a bare end tube. You will also need to use a reference pressure reservoir (see Chapter 4).

If you are making modifications to achieve better separation, it's rather hard to change the shape of the panel. Instead, separation edges are devices that are best added to the surface of the panel. One easy approach is to add straight extensions, rather like the wake-reducing extension I covered earlier for the Skoda. As with that design, all your separation edges can be angled inward to reduce the size of the wake, as well as give the required clean separation. If the use of straight rearward

REDUCING DRAG

Figure 6-46: This Mercedes has beautifully clean separation on the top of the boot lid, but, to promote side separation, it lacks even the sharp edge on the tail-lights, let alone the bumper. Dust signatures are an excellent way of looking at separation.

Photo 6-47: The dust signature on the rear of this squareback vehicle shows the lack of a clean vertical separation point. Note how the airflow wraps around the rear differently at the rear window, above and on the tail-light, and on the bumper. You can also see the dust deposition in the separated flow immediately behind the wheelarch lip.

Figure 6-48: Left – using tuft testing to see the flow at the rear side separation point (circled). Note how the airflow wraps around the bumper curve, creating a low pressure and so drag. Right – trialling a rear separation edge made from plastic sheet. Note how the tufts now show clean separation.

139

SPEEDPRO SERIES

Figure 6-49: Left picture, wool tuft testing on a 1973 Beetle showed attached flow over the cambered roof and down over the rear window. Right, owner Gerrelt Molhoek then made this neat spoiler that causes the flow to separate at his desired point. (Both photos courtesy Gerrelt Molhoek)

extensions is a bit visually daunting, you can add more subtle rubber strips that, if placed carefully, will not increase the area of the wake but will still cause clean separation. When clean separation occurs, wool tufts will more likely stream straight backward from the separation edge, rather than wrapping around into the wake. Dust testing of successful separation edges will show all the dust deposited inside the edge – on one side there should be dust (ie in the wake), and on the other side, none (ie out of the wake).

One older car on which the separation point was changed was a 1973 Volkswagen Beetle owned by Gerrelt Molhoek of the Netherlands. Wool tuft testing showed attached flow over the cambered roof and down over the rear window. While the Beetle deserves credit for being the first streamlined car to be a commercial success (and was it ever!), it also has high drag. Gerrelt wondered if a spoiler mounted above the rear window could improve flow separation. He made a mock-up from cardboard, and then after a few unsuccessful attempts, made a neat spoiler from glass fibre and polyester resin. The spoiler is angled 12° downward from the horizontal and is held in place by the window rubbers and some double-sided tape. Wool tuft testing after fitment clearly showed that separation now occurs at the spoiler.

I am not aware of any testing that specifically shows the drag reduction caused by adding separation edges to a vehicle that had originally none, but based on the Commodore example described above, and the fact that these edges have been adopted by almost all new car manufacturers, I'd guess the drag reduction potential of using side and top separation edges, where currently none exist, at perhaps 3 or 4 per cent.

REDUCING SEPARATION BUBBLES

Separation bubbles occur where flow separates and then reattaches further along the car. They do not increase the size of the wake, but they reduce the pressures that are being brought to bear on what are often rear-facing surfaces. In some situations (eg separation around an A pillar) they can cause vortices that then cause problems downstream. I'll call a separation that results in a separation bubble 'temporary separation' to differentiate it from separation that occurs at the end of the car.

Changing the car's shape
Temporary separation is caused by too sharp a turn for the airflow to negotiate. Rather like an understeering car just keeps going straight on rather than following the corner, so the airflow keeps 'going straight on.' Therefore, the first modification that can be undertaken to reduce temporary separation is to reduce the sharpness of the corner. Because in this book we're not designing cars from scratch but instead looking at how you can make changes to an existing car, there's not a lot of point in saying that you should remodel the transition curve from the roof to the rear window, or add camber to roof profile. Instead you will need to use attachments that can achieve similar outcomes.

REDUCING DRAG

THE EXCELLE GT

The 2015 Buick Excelle GT was tested in the aerodynamics wind tunnel of the Shanghai Automotive Wind Tunnel Center. The results of the testing are shown in SAE paper 2017-01-1512. Change in both C_D and C_L were measured with the car in 12 different configurations. The testing is fascinating because it shows the magnitude of the changes in drag and lift that are possible with simple changes.

Configuration 1 was with the car in standard form. Configuration 2, removing the front air dam (spoiler), caused a 2.4 per cent increase in C_D (implying the presence of a rough underfloor) and a very large 48 per cent increase in C_L. Additionally, removing the front wheel deflectors (Configuration 3) caused a further increase in C_D as more of the airflow impinged on the front tyres, but a slight reduction in C_L over removing the front air dam alone. The reduction in lift implies less build-up of pressure under the front of the car with the removal of the tyre deflectors.

Configuration 4, raising the boot lid by 15mm, reduced C_D fractionally and reduced C_L by a small amount. Both are likely related to better flow attachment over the rear of the car. Adding a separation edge to the rear lights (Configuration 5) reduced C_D fractionally (as expected with the cleaner separation) but, oddly, also reduced lift by nearly 13 per cent! Configuration 6, adding a rear roof separation edge, increased wake size (drag up by 2.7 per cent) and reduced pressures on rear upper surfaces (lift up by 6.8 per cent).

Configuration 7, removing the external rear vision mirrors, reduces C_D by nearly 4 per cent (and, with the reduction in projected frontal area, would reduce overall drag by a larger amount than this) and increases lift by a whopping 22 per cent! (Presumably, the airflow over other parts of the car is changed by the removal of the mirrors.) Smoothing the A pillar (Configuration 8) decreases C_D fractionally but has a bigger impact on C_L, increasing it by 13.6 per cent.

Configuration 9, adding large front wheel deflectors, dropped C_D a little but increased C_L as the air ran into the

deflectors and so pressure was built up ahead of them. However, using saw-tooth deflectors (Configuration 10) had the same impact on C_D as the conventional deflectors, but dramatically decreased lift (down by 43 per cent). Unfortunately, the paper does not have dimensions of these deflectors, so the saw-tooth deflectors were not necessarily the same size as the conventional deflectors.

The effectiveness of a rear spoiler was heavily dependent on its height. Configuration 11, that used a 30mm high spoiler, increased both C_D and C_L, the latter being significant. However, increasing spoiler height to 48mm resulted in a relatively minor increase in C_D but a major reduction in C_L – lift was down by 63 per cent.

Configuration		Change in C_D (per cent)	Change in C_L (per cent)
1	Baseline	0	0
2	Air dam removed	+2.4	+47.9
3	Air dam removed, wheel deflectors removed	+3.8	+41.5
4	Boot lid raised by 15mm	-0.3	-3.1
5	Tail lamp separation edge added	-0.3	-12.7
6	Rear roof separation edge added	+2.7	+6.8
7	Mirrors removed	-3.8	+22.2
8	A pillar step smoothed	-0.3	+13.6
9	Big front wheel deflectors added	-1.7	+4.7
10	Saw-tooth front wheel deflectors added	-1.7	-43.1
11	30mm high rear spoiler added	+0.7	+10.1
12	48mm high rear spoiler added	+1.4	-63.2

SPEEDPRO SERIES

Figure 6-50: The Mercedes-Benz 190 E 2.5-16 Evolution II (W 201) used a panel that fitted at the top of the rear window to give a gentler transitional radius, so resulting in better flow attachment. (Courtesy Mercedes)

One of the earliest cars I've ever wool-tuft tested was a 1970s Ford Escort. (You can see a picture of it being tested on page 60.) Unlike every other car I have ever tested, the Escort had flow separation between the headlight and the bonnet and between the headlight and the guard (fender). In this case, a potential solution would have been to fit slightly bulbous clear headlight covers – this would have reduced the effective radius of the corners the airflow had to negotiate.

The Mercedes-Benz 190 E 2.5-16 Evolution II (W 201) from 1990 used a wild aero kit. It included a panel that fitted at the top of the rear window that gave a gentler transitional radius, so resulting in better flow attachment (and presumably improving flow to the rear wing). The A-pillars on older cars had sharp transitions that can be smoothed by having a shaped plastic moulding placed over the top. This often done to reduce noise

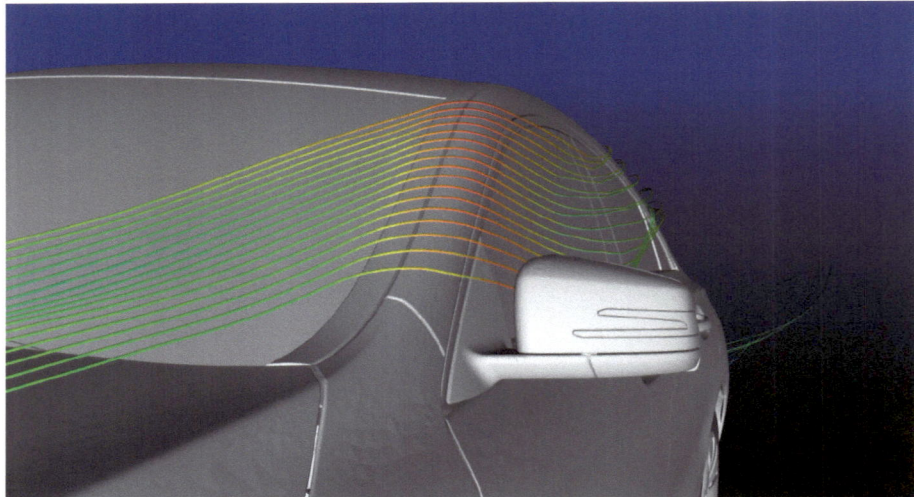

Figure 6-51: CFD image of attached flow around A and C pillars. Flow separation around these pillars – especially the A pillar that is normally a tight radius) can be reduced by the use of overlay mouldings. (Courtesy Mercedes)

– more on this approach in Chapter 10.

Vortex generators
The other way of addressing separation bubbles is to look at the underlying cause. Why is the airflow unable to follow the change in direction – especially when further upstream, it will have no problem in doing so? The answer is that the boundary layer – the blanket of air nearest the car's body that is not moving at the speed of the

REDUCING DRAG

free airflow – has thickened. If we can put some energy back into the boundary layer, it will again better follow corners. One way that this can be done is by the use of vortex generators.

Vortex generators really came to the attention of general car enthusiasts with the release of the Mitsubishi Evolution IX Lancer in 2004. Not only were these odd-shaped things across the rear of the roof unusual, but Mitsubishi also released a detailed technical paper on how they worked. That paper was called *Research on Aerodynamic Drag Reduction by Vortex Generators*, and it was published in the Mitsubishi Motors Technical Review, 2004, No 16.

As with many booted sedans, on the Lancer the airflow tended to separate from the body at the trailing end of the roof. The result was a larger wake of disturbed air and a lack of pressure recovery on the rear window. Mitsubishi engineers then experimented with vortex generators placed just ahead of the separation point. These performed as advertised, energising the boundary layer. However, the vortex generators themselves developed some drag so the end result was the balance of the decreased car body drag minus the increased vortex generators' drag.

Mitsubishi did extensive testing of the Evo vortex generators in its full-sized wind tunnel. The test speed was 180km/h (112mph). As mentioned, it's possible for the vortex generators – even when functioning correctly – to add an excessive amount of drag by their very presence. A rule of thumb is that the height of the vortex generators should therefore not be much greater than the thickness of the boundary layer. In the case of the Lancer, the boundary layer 100mm (4in) upstream of the rear

Figure 6-52: The Mitsubishi Evo IX brought vortex generators to the attention of car modifiers. Here they can be seen as the line of spikes above the rear window. (Courtesy Mitsubishi)

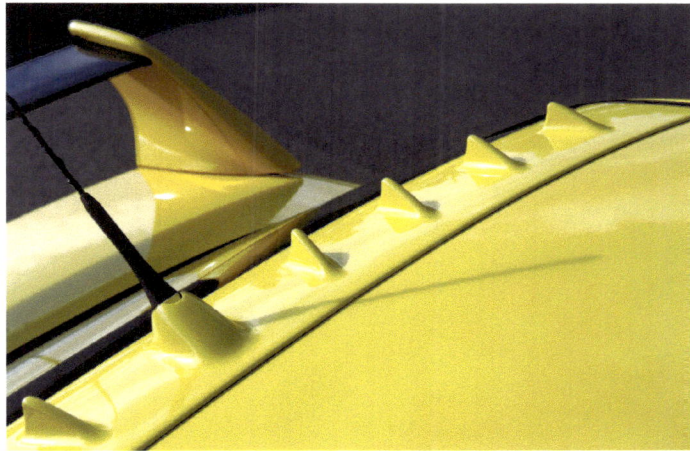

Figure 6-53: A close-up view of the Mitsubishi vortex generators. (Courtesy Mitsubishi)

window was 30mm (about 1¼in) thick. (Although later in the paper the engineers said it was 15-25mm thick.) In the technical paper, the vortex generators on the roof of the Lancer were placed immediately upstream of the flow separation point, a position which was 100mm (4in) ahead of the rear window (although this appears to have changed on the production model).

Two completely different designs of vortex generators were trialled by Mitsubishi. The first was a bump-shaped device that looked a bit like an upside-down spoon. However, at the rear of the 'spoon' a flat surface was placed with a rear slope angle of 25-30°. The other design was modelled on an aircraft's delta wing (although it looked more like a wedge doorstop than a wing).

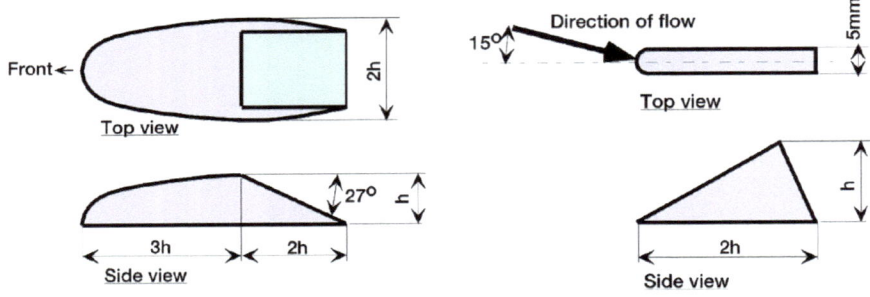

Figure 6-54: The two different types of vortex generator trialled by Mitsubishi on the Lancer. (Courtesy Mitsubishi)

The shape was not mounted in-line with the airflow but instead at a 15° angle to it. (This means that the airflow direction must be carefully mapped right across the roof if vortex generators of this type are to be angled correctly.)

Three different height vortex generating 'bumps' were tested – 15mm, 20mm and 25mm. Looking firstly at the change in C_D, the 15mm high bumps caused a decrease in overall drag of about 0.001, and the 20mm and 25mm bumps a decrease of about 0.003. Changes in coefficient of lift showed that the higher the bumps, the greater reduction in the coefficient of lift to a maximum of a 0.005 decrease for the 30mm bumps. Figure 6-55 shows these results.

Three different delta-shaped vortex generators were also trialled. The tested height of these vortex generators were again 15, 20 and 25mm. The test results showed a similar decrease in drag and lift coefficients of 0.006 for 15 and 20mm delta vortex generators. For 25mm high deltas, the drag decrease remained much the same but the coefficient of lift decreased by 0.007. Figure 6-56 shows these results.

The delta-shaped vortex generators therefore gave clearly better results than the bumps. However, Mitsubishi did not quote drag or lift coefficients for the car, so assessing the percentage changes are difficult. If C_D 0.35 is assumed, the drag reduction with the delta-shaped vortex generators is 1.7 per cent.

Mitsubishi performed a number of other tests to see the effect of the vortex generators. These included CFD simulations and actual pressure measurements on the surface of the vehicle body. With the vortex generators fitted, the calculated velocity past the rear wing was increased. The measured pressure distribution on the boot lid gave an even clearer indication of the positive changes. A greater area of high pressure could be clearly seen on the boot lid and the rear window. However, there was also a lower pressure behind the vortex generators themselves, indicative of the drag being created by their presence. These changes are shown in Figure 6-57.

In its technical paper, Mitsubishi's engineers drew the following conclusions:

"(1) Vortex generators (VGs) were installed immediately upstream of the flow separation point in order to control separation of airflow above the sedan's rear window and improve the aerodynamic characteristics. It was found that the optimum height of the VGs is almost equivalent to the thickness of the boundary layer (15-25mm) and the optimum method of placement is to arrange them in a row in the lateral direction 100mm upstream of the roof-end at intervals of 100mm. The VGs are not highly sensitive to these parameters and their optimum value ranges are wide. Better effects are obtained from delta-wing-shaped VGs than from bump-shaped VGs.
(2) Application of the VGs of the optimum shape showed a 0.006 reduction in both the drag coefficient

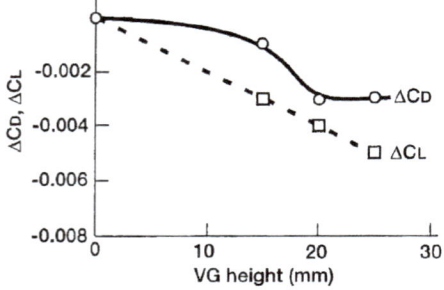

Figure 6-55: Testing of 'bump-shaped' vortex generators showed a linear reduction in lift coefficient with increasing vortex generator height. However, drag dropped before then remaining constant. (Courtesy Mitsubishi)

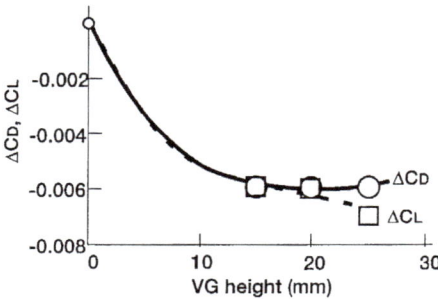

Figure 6-56: Delta-shaped vortex generators gave better results than the bump-shaped design, with greater decreases in both drag and lift. (Courtesy Mitsubishi)

REDUCING DRAG

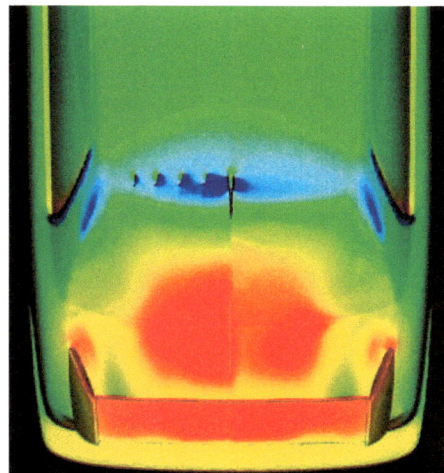

Figure 6-57: With the vortex generators fitted (left of image), higher pressures were recorded on the boot lid and rear window. However, the drag created by the presence of the vortex generators themselves can also be seen (increased area of dark blue). (Courtesy Mitsubishi)

and lift coefficient of the Mitsubishi Lancer Evolution.
(3) It is confirmed that VGs create streamwise vortices, the vortices mix higher and lower layers of boundary layer and the mixture causes the flow separation point to shift downstream, consequently the separation region is narrowed. From this, we could predict that VGs cause the pressure of the vehicle's entire rear surface to increase, therefore decreasing drag, also the velocity around the rear spoiler to increase, and the lift to decrease."

Mitsubishi found the vortex generators to be effective, but not all research suggests that vortex generators will be successful – it depends very much on how they are used.

In a paper published in the *Journal of Automotive Engineering* – "Comparative effects of vortex generators on Ahmed's squareback and minivan car models" – researchers analysed in a wind tunnel the performance of two different types of vortex generators, attached to both a simple shape and a real car – a Peugeot 308. The vortex generators were placed on the trailing edge of the roof.

One type of vortex generator was formed from folded sheet metal and comprised two triangular-shaped vertical wings splayed apart – rather like vortex generators used on aircraft wings. The other type of vortex generator comprised simple vertical cylinders, formed by stacking magnets. The sheet metal vortex generators were 60mm high, 160mm long and had a front gap between the 'vertical wings' of 25mm and a rear gap of 100mm. The 'wings' were highest at their rear. The magnet cylinders were 20mm in diameter and, when tested on the car, 30mm in height. The sheet metal vortex generators were spaced 280mm apart (on the car six were used) and the cylinders were spaced 70mm apart (on the car 12 were used).

And the results? Terrible! On the Peugeot, the use of the cylinders increased drag. There was a variation in the amount of drag increase with the position of the cylinders on the roof, but in all cases, the result was worse than not having them. And the sheet metal vortex generators? The result was even poorer, with increases in drag and lift, and a reduction in base pressure, caused by major changes in the wake pattern. In fact, with the sheet metal vortex generators fitted, drag increased by 17 per cent and rear lift by a staggering 102 per cent. Importantly, on the simple (non-car) shape that was also tested, the results were more positive, with a slight increase in base pressure – however, there was still an overall increase in drag.

So what can we make of this? As we will also see in a moment, using vortex generators to improve flow attachment over changes in car body shape (eg to improve flow attachment on a sedan's rear window) can be quite effective. However, using vortex generators to alter the patterns of airflow in the wake appears to increase drag and lift.

AirTabs

So what vortex generators are available off the shelf? The best that I have found are AirTabs. AirTabs are made in the US and are widely available. They can be bought directly from the manufacturer quite cheaply, and are available in black or white. AirTabs are made from

Figure 6-58: A Peugeot 308. Vortex generators of different types were trialled on the trailing edge of the roof. The results ranged from poor to terrible! (Courtesy Peugeot)

145

SPEEDPRO SERIES

Figure 6-59: Commercially available AirTab vortex generators. They are available in either black or clear/white and are quite cheap. In use, the open mouth faces into the airflow.

ABS plastic and they come with pressure-sensitive adhesive tape on their base. They are 120mm long, 82mm wide and 25mm high (4¾ x 3¼ x 1in). Note that in use, the open mouth faces into the airflow. Let's now look at some applications I have made of these AirTab vortex generators.

Toyota Prius

I carried out testing on an NHW10 Toyota Prius – a sedan, unlike later Prius shapes. The first step was to track the airflow pattern over the rear window, using wool-tufts. This testing showed that there was attached flow across the transition from roof to rear window. The attached flow continued down the window at both ends of the rear glass, however, in the lower-middle area there was separated flow. In other words, a separation bubble formed at the middle/base of the rear window which would adversely affect the flow onto the boot-lid.

To see if the separation bubble at the base of the rear glass could be eradicated, four AirTab vortex generators were centred at the trailing edge of the roof. With these in place, the difference in airflow

Figure 6-60: The NHW10 Prius had a C_D of 0.29 – excellent for the time. However, could flow over the rear window be improved? (Courtesy Toyota)

Figure 6-61: Left: Tuft testing of the standard Prius showed that there was attached flow from the roof to rear window. The attached flow continued down the window at both ends of the rear glass, however a separation bubble formed at the middle/base of the rear window (ringed). Centre: The car with four AirTab vortex generators positioned across the roof just ahead of the rear window. (They are hard to see because of the glare on the roof). With these in place, the airflow down the middle of the rear window now stays attached. However, either side of the AirTabs' path of influence, the separation remains.
Right: Six vortex generators are now in place. The airflow pattern is transformed, with no separation bubble forming at all. (However, there is still a little separation near the base of the pillars – probably trailing vortices.)

REDUCING DRAG

was immediately apparent. This time, the airflow down the middle of the rear window remained attached to the glass. This change in flow pattern was directly downstream of the vortex generators. However, either side of this path of influence, the separation remained.

Another two vortex generators were then added, giving a total of six centred on the trailing edge of the roof. Again, the difference was obvious. The airflow pattern was completely transformed, with no separation bubble forming at all. However, with such good airflow, any separation became more visible and some could be seen at the base of the window at each extreme end. Would fitting another two vortex generators (so extending the line across the whole width of the roof) fix this problem?

The answer was 'no.' With eight vortex generators placed on the roof, the separation at the lower edges of the rear glass remained – perhaps caused by airflow wrapping around the C-pillars and causing vortices to be formed. I decided then to go back to six vortex generators and run with the small edge separation remaining.

So, was there a fuel economy gain (indicative of decrease in drag) from fitting the six vortex generators? Unfortunately, I don't know. The NHW10 Prius doesn't run a trip fuel consumption display, and I simply didn't think that filling the tank each time was a sufficiently accurate way of checking the fuel consumption – not when we're talking a car where just a tiny difference in tank fill volume would dramatically change the results. This was one case where fuel consumption records would need to be accurately kept over a long period if valid data were to be gained.

However, one thing is clear from this test – the AirTab vortex generators can certainly energise the boundary layer, so promoting attached flow where previously there was separation.

Honda Insight

An additional idea promoted by those selling vortex generators is that vortex generators can pull extra air into the low-pressure wake, increasing its pressure and so decreasing drag. As it happened, at that time I had available a vehicle and a daily route that was perfect for checking out this 'filling-the-wake' idea. The test car was a standard Gen 1 Honda Insight. Every day for four weeks it travelled a long freeway route, giving very consistent fuel consumption and allowing a good A/B comparison.

Five AirTab vortex generators were placed across the upper trailing edge of the rear hatch. After fitting the vortex generators, the measured trip fuel consumption immediately rose 3-7 per cent. No change in the feel of the car could be felt – there was no apparent improvement in stability, reduction in wake noise or any other positives. Simply put: the vortex generators made things worse, so I took them off. (This result is similar to the outcome of the test described earlier on the Peugeot 308.)

A further test was performed on the Honda. As standard, the Honda runs a short front undertray with a slightly raised section in the front bumper flowing air to it. Further back under the car there are some in-fill panels, but the standard underside is certainly not a smooth,

Figure 6-62: It is sometimes suggested that vortex generators can aid airflow into the wake, increasing base pressure and so lowering drag. However, testing of them in this configuration on a Honda Insight showed increased fuel consumption.

SPEEDPRO SERIES

Figure 6-63: Trialling of vortex generators on the leading edge of the front bumper underside showed an improvement in stability with no increase in fuel consumption.

flat and continuous surface. I thought that if the boundary layer of air under the car was energised, the flow might better 'jump the gaps' on the underside. Additionally, it might accelerate air past the frontal undertray, reducing lift. Or, to be honest, fitting the vortex generators to the undertray might do *something* – I doubt if anyone in the world could say what the outcome would be without first trying it!

Four AirTab vortex generators were placed at the leading edge of the undertray. Immediately noticeable was improved aerodynamic stability. In standard form the Insight is not an aerodynamically stable car at speed – it is discombobulated by, especially, the bow waves of trucks. This can be felt when slowly passing a truck heading in the same direction on a multi-lane road. As the Honda draws adjacent to the front of the truck, the very light car is pushed away from the truck. (This effect, once recognised, can be felt to a degree in lots of cars – but it's quite clear in the Honda.) With the undertray vortex generators in place, the effect of the truck bow waves was diminished. The bow wave effect could still be felt, but it needed less steering correction.

But would there be a trade-off in extra drag, resulting in increased fuel consumption? The answer to that is – no. With the undertray vortex generators in place, the fuel consumption did not change.

Honda Legend

At one stage I owned a 2006 model Honda Legend. The Legend has, for its time, a good C_D of 0.29. However, driving the car showed evidence of a very interesting airflow pattern on the rear window. After overnight rain, when the ungaraged car had been thoroughly soaked, driving the car resulted in a progressive clearing of the water film from the rear window. This clearing occurred as a result of airflow over the window. (Note that I am not talking about discrete droplets but an overall film of water.) Watching the water film in the rear vision mirror showed that, after about 30km (~20mi) at 100km/h (62mph), the film down each side of the rear glass was gone, but a patch remained in the middle. (This is similar flow visualisation to the 'eroding clay' technique described in Chapter 3).

A patch of water film remaining the middle of the rear glass implied that the airflow was not attached down this portion of the glass. As we have already seen, this is not uncommon – flow that wraps around the C pillars will often contribute to the air that moves on the rear window, so it's the middle that's the hardest to get right.

Figure 6-64: Watching the pattern of water clearance on the rear window of this Honda Legend indicated that a separation bubble may have been forming at the centre of the rear window.

REDUCING DRAG

Figure 6-65: (a) Tuft testing of the Honda Legend showed that there was better flow from the roof onto the central part of the glass that I had thought might be the case, but that there was a clear separation bubble located at the base of the glass in the middle, and in an adjacent area of the boot lid.
(b) Three Airtab vortex generators placed at the top of the rear window did not cause any change to the flow pattern.
(c) The vortex generators were then moved down the glass, so that they were located just ahead of where the separation bubble was occurring. Doing this resulted in the flow separation on the lower part of the window being a little improved, as was the airflow on the forward part of the boot lid. However, there remained flow separation to the left and right.
(d) With five AirTabs in place, the flows on the lower part of the window, and the front part of the boot lid were transformed.

It was time to do some wool-tufting! This showed that there was actually better flow from the roof onto the central part of the glass than I had thought might be the case, but that there was a clear separation bubble located at the base of the glass in the middle, and in an adjacent area of the boot lid.

I then placed three AirTab vortex generators at the top of the rear window. However, the flow pattern did not change. Flow separation continued to occur at the lower part of the window, and there was a very clear separation bubble on the forward edge of the boot lid.

So, what if the AirTabs were moved further down the glass – so that they were located just ahead of where the separation bubble was occurring? Doing this resulted in the flow separation on the lower part of the window being a little improved, as was the airflow on the forward part of the boot lid. However, it looked very much as if three AirTabs weren't enough – there remained flow separation to the left and right.

With four AirTabs in place, the flows on the lower part of the window, and the front part of the boot lid, were excellent. Not able to be seen in the photos shown here, most of which were taken with a shutter speed of $1/1000$ of a second or more, is that the tufts were also far more stable in their movement than prior to fitting the AirTabs. (And rear vision with the AirTabs in place? It was actually not at all impaired – a serendipitous outcome resulting from a combination of the heights of the seat height, my height and interior rear vision mirror location.)

So the airflow looked better attached, but was there a reduction in drag? Fuel economy on my regular long trip improved by about 3 per cent. Looking back at this from a distance of a few years, I think that this figure is probably a bit high, but there's absolutely no doubt that the flow over the rear of the car was much improved. It would have been interesting to measure pressures in the middle of the rear window, with and without the vortex generators in place.

SPEEDPRO SERIES

SAFETY UNDER THE CAR

There's an important safety aspect that I want to cover before going any further. When you are working under a car, you *must* have jack-stands in place. Jacks are for lifting cars; jack-stands are for holding them in that elevated position. The following approach is advised.

1. Place the car on a firm and level surface.
2. Chock the wheels so the car cannot roll forward or backward.
3. Have the handbrake (e-brake) on.
4. Jack one corner of the car.
5. Place a jack-stand under that corner and then lower the car until the weight is on the jack-stand.
6. Lower the jack and then give the car a hearty wriggle. Ensure it doesn't then fall off the jack-stand.
7. Do the same for the other corners.

There are many reports each year of people working on their cars being killed or injured by cars falling on them. I am too imaginative: I am terrified of that idea, and so I often double-up on jack-stands, with a second set positioned to catch the car should it fall. I have also seen a photo that sticks in my mind: a jack-stand, positioned under an older car, simply punched its way through the floor when the rusty body's weight was placed on the stand. A simple trick: if the wheels are off the car, place them flat under the body – they will give you breathing space should the car fall, and their greater area reduces localised pressures on the car body. I live in an earthquake zone, and so I also cater for the fact the ground may be shaking. In my book *Setting up a Home Car Workshop* (also published by Veloce) I show how to build a set of full-length car ramps. I actually built them specifically for installing aerodynamic undertrays on my cars. So please take care – it's not worth losing your life.

SKODA ROOMSTER – VORTEX GENERATORS ON A FRONT UNDERTRAY

When I had the standard front undertray off the 1.9-litre diesel Skoda Roomster I then owned, I decided to add some aero enhancements. Across the full width of the undertray I glued a line of Airtab vortex generators. The idea was to encourage the boundary layer to stay attached to the undertray, so better drawing-out air from the engine bay as the flow moved past the end of the tray. I'd just fitted a huge front-mounted intercooler and so this seemed like a good idea. I already had the AirTabs, and gluing them in place was only a five-minute job – so if they worked, great; and if they didn't, not much time or money lost. But when I drove down the road for the first time, the AirTabs were furthest from my mind. Because what I could feel was a vibration – a vibration through the floor, gearlever and steering wheel.

Being an older design diesel, the engine in the Skoda Roomster is quite coarse, and so my first thought was that the new intercooler plumbing was too firmly mounted, so transferring engine vibration to the bodywork. I checked under the bonnet to ensure that the pipework wasn't banging against anything – but it looked fine. I idled the engine and physically felt the pipes – and yes, there was quite a lot of vibration occurring in them (perhaps also because of internal pressure waves – diesels breathe a lot of air, even at idle). But then again, that had been the case with the original plumbing, where driving vibration wasn't an issue.

I went for a longer drive at highway speeds and the vibration was so bad that I knew something had to be done. And it was more than vibration – the car was also noisier. This was terrible – even in standard form, the Roomster diesel is no paragon of NVH (noise, vibration, harshness), and I'd made it a lot worse. So how much of the noise and vibration was coming from the engine? I drove along at 100km/h (62mph) and then selected neutral, letting the revs drop back to idle. *And most of the vibration and noise remained!* So what was going on?

Then I remembered the vortex generators. Surely, *surely* they couldn't be causing these problems? There was only one way to find out – off they came. Incredibly, the noise and vibration disappeared.

The AirTabs must have been causing vortices, perhaps to impinge on the floor near the firewall, shaking the car and generating noise. It seems implausible, but there's no other possible explanation.

REDUCING DRAG

UNDERTRAYS

Changing the aerodynamics that are at work beneath your car is one of the most important areas of aerodynamic modification. This is especially the case if you do not have a late model, slippery car where a lot of attention has already been paid to this area. In an older car, modifications under the car can reduce drag and lift (or even give downforce – more on this in the next chapter).

There are two major benefits in optimising under-car aero:

- you can achieve major changes in aerodynamic performance
- the work you do is hidden from view, so unlike modifications such as spoilers, vortex generators and separation edges, there's no need to worry about aesthetics

So, how do undertrays (sometimes called belly pans) work? In short, they smooth the path of airflow. In most older cars, the underside is very rough. One reference describes it as having a surface roughness of ±100mm (±4 inches) – which is an interesting way of putting it. As a result, the airflow trying to pass over this surface roughness cannot remain attached, and so becomes turbulent. This has two results: the undercar airflow acts as if it is wake (ie the wake grows in size), and it develops an average pressure that is similar to ambient. (It may be higher at the front of the car and lower at the back of the car, but overall, it's not much different to ambient air pressure.)

And what benefits does smoother airflow have? Firstly, its average velocity is higher – and higher velocities equal lower pressures. But the very fact that airflow can smoothly pass under the car also prevents the frontal area between the wheels being part of the mechanism that generates the wake. (I will explore this in more detail shortly, but at this stage I want to paint a broad picture.) Thus, for low drag, a smooth underfloor is a must.

And the possible reductions in drag? References vary widely in the suggested gains. Later in this chapter, I show a table that suggests a 15 per cent drag

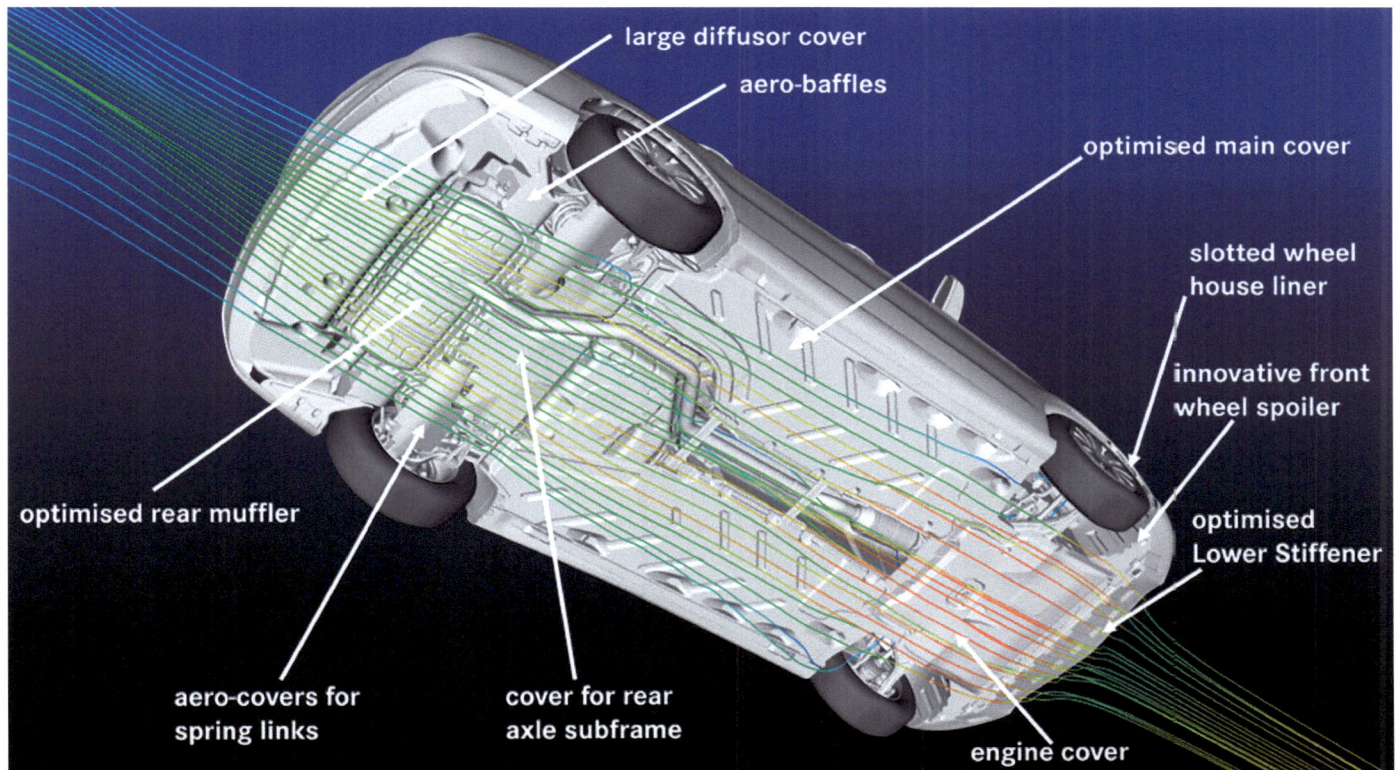

Figure 6-66: The underside of a Mercedes CLA class showing the attention to detail now being paid to reducing drag. The rear of the car is toward the left. Note the covers under the rear suspension arms, the flat bottom of the transversely-located muffler, and the diffuser. Importantly, for cooling purposes the exhaust is left exposed. (Courtesy Mercedes)

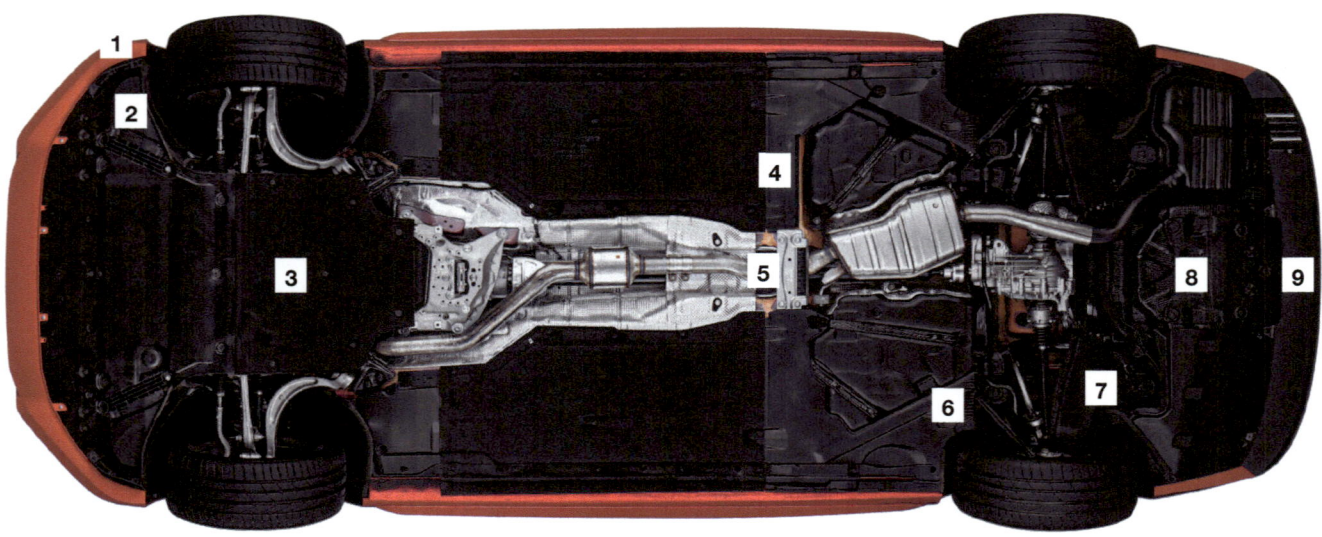

Figure 6-67: Top picture – The Audi A4 Avant has a C_D of 0.26. The slippery underbody makes a major contribution to this. Lower picture – (1) wheelarch pre-spoiler, (2) wheel spoiler, (3) engine undertray, (4) spoiler, (5) spoiler, (6) wheel spoiler rear, (7) suspension arm cladding, (8) spare wheel well cover (9) spoiler spare wheel well. (Courtesy Audi)

reduction is possible. SAE paper 2016-01-1613 (the Canadian study already mentioned several times in this chapter) is much more conservative. It tested 2013-2015 model cars in a full-size wind tunnel, completely covering the underside of the vehicles and comparing the drag with vehicles devoid of any undertrays. It recorded reductions in drag of 1.5-4.5 per cent. Obviously, the possible gain depends very much on how poor the underside is in standard form.

I've been lucky enough to visit some of the great car museums of the world, and the one thing I consistently do is lie on the floor and look under cars. Not *one* of the slippery cars – either past or current – has a poor underfloor treatment. Some cars, like early Tatras and Porsches, the General Motors EV1 and the Tesla Model S, are really quite exceptional in this regard – see Chapter 2 for more on these cars.

Undertray practical design points

In addition to the aerodynamic shape, when finalising an undertray design, there are some important points to keep in mind.

Unless you are intentionally reducing flow through heat exchangers, be careful to ensure

REDUCING DRAG

that exits for airflow from the radiator and other heat exchangers are not blocked by your undertray. Remember, if all the air exits are blocked, no air will flow through the heat exchangers! When considering this aspect on a traditional front-engine car, look at four possible air exits for this heated air:

- past the rear of the undertray to the underside of the car
- through a gap left in the undertray to provide clearance to the exhaust, and then to the underside of the car
- through gaps in the inner guards (fenders) and then into the wheel wells
- through specific vents in the bonnet

You can normally close off excess openings so that you have only one of the above approaches remaining. But after fitting an undertray, monitor temperatures carefully – for example, often engine oil temperatures will rise without the cooling flow of air past the sump.

There are three clearances you need to be careful of when designing and making undertrays. You need to have adequate clearance to the ground on full suspension bump, to clear moving parts like engine accessory drive pulleys, and to leave an adequate gap to hot parts like the engine sump and exhaust. Additionally, with a front undertray, ensure clearance for the wheels on full steering lock.

Just a further word on clearance to exhausts. I have seen people install trays that cover the *complete* underside of the car. Often these trays are made from plastic that has been glued or riveted in place. However, if you look beneath even the latest (non-electric) cars having low drag, you'll find that invariably

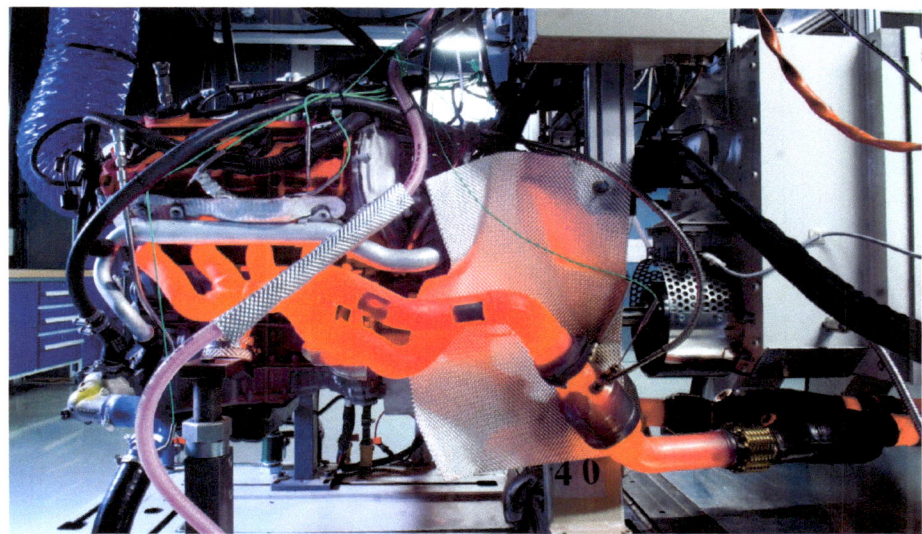

Figure 6-68: If you're tempted to cover the complete underside of your car (including the exhaust) have a long and hard look at this photo! Exhausts really do glow red-hot when the engine is being worked very hard, and so the exhaust needs ventilation flow, especially over the first half. (Courtesy Audi)

the exhaust pipe is *not* enclosed – especially the first section immediately after the engine. This is because, at high loads, this section of the exhaust can become hot enough to glow red! The radiant heat emitted from such an exhaust pipe could set fire to a plastic undertray placed close to it. You would then have a fire under the car – and fuel lines are normally routed nearby… When installing an undertray, look carefully at the heat shielding installed by the manufacturer, and take your cues as to the required clearances from that.

Figure 6-69: It's easy to forget how poor the undersides of many cars are. This 2017 Mustang cries out for some underfloor aero! (Courtesy Georgina Edgar)

SPEEDPRO SERIES

Note that many cars that do have all-enclosing undertrays (eg exotic supercars) have mid-mounted engines, and so there are no exhaust pipes running from the front of the car to the rear. In these cases, it's of course fine to have undertrays that completely cover the front and centre portions of the underside. Incidentally, and worth exploring if you do need some cooling air directed appropriately, many of these exotics use NACA ducts in their undertrays to achieve cooling.

When you are designing and installing an undertray, ensure that you do a workmanlike (no gender implied) job. An undertray that is glued into place will probably need to be destroyed if maintenance needs to be carried out, and all cars with conventional engines need to have regular oil and filter changes. You also do not want the undertray coming loose at speed – as described in more detail in the next chapter, that undertray may be providing downforce. A sudden loss of downforce, as for example if the undertray is torn off over a bump, could be quite dangerous – yes, even in a road car.

As described in more detail below, I like to use ABS plastic sheet to form undertrays. I also prefer to hold them in place using rivnuts and bolts. Rivnuts (sometimes also known nutserts, blind rivet nuts and similar) can be an absolute lifesaver when working on cars. They allow you to put special threaded inserts into blind holes, so letting you screw bolts into the holes. To put this another way, they turn empty holes into tapped holes – and unlike using a tap to make a thread, they will work in thin sheet. This allows you to drill into underside sheet metal (*not* the fuel tank!), insert a rivnut and then expand it into place with a special tool. You then have threaded

Figure 6-70: An undertray fitted to a 1998 Toyota T-100 Xtra cab SR5 truck. Note that this undertray is positioned substantially lower than the exhaust and so can cover the complete underside of the truck. The undertray is made from aluminium plate. (Courtesy Earl Poon, EcoModder.com)

holes that you can use to take the bolts holding the undertray into place.

Don't skimp on the number of bolts you use to hold the undertray in place. I try to use at least one every 300-400mm (12-16in) in every direction. Generally, I do not use push-in plastic clips of the sort often used as original equipment. If no underside body panels are available in which to insert a rivnut, you may need to develop a frame or some other type of support. If you are a high-performance driver, consider integrating the undertray supports with added chassis braces that provide better body rigidity.

Toyota Prius undertray

I decided to fit a new front undertray to an NHW10 Prius that I then owned. The Prius upper body is quite low-drag in design, but when you look under the car, the picture dramatically changes. Especially under the front, it's an aerodynamic dog's breakfast. A plastic moulding covers the front of the lower engine and power split device (ie gearbox), and short deflectors are positioned ahead of each front wheel. But there's no continuity in line backward from the lower edge of the bumper, and bits and pieces are hanging down into the flow everywhere. About the only clear evidence of underfloor aero treatment is an infill panel positioned near the rear of the car.

Looking at the front underside of the Prius, there appeared to be a few undercar options.

1. Increase the size of the existing front wheel deflector plates. But this would likely result in more drag and more lift – not wanted.
2. Remove the deflectors, install an

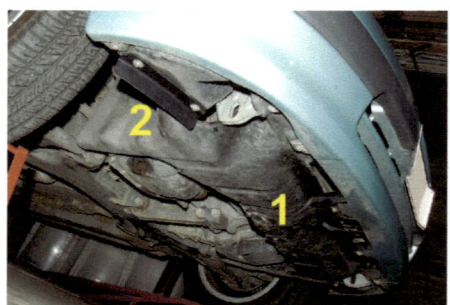

Figure 6-71: The view under the front of a standard NHW10 Prius. A plastic moulding (1) covers the front of the lower engine and power split device (ie gearbox), and short deflectors (2) are positioned ahead of each front wheel.

REDUCING DRAG

undertray across the full width of the car ahead of the front wheels, and then reinstall the deflectors on the new undertray.
3. Install a full-width undertray ahead of the front wheels that curved under the existing deflectors.

I decided on the last of the three options, firstly building a quick and simple prototype undertray to see if this approach worked. I sourced some thin plastic that had originally formed a sign. It was cut to the right shape and the rear edge reinforced with aluminium extrusion. Two pieces of sign needed to be used, and these and the aluminium strengthening piece were held together with high quality plastic adhesive tape. More tape was used to hold the trial undertray in place. On-road testing comprised two aspects:

1. Did the car feel more stable, less stable, or the same in freeway conditions?
2. Was there any measurable change in fuel consumption?

On a multi-lane freeway – always a good test of car stability with its relatively high speeds, air disturbance from other vehicles, and open space for crosswinds to impact – the Prius felt just the same. Or at least, I couldn't tell any difference. If I had to guess, I'd say that the car felt a fraction more stable – but in the real world, increasing the weight in the electric power steering (a modification previously undertaken) had improved things much more dramatically than the effect of the undertray.

However, indicative of lower drag, there was a clear and measurable improvement in fuel economy. The Prius uses a colour

Figure 6-72: A trial front undertray was made from plastic sign material and tape. Note how it covers the original front-wheel deflectors. This trial tray worked well.

Figure 6-73: The final undertray was made from ABS plastic sheet. Here is the starting point of the process, with the plastic sheet held against the car underside, allowing the front cut line to be marked.

LCD to show average fuel economy in 5-minute intervals. That is, each 5 minutes, the screen updates to show by means of a bar graph the fuel economy over the last 5 minutes of driving. This Japanese domestic market car showed the fuel economy in kilometres per litre. (100 divided by km/l = litres/100km.)

On a flat road at 100km/h (62mph), in standard form the Prius invariably turned in a 5-minute average fuel economy reading of 18 kilometres/litre (5.6 litres/100km). That figure was achieved in cruise conditions on flat roads over literally thousands of kilometres of testing. Just occasionally, 20km/litre was achieved – but rarely.

However, with the trial undertray in place, 20km/l became the normal 100km/h cruise fuel consumption. In this configuration, it was 22km/l that was the occasional best result. In other words, 100km/h cruise economy improved from 5.6

SPEEDPRO SERIES

Figure 6-74: Cutting out the ABS sheet with an electric jigsaw. The sheet is easily cut with a wood-cutting blade.

Figure 6-75: Side folds have now been made and the material cut away to allow wheel clearance. I had intended heating and bending along both marked lines but actually did only the rearmost.

Figure 6-76: This aluminium angle was needed to stiffen the rear edge of the undertray.

litres/100km to 5 litres/100km – and economy in the Fours was now occasionally occurring. (At 62mph, the improvement was from ~42-47 US mpg.) In a highly developed car like the Prius, to improve open road cruise fuel economy by about 10 per cent is a startling result – far better than I had hoped for, especially with no noticeable downside in stability.

While the prototype undertray appeared to work very well, it did have a downside – being made of thin plastic and adhesive tape, it wasn't going to last very well … or look too good either! A new one was needed.

I decided to use 4mm grained ABS plastic sheet to form the final undertray. ABS is tough (it can be bent, hit with a hammer, etc, without shattering) and can be cut, filed and sanded with normal woodworking tools. After it has been heated, it can be bent into shape. Finally, unlike plywood or painted aluminium, ABS doesn't show the results of scrapes on the road – it's the same colour right through the thickness of the sheet.

Each end of the undertray needed to be bent upward, allowing it to fit around the existing tyre deflectors (the deflectors were retained but completely covered by the new undertray). The bending was achieved by marking a bend line with tape and then using a heat-gun to heat the plastic along that line. When pliable, the undertray could be bent into position. With the sides bent into approximate position, the undertray could be removed and the front edge cut to shape. Cut-outs were also made on the trailing edge to give the standard tyre clearance – ie the rear cut-outs were within the wheelarches.

A slight rear upturn was bent in the undertray, however, when fitted, the rear edge of the undertray didn't have the stiffness that it was thought the bend would give it. To provide this rigidity, a length of aluminium angle was bolted to the upper surface of the trailing edge. In retrospect, this trailing edge should have been rounded.

The Prius undertray was held in place with screws inserted around the leading edge, screwing into the existing bumper underside. Very coarse thread, short woodscrews were used. The heads of the screws were sanded back to provide a flat-head fastener.

The final undertray varied from the prototype plastic prototype in four respects:

- Directly in front of the tyres the final undertray was not quite the same shape as the prototype. This was because the plastic prototype had major gaps covered in tape, while the final design needed to have its plastic shaped to cover these gaps.
- The final version was slightly longer – it extended about 40mm (~1½in) further rearward between

REDUCING DRAG

Figure 6-77: The finished product. Unlike surface body aerodynamics, undercar aero is largely hidden from the casual glance.

the wheelarch openings, resulting in an approximate total size of about 1600 x 440mm (63 x 17in).
- The final undertray was slightly curved from front to back. It was hoped that this curve might reduce lift.
- The final undertray was made from much thicker material than the plastic prototype, and so held its shape better under aerodynamic forces.

The first task on the road was to assess whether the fuel economy had continued to be better than standard. This proved to be the case. Second, was there now any noticeable aero instability at speed? Certainly, the final version of the undertray was no worse than the plastic prototype– and may well have been *slightly* better. It was similar to what I said above – if I had to guess, I'd say stability might have been a fraction better again than standard. Indications being used were that the car felt less unstable when being passed by trucks, and the steering might have been a fraction heavier in bends taken at speed. However, no particular claims are being made in this area, other than the car was definitely not worse in stability than standard.

Finally, an unexpected outcome of the undertray was that the car became quieter. Whether that was a reduction in undercar aerodynamic noise, or simply the blocking of a path for engine noise transmission, I am not sure. But what was noticeable was that tyre whine and A-pillar aero rustles could be heard – previously, they were drowned out.

And the downsides of the undertray? First, the undertray needed to be removed each time an oil filter change was carried out. Second, the front tow/tie-down hooks were no longer accessible with the undertray in place.

WHEELS, TYRES AND WHEELARCHES

For many years, the drag created by wheels – and the openings in which they rotate – was ignored. In part, this was because wind tunnels used either model or full-size cars that did not have their wheels turning. However, in more recent times, a lot of attention has turned to reducing drag created by the wheels. Many wind tunnels now use a five-belt system, where each wheel is equipped with a belt to provide wheel rotation, and a fifth belt moves under the car's body itself. Some motorsport car manufacturers use a single, full-width belt.

Barnard (*Road Vehicle Aerodynamic Design*) suggests the change in

Figure 6-78: Wheel drag can be divided into ventilation resistance (the drag caused by air movement generated by the wheel) and aerodynamic drag force (the drag caused by the external airflow passing over the spinning wheel). (Courtesy Mercedes)

157

SPEEDPRO SERIES

FRONT DEFLECTORS?

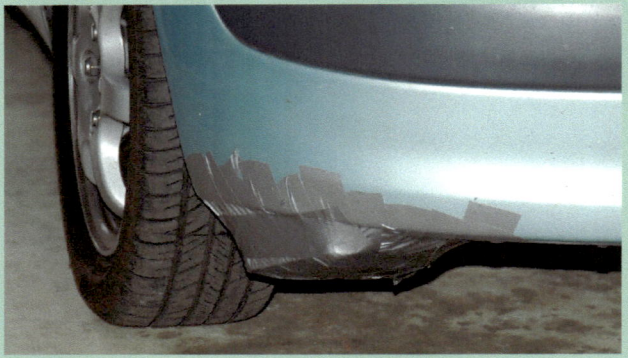

With the front undertray on the NHW10 Prius such a success, what about trying curved front wheel deflectors?

As I did with the undertray, the first step was to do some quick-and-dirty prototyping with lots of duct tape and whatever other materials came to hand. The intention of the add-ons was to deflect air around the high-drag area of the front wheels without using flat-plate deflectors (as commonly used in production cars) which can cause a wider wake and stall the air, creating high drag areas. In addition, flat plate deflectors invariably cause lift as the air pressure bears upward on the undertray positioned ahead of them.

The first step was to use an electric carving knife to approximate the shape of the deflector in expanded polystyrene. High quality duct tape was then used to hold the foam blocks in position, with each attached to the new undertray ahead of the front tyres. The gaps were then filled with tape to give a smooth fairing for each front tyre. The deflectors didn't cover the full width of the tyre but the intention was to move at least some of the oncoming air away from the whirlwind associated with the spinning wheel, and also shield the front suspension arms. It's important to note that this type of prototyping takes little time but can be very effective in showing whether or not you're heading in the right aero direction.

So… was it the right direction? Well, yes and no. About 300km (190mi) of freeway driving at 100km/h (62mph) testing was undertaken with the two prototype deflectors in place. And the results? Firstly, fuel economy slightly improved, indicative of the drag again being decreased. The gain wasn't nearly as great as achieved by the new undertray, but the best-ever 100km/h freeway 5-minute fuel consumption was achieved – 24km/l (4.2 litres/100km), and that was with the air conditioner on! Secondly, 5-minute fuel consumptions of 22km/litre were also more frequently achieved with the deflectors in place.

However, the aero stability of the car *was clearly poorer than standard*. While I stated above that the new undertray may have resulted in a stability improvement, with the deflectors in place, stability was without a doubt inferior to standard. The reason that stability had declined can be sheeted home to an aerodynamic pressure build-up on the undertray ahead of the new deflectors. With the deflectors in place, steering corrections were more frequently needed and the car was more susceptible to the bow waves and wakes of cars in adjoining lanes.

I'd already bought the high-density foam rubber from which I'd intended to make the final versions of the deflectors, but after testing the quickie foam-and-duct-tape prototypes, I decided not to go ahead. Here was a clear case of deciding to either further reduce drag and have lower stability – or to have higher drag and higher stability. I chose to maintain the stability!

C_D caused by the presence of wheels on a normal car is about 0.08. Assuming a car C_D of 0.30, that indicates that wheels make up about 26 per cent of total drag – a significant amount.

The drag caused by air being moved by the wheel itself is termed 'ventilation resistance.' The drag caused by the external airflow passing over the spinning wheel is called its aerodynamic drag force. At speeds over 60km/h (37mph), the aerodynamic drag of the wheels absorbs greater power than ventilation resistance.

In a thesis by Alexey Vdovin at the Chalmers University of Technology, Gothenburg, Sweden (*Investigation of Aerodynamic Resistance of Rotating Wheels on Passenger Cars*), the drag created by 17 different wheel designs were wind tunnel tested. The starting point was an alloy, five-spoke rim with relatively narrow spokes. By the use of add-on parts, this rim was able to be modified to produce wheels with different aerodynamic characteristics. Wheel designs tested included those with twisted 'fan-type' blades (either orientated to flow air inward or outward), thick flat spokes, a high-drag configuration (with flat paddles added to the spokes), and a fully-covered rim (using a flat hub-cap disc).

The design that gave the lowest ventilation resistance was a wheel that retained the spokes in thick form but covered the outer radius of the wheel with a cover. This wheel decreased ventilation drag by about 20 per cent over the worst wheel – the high-drag configuration with flat paddles. And how did the fully-covered wheel perform? It dropped drag by only about 10 per cent over the high-drag design.

Regarding the winning design, the author states that:

REDUCING DRAG

"A possible reason for this [low drag] may be that the exposed parts of spokes in this configuration have lower relative speed, since the spoke length was shorter and they are positioned closer to the centre of rotation of the wheel. This meant that the leading edge of the spoke was subjected to a lower pressure."

And the relatively poor performance of the completely covered wheel?

"Firstly, having no openings in the rim permits an attached flow on the outer side of the rim; this can result in increased surface friction. Secondly, with such a configuration the air cannot pass through the rim and the pressure inside the wheelhouse may be affected."

Interestingly, the 'fan' wheel design that moved air out of the wheel housing also gave good results in the ventilation resistance test.

In the aerodynamic resistance testing, the fully-covered rim performed best, with a drag reduction decrease over the 'high drag' wheel of about 5.5 per cent. However, the wheel that retained the spokes in thick form but covered the outer radius of the wheel with a cover (the winner in the ventilation resistance test), did very nearly as well as the fully-covered wheel.

Putting both tests together?

"The best result was shown by the thick outer radius cover, since it produced the lowest ventilation moment and performed rather well in terms of aerodynamic drag. This configuration was also better than the fully covered rim from a brake cooling point of view, since it allowed air exchange through the rim."

The 'fan blade out' design came in at about middle place in the final figures.

SAE paper 2007-01-1047 (*The Influence of Rotating Wheels on Total Road Load*) adds some further information. Ventilation resistance and aerodynamic drag are measured for a number of wheels, including those of different diameters.

As found by Vdovin in the previously cited thesis, a wheel with only small openings (in this case, a standard Audi alloy wheel, with holes around the periphery) had effectively the same ventilation drag as a wheel completely covered with a disc. Conversely, a wheel with lots of open area – especially if it was bigger in diameter than the standard rim – had increased ventilation drag. Why the change with the bigger rim? The paper suggests that it was because the increased wheel diameter reduced the area of the wheel that was blocked by the brake disc. Even sealing the larger rim (tape was used in the testing, but a hub cap would do the same) did not reduce ventilation drag of the larger rim to the level achieved with the smaller wheels. So from a ventilation drag perspective, the worst scenario is a larger than standard wheel with lots of open area in its design.

Overall, the best results in this study came from small diameter wheels (that is, as small as could be fitted over the brakes), equipped with complete outer covers. However, as Vdovin pointed out, for brake cooling, there should be some openings in the wheel, and if these are small, the total drag contribution is almost identical to completely covered wheels.

SAE paper 2011-01-0165 (*Influences of Different Front and Rear Wheel Designs on Aerodynamic Drag of a Sedan Type Passenger Car*) explored the total impact on drag of a car with different wheel designs. The total lowest drag was achieved by the use of a 30mm outer radial cover for the front wheels, and fully-covered rear wheels. The success of the fully-covered rear wheels was easily explained: there were significantly smaller rear wheel wakes and reduced crossflow through the rear wheels and rear wheel housings. However, the behaviour of the front wheels was much more complex, with effects on upstream and engine bay flows, and downstream effects on the rear wheels, vehicle base and rear end underbody. For example, covering the front wheels resulted in increased drag at the rear wheels, as the rear wheels were exposed to faster flows from the front wheel housings.

Tesla engineers produced a full technical paper (SAE 2012-01-0178 – *The Aerodynamic Development of the Tesla Model S – Part 2: Wheel Design Optimization*) on the development of the 'Aero' wheel option for the Model S. They too found that completely sealing the front wheels caused higher drag. This was primarily because the airflow under the front of car had a major yaw component, diverging from the centreline by about 20°. This flow impacted the exposed inner face of the wheels, and where the wheel face was completely sealed, created higher pressures in the wheelhouse. This resulted in not only higher drag but also higher front lift. The Tesla engineers settled on a wheel that left a gap around its periphery, with lowest drag recorded when the gap was within the 15-20mm range. The production Aero wheel uses a gap of 20mm. This design reduced the Tesla Model S C_D from 0.247 to 0.225, a 9 per cent decrease – astonishingly good on a car already so low in drag.

SPEEDPRO SERIES

Figure 6-79: The Aero wheel on the Tesla Model S was developed after extensive research on low-drag wheel designs. It performs better than a completely covered wheel. (Courtesy Dr Gary M Guest)

and after the rear wheels, and in front of the front wheels. A drag reduction of 5 per cent was claimed. The SAE 2004-01-1307 paper by Volvo models the use of front wheel deflectors and finds that those positioned and sized effectively can reduce C_D by about 0.01. A higher figure (0.016) was the reported benefit of front wheel deflectors by Jaguar Land Rover in the development of the recent Range Rover Evoque, where care was also taken to shield the front suspension arms. Wheel deflectors can be seen under nearly all current cars – it's a good area to look and learn.

Air curtains are now integrated into many front-end designs. In this approach, air enters a front opening positioned in an area of high pressure. The air is then fed through a narrow duct around the front of the wheel housing, exiting from a vertical slot positioned just inside the front edge of the wheel housing. In this way, a stream of air is fed across the face of the wheel, giving better flow reattachment on the panels behind the wheelarch and so reducing the width of the wake created by the front wheels. (In a way, this is rather

The aerodynamic drag (and perhaps also some of the ventilation resistance) can be decreased if the outside of the wheel is no longer fully exposed. This is achieved by enclosing the upper part of the wheel by a spat, normally able to be detached for changing the wheel. This approach is most often used on the rear wheels, as these do not provide any major steering input. However, in some cases, the front (steering) wheels have also been enclosed. If this was the designer's intention from the outset, the wheels can be recessed sufficiently far to give turning clearance. Alternatively, flexible outer covers can be used that deflect as required for steering.

The Ford Probe IV concept car took an innovative approach. The wheels used an inner front wheel mudguard (fender), rather like the cycle guards used on Lotus 7-type cars. This inner mudguard pushed on the outer, flexible spat when steering was required. Thus the rotating wheel never came into contact with the flexible spat. A 9 per cent reduction in drag was claimed for this approach. Figure 6-80 shows the configuration.

A key aspect of the drag created by tyres is the frontal area exposed to the oncoming airflow. The Ford Probe IV used fairings placed before

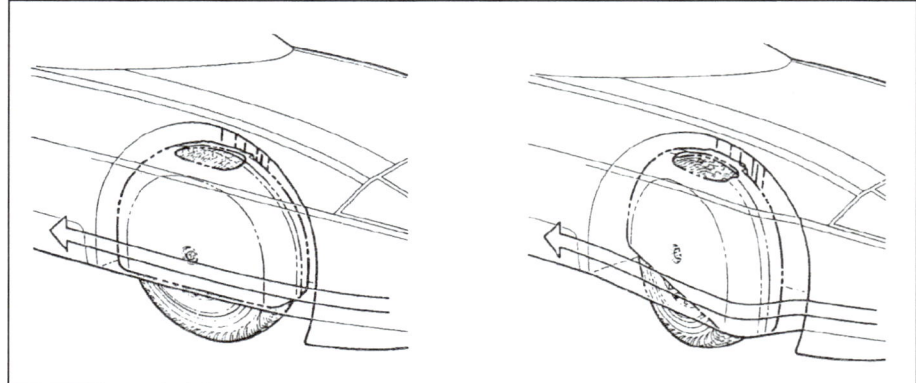

Figure 6-80: The Ford Probe IV concept car used an interesting approach to reducing drag from the front wheels. A flexible spat covered the wheel, and the wheel equipped with an internal cycle-type guard (left). When large steering angles were required, the inner guard pushed on the flexible spat (right), thus avoiding the rotating wheel contacting the spat. (Courtesy Ford)

REDUCING DRAG

Figure 6-81: Many cars now use air curtains like this one on a Ford Mustang. Air is directed from a front intake through a slot positioned in the wheel house ahead of the forward, outside edge of the tyre. (Courtesy Georgina Edgar)

like the turning vanes covered in Chapter 8.) At the time of writing, air curtains were also being introduced for rear wheels, eg Lexus LC500. To feed the rear curtain, a side-mounted duct is required, so this works better on cars with a sporty appearance.

Air curtains are built into the inner guard (fender) liner and front bumper, but is it possible to create the same effect with an external add-on? I experimented using curved wings, placed vertically and mounted directly ahead of the front wheels on a small car. A small gap was left between the bumper cover

Figure 6-82: This simulation shows the air curtain in action. Better flow reattachment occurs on the side panels behind the wheelarch. (Courtesy Ford)

161

SPEEDPRO SERIES

Figure 6-83: Ahead of the rear wheelarch is the side duct that feeds the rear air curtain on the Lexus LC500. (Courtesy Lexus)

and the inside surface of the wing section. The wings were made from GOE222 profile aluminium extrusions (more on these in the next chapter). The curvature of the wing largely matched the curvature in plan form of the bumper, this creating a slot between the wing and bumper. This in turn created a low drag, external air curtain for the front wheels.

And did it work? I fitted the vertical winglets to my Gen 1 Honda Insight and did extensive wool tuft testing with the car standard, and with the winglets located at three different positions. The optimal position was with the trailing edge of the wing just ahead of the wheel opening, and with the forward opening of the slot greater in area than the rear opening. In this position, better flow was observed across the face of the wheel, and fractionally better flow attachment could be seen immediately behind the wheel. However, in the case of the Honda, the car is already good in these areas, and so I didn't proceed with making the fittings permanent. However, I think such an approach could be effective in other cars, especially those with open design, spoked rims.

In summary, to give lowest drag:

- use the smallest wheels that fit over the brakes
- completely cover the rear wheels
- leave small ventilation gaps in the front wheels
- fit tyre deflectors or fairings, front and back
- in cars without air curtains, consider fitting them

TRAILING VORTICES

Trailing vortices occur where there is lift-induced drag. Reducing this lift in turn reduces the strength of the vortices, and is therefore a good indicator of reduced drag. One of the best descriptions of the behaviour of trailing vortices on a three-box (sedan) can be found in Honda's paper on the development of the 2013 model Accord Hybrid (SAE 2015-01-1434). It starts by looking at the vortices developed by the A-pillars.

A-pillar vortices travel across the roof and cause a decrease in pressure at the rear of the vehicle. In turn, this creates more wraparound of airflow from the sides of the car onto the rear

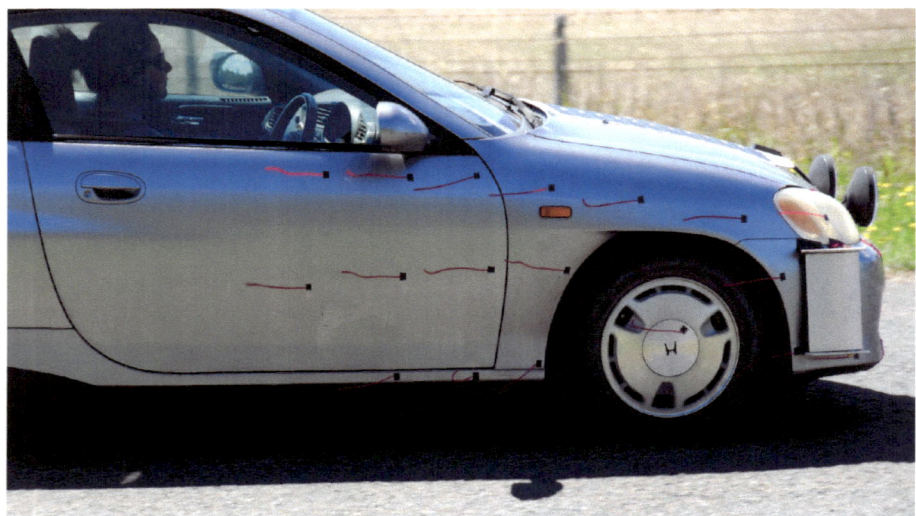

Figure 6-84: Tuft testing of an external air curtain, created by using a GOE222 profile wing extrusion mounted vertically just ahead of the wheelarch opening. Note the excellent flow across the face of the rim.

REDUCING DRAG

Figure 6-85: Extensive optimisation for decreased drag was carried out on the rear of the Honda Accord hybrid. (Courtesy Honda)

window, so creating stronger vortices off the C-pillars.

So, how to weaken the A-pillar vortices? The paper suggests three approaches:

1. decreasing the angle between the windscreen and side glass
2. expanding the A-pillar width to reduce the curvature change
3. reducing the drip moulding (gutter) height to lower the step in the path of the flow

Note that approaches 2 and 3 can be achieved by the addition of A-pillar covers on existing cars – a method used in Chapter 10 to reduce A-pillar noise. In the case of the Accord, the third of the above approaches was used.

Now, what about at the C-pillars? As we've seen, maintaining attached flow to the trailing edge of the boot on a sedan reduces the cross-sectional area of the wake. However, associated with this is a downwash of air as it passes from the roof to the rear window. This pressure drops, and strengthens the longitudinal vortices being shed from the C-pillars. This is because the side airflow more readily travels around the pillars onto the rear glass. Increasing the height of the boot lid will reduce the drop in pressure occurring on the rear window (so reducing C-pillar vortex formation), but in doing this, another problem is created. The taller vertical trailing edges each side of the boot sides create their own vortices! There is, therefore, an optimal boot lid height to reduce the severity of the C-pillar vortices without creating newly-strengthened vortices off the boot sides.

The Honda engineers took three approaches to solving this problem. Firstly, the height of the rear portion of the roof was lowered. Secondly, a 25mm reduction in wheelbase of the new model gave the effect of extending the trailing edge of boot lid rearward. Finally, a 'ducktail' (rising edge) was added to the lid, giving better pressure recovery on the upper surfaces.

Part of the improvement gained was quantified by measuring the pressure difference between the boot trailing edge and the side panels. The lower this difference, the weaker the developed trailing vortices off the C pillars. This is interesting because, even if we cannot see the trailing vortices, we can make similar pressure measurements using the techniques described in Chapter 4.

SPEEDPRO SERIES

ADD-ON MODIFICATIONS

SAE paper 2016-36-0203 (*Aerodynamic Enablers Review for Automotive Applications*) attempts to summarise the reductions in drag available from various simple aerodynamic changes. These are shown in the following table, with the percentage changes referenced against a C_D of 0.35.

Aerodynamic part	C_D reduction	Percentage reduction
Active grille shutter	0.020	5.7
External mirror optimisation	0.008	2.3
Cameras instead of mirrors	0.014	4.0
Wheel air curtains	0.010	2.9
Wheel design	0.007	2.0
Active rear spoiler	0.007	2.0
Rear diffuser	0.014	4.0
Full underbody cover	0.052	15.0
Front underbody cover	0.032	9.0
Tyre deflectors	0.008	2.3

Two important points need to be made. First, as described earlier, do not assume that making multiple modifications will result in a change in drag found by adding the percentage changes! Second, the actual effect of each change will depend heavily on the car in which you are making the modifications – the table is indicative only.

NISSAN LEAF

The Nissan Leaf has a C_D of 0.29. The front wheels are close to the front of the car. To give a longer stretch of attached flow on the bumper corner, and so give smother flow past the front wheels, a sharp edge has been incorporated into the bumper shape. The bulbous headlight splits the airflow so that it passes either side of the side mirrors, decreasing mirror noise by 3dB.

Clean separation is provided by the roof extension, while the sharp corner on the rear guard (fender) creates separation at the lower part of the side of the car. The separation at the upper part of the side is provided for by the edge incorporated into the rear light.

The Leaf has extensive undercar aerodynamics, including this genuine rear diffuser. In addition to the diffuser, there is a large flat floor cover under the middle of the car and a specially-shaped cover under the motor compartment. The motor compartment undercover incorporates a convex depression to accelerate the airflow and so reduce pressure, preventing the oncoming flow diverging into the suspension arms and tyres.

(All photos courtesy Nissan)

REDUCING DRAG

TESTING MULTIPLE DRAG-REDUCING MODIFICATIONS

As described in Chapter 5, measuring changes in drag can be difficult. That's even more the case when trying to assess the worth of individual drag-reducing modifications. After all, each aerodynamic modification has the ability to influence how other modifications behave, so it's a bit like trying to judge the taste of a cake by sampling the individual ingredients – before the mix is cooked!

To address this, Darin Cosgrove of the EcoModder.com forum decided to perform six drag reduction modifications. He then tested the results achieved from the combination of all of these modifications against the standard car. However, note that in addition to making the aerodynamic changes, Darin increased the tyre pressures, going from 230kPa (33psi) to 310kPa (45psi). This change would have potentially reduced the fuel consumption by perhaps 5 per cent.

The car was a Canadian-spec 2015 Nissan Micra 1.6L with manual transmission. In standard form, the Micra has a C_D of 0.315.

All the modifications were temporary.

- A partial grille blocker was made from plastic sheet and held in place with wire ties. The blocker covered the complete upper grille, and the lower grille in the areas not directly in front of the radiator.
- A front air dam (spoiler lip) was made from 100mm (4in) lawn edging.
- A large rear roof and side hatch extension, similar to the one shown earlier for the Skoda Roomster, was installed. The 'roof' part of the extension sloped downward at about 10°.
- Smooth wheel covers were made from plastic sheet.
- The top halves of the rear wheelarches were covered with cardboard and plastic.
- Small deflectors were placed in front of the rear tyres.

Testing of changes in fuel economy (mileage) was carried out at 90km/h (56mph) on a flat, straight stretch of two-lane highway. The ambient temperature was 16°C (61°F). Four runs were made in each direction with the car stock, and four runs were made in each direction with the aero modifications (and increased tyre pressures).

In standard form, the car returned an average of 5.0/100km (47.0 US mpg). In modified form this improved to 4.5 litres/100km (52.5 US mpg), an 11.6 per cent gain.

Figure 6-86: The front of the Micra shows the partial grille blockers and spoiler. The passenger side rear vision mirror was later removed.

REDUCING DRAG – DEVELOPMENT OF A NEW CAR

When Land Rover developed the 2017 Discovery, they were replacing the 4th generation Discovery – which in turn was closely based on the 3rd

Figure 6-87: The Micra was also treated to a rear extension, flush wheel covers, rear wheelarch closures and small deflectors placed in front of the rear wheels. (Both photos courtesy Darin Cosgrove, EcoModder.com)

SPEEDPRO SERIES

Figure 6-88: The 2017 Land Rover Discovery had substantial aerodynamic development, aimed at lowering drag and rear window soiling.

generation released in 2004. That car had a frontal area of 3.15m² and a C_D on launch of 0.41, which dropped to a little less than 0.40 over subsequent model years. The new Discovery had to maintain many of the previous models' attributes, including off-road performance and seating for seven adults. This meant that reducing frontal area would be difficult, so overall drag needed to be reduced by decreasing C_D.

The CFD software package *EXA PowerFLOW* was used extensively during early development, and testing was also carried out at the UK's Horiba-MIRA full-scale wind tunnel and the Aeroacoustic Wind Tunnel at the FKFS research institute, part of Stuttgart University.

An early step was to develop a 'road map' that targeted potential drag reductions. Starting with the Discovery 4's C_D of just under 0.4, the initial design challenge was to reduce this to around 0.36 by design (styling) decisions. The following points were concentrated on:

- front bumper corner
- rear tapering and sharp separation edges
- roof spoiler
- roof header and line
- front bumper chin

Engineering optimisation could then take over, aiming at a C_D of 0.330 by:

Figure 6-89: Visible in this schematic are the front wheelarch air curtains and the rear slotted roof spoiler.

REDUCING DRAG

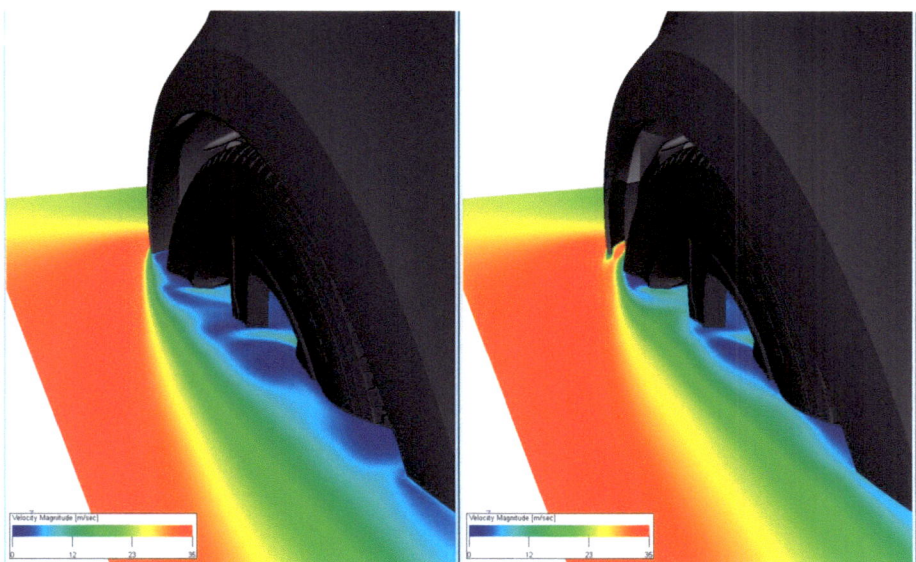

Figure 6-90: Simulation of the airflow velocities without the air curtains (left) and with them (right). Note the faster-flowing air across the wheels with the air curtains present.

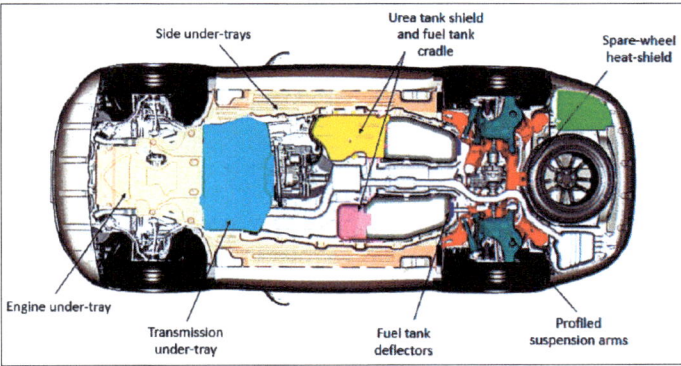

Figure 6-91: Much of the low-drag development of the Land Rover occurred under the car. Note that strategically-placed deflectors, while often small, reduced drag.

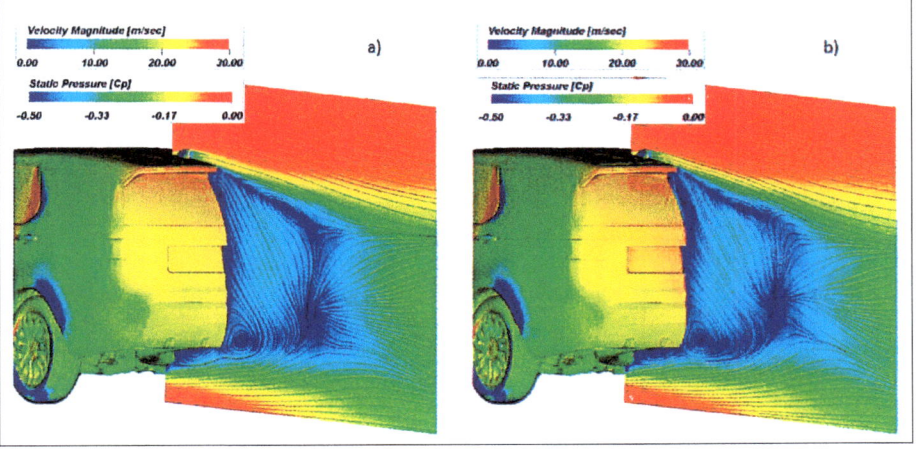

- active grille shutters
- lowering of suspension at speed (-13mm, ½in)
- front wheel deflector optimisation
- transmission undertray
- fuel tank front fairing
- fuel tank deflectors

Following that, there were some additional opportunities that could be pursued for small gains:

- aero wheels
- spare wheel cover

In the early design stage, the front corners were identified as one the most promising areas for drag reduction, with an overall potential of about 15 counts (1 count = 0.001 C_D). Improvement was achieved through a combination of conventional shape optimisation and the development of an air curtain. The air curtain feeds air from front bumper openings around and onto the front wheels. This reduces the disturbance to the airflow caused by the front wheelarches, increasing flow attachment on the tyre sidewall. The net effect is to reduce momentum losses in the flow over the wheelarch and downstream body side.

A key area of effort was directed at improving underfloor aero. The engine and side under-trays are the same as other vehicles sharing the same architecture. However, the large transmission undertray has been clad in a smoothed skin, giving improved NVH (noise, vibration, harshness)

Figure 6-92: The simulated airflow velocities and pressures without the slotted rear spoiler (left) and with it present (right). The differences are quite subtle but rear window soiling was reduced by about 20 per cent with the slotted design.
(All photos courtesy Land Rover)

performance and decreasing drag by three counts. The trailing edge of this undertray has been given a turned-down edge, resulting in a stable separation bubble over the exposed rear part of the transfer case. Fascinatingly, the flow reattaches on three provided flat surfaces: the urea tank shield, the Selective Catalytic Reduction (SCR) exhaust device (it looks like a muffler but isn't), and the fairing that has been added to the front of the fuel tank.

However, with the flow reattached, another problem appeared. Large radii are used on the trailing surfaces of the fuel tank and these were causing the flow to turn upward in an unsteady manner. (This is a bit like flow wrapping around the trailing edge of a boot – except upward rather than downward.) The solution was similar – add two small spoilers to provide clean separation and guide the flow so that it misses the subframe and suspension arms. This approach resulted in a drag reduction of two counts. The suspension arms were also profiled to give low drag.

The spare wheel, located horizontally under the rear of the vehicle, is equipped with a heat shield, and this was optimised to give the required thermal protection without a drag penalty. For powertrains requiring only one rear muffler, an infill panel was developed to cover the space created by the missing muffler. Underside clean separation was ensured by the spare wheel heat shield location, which gives separation irrespective of the presence or otherwise of the spare wheel. The profiled rear suspension arms incorporate appropriate trailing edges to give flow separation to the airflow passing outboard of the spare wheel.

Another device aerodynamically finessed was the rear roof spoiler. This was equipped with slots sufficient to give some airflow over the rear window (reducing soiling by about 20 per cent), but not so much that drag suffered.

Six different designs of wheels were trialled. In addition to achieving low drag, the designs were also assessed for their impact on brake cooling and the change in ventilation resistance. From a baseline of the suitably scaled wheel used on Range Rover hybrids, a decrease in drag by eight counts was achieved while still meeting brake cooling requirements.

The use of air suspension in some versions allows ride height to be easily altered. Lowering the ride height in this way decreases drag by another six counts.

The final C_D is 0.33, a 17 per cent improvement over the outgoing model.

Note: The above content is based on material drawn from *The Aerodynamics Development of the New Land Rover Discovery* – see the references at the end of the book for more detail on this paper.

Figure 6-93: Arto Viinanen has made a variety of modifications to his Audi A2 to reduce drag. The modifications include a front grille block (since altered over the pictured version as the engine overheated), flush wheel covers and undertrays.

Figure 6-94: The flush wheel covers are made from 3mm thick, vacuum-formed plastic by Tuneko. They are held in place by a single middle fastener that incorporates the Audi badge.

REDUCING DRAG

REDUCING DRAG – MODIFYING AN EXISTING CAR

Arto Viinanen of Finland has made a variety of modifications to his Audi A2 to reduce drag. The car, a 2001 model 1.2 TDI, improved in fuel consumption from about 3 litres/100km (78.4 US mpg) to 2.6 litres/100km (90.5 US mpg). One thing that makes Arto's car stand-out is that the modifications have been so well executed that they look standard. His changes include flush wheel covers held in place with a single central fastener, a full-length undertray, and front and rear tyre deflectors.

Figure 6-95: The full-length undertray is made from aluminium sheet and uses aluminium extrusion to stiffen it.

Figure 6-96: The front wheel deflectors cleverly reduce the open area of the wheelarch in addition to deflecting air around the tyres.

Figure 6-97: The rear wheelarches are enclosed in removable spats. The continuation of the trim blackout makes them look very integrated.

Figure 6-98: Note the deflectors placed ahead of the rear tyres. Arto's car shows a stylish mix of low-drag modifications – if you didn't know the A2's original form, you could think many of the changes are standard. (All photos courtesy Arto Viinanen)

169

SPEEDPRO SERIES

PHILIP'S TRUCK

Philip Knox has a radically aero-modified 1994 Toyota T-100, ½-ton, 2-wheel drive pickup truck. The truck is fitted with a standard 2.7-litre DOHC, 16-valve, 150hp engine with manual 5-speed transmission. The truck is constantly evolving, but in the form shown here had:

- a quasi-semicircular plan-view, 300mm (12in) nose extension with integral flexible air dam, reduced area, radiused cooling/combustion air inlet with diverging ductwork and bullet-valve (grille-block), and Plexiglass headlamp and turn signal covers
- composite hood blister
- composite and sheet aluminium belly pan (although missing the centre section), then diffuser (destroyed by an underwater hazard at Bonneville International Speedway)
- 2.8-degree up-swept diffuser
- rocker panel extensions
- urethane rubber, ablative, tyre gap-fillers
- articulated front wheel skirts
- fixed, partial coverage, rear wheel skirts
- General Motors Oldsmobile, 15-inch, aspirated, full-coverage wheel covers
- composite rear load area cover
- 1.2m (48in) composite boat-tail extension
- Plexiglass tail lamp fenestration fairings

In standard form, the truck achieved 9.36 litres/100km (25.14 US mpg) at 105km/h (65mph). In the pictured form, this improved to 6.88 litres/100km (34.18 US mpg), an improvement of 26.5 per cent.

(Courtesy Philip Knox, EcoModder.com)

Chapter 7
Reducing lift and improving stability

- **Causes of lift**
- **Stability**
- **Determining the centres of pressure and gravity**
- **Forces affecting stability**
- **Undertrays and diffusers**
- **Spoilers**
- **Wings**
- **Downforce in a road car**
- **Active aerodynamics**

Aerodynamic modification of road cars can reduce lift and improve stability. In fact, you can make significant changes to the amount of lift your car is subjected to, and in some cases, create downforce at road-going speeds – even in countries with low speed limits. Reducing lift or gaining downforce can have a dramatic impact on tyre grip. And you can also improve stability, something that few people seem to have in their sights when making aerodynamic modifications.

LIFT

Imagine a car-shaped body that is sitting directly on the road, without any gap underneath. All air that passes the car will have to flow around it or over it, and the air that flows over the car will generate lift. Why? Because the air will need to accelerate and, as we know, faster-travelling air generates lower pressures – and so, in this case, lift.

Now let's raise the car body off the ground a little, as it is when standing on its wheels. Air can now pass underneath. If that air runs into obstructions, the airflow will be modified. Any resulting increase in pressure on the underside of the car will generate a contribution to lift. You'd therefore think that it would be best to prevent air going under the car: for example by using a deep front spoiler (air dam). In fact, that is very much the approach that used to be taken with road cars.

But what if we have a smooth underside to the car? If we can get that airflow passing at relatively high speed over a smooth surface under the car, we will generate a low pressure and so reduce lift (or even create downforce), as well as create less underside friction drag.

Therefore, to reduce lift, or create downforce, we need to do this: increase the average pressure acting downwards on the upper body, or reduce average pressures acting under the car, or both.

It's important that the idea of *average* pressure is understood. In Chapter 1 we looked at the pressure distribution along the centreline of a vehicle, and in Chapter 4 I described how you can measure this yourself. You'll remember that the front half of the car had two zones where low pressure was generated – where the airflow wraps around the leading curve of the bonnet (hood), and

Figure 7-1: In the past, some cars generated a lot of lift. This 1965 Ford Galaxie, at 160km/h (100mph), had about 163kg (360lb) of lift, with 145kg (320lb) of that on the front!

in mind that most of the time, there is at least some yaw component in the airstream passing over the car. Most vehicle shapes develop more lift as yaw angles increase. Since that's when there is a cross-wind, you gain more lift force at precisely the time you least want it! (More on stability later.)

The amount of lift, especially in older cars, could be quite staggering. A paper published in 1968 (*Problems of Ground Simulation in Automotive Aerodynamics* – SAE 680121) had some fascinating figures for the actual, on-road measured lift of a 1965 Ford Galaxie. This car, that had a C_L of 0.552(!), developed about 68kg (150lb) of lift at 100km/h (~60mph) – nearly all of it on the front wheels. At 160km/h (100mph), lift had risen to about 163kg (360lb), of which 145kg (320lb) was on the front! (Incidentally, this car had a measured C_D of 0.54.)

where it passes from the windscreen onto the roof. These are small but intense areas of low pressure. If we multiply the pressures at these zones by the areas affected, we get a resulting force – a lot of which is upward.

However, low pressures also exist on other parts of the top surfaces of the car. They're not as intense as the two pressures just described, but they occur over much larger areas. Because of their large areas, these rather innocuous-looking areas of low pressure, if they are widespread, can also generate substantial lift. (And of course, the opposite applies too. If you can develop a higher than ambient pressure – even only a little higher – but it is over a large area on the upper surface of the car, then you will develop a good downward force.)

The majority of passenger cars generate lift, and these lift forces can be quite high. Lift reduces the grip that your tyres can develop and makes the car less stable at speed.

When discussing lift, also keep

Figure 7-2: Airflow wrapping over upper surfaces creates lift. The pictured Mercedes-Benz 300 SL (W 194 series, 1952) had a C_{Lf} of 0.318 and a C_{Lr} of 0.140. (Courtesy Mercedes)

REDUCING LIFT AND IMPROVING STABILITY

Hucho (*Aerodynamics of Road Vehicles*) describes an early Porsche 911 Carrera developing a more modest 25kg (55lb) of front lift at 160km/h (100mph), and a much higher 55kg (120lb) of rear lift at the same speed. Fitting front and rear spoilers to the car reduced this to a tiny fraction of these figures, but you can see why the attachments were needed. (The more pronounced rear lift in the case of the Porsche reflects the airflow generating low pressures over the rear half of the car as the air wraps over the long curve.)

In the development of the 2006 C6 Z06 Chevrolet Corvette, GM engineers spent most of their time trying to reduce the lift generated by the previous version of the car. They were happy when they had reduced the previous car's lift by an astonishing 607.5kg (1336.5lb) – although that was at 300km/h (186mph).

Frere (*Porsche 911 Story*) has interesting data on the 911 series cars – the 964 (sold between 1989-1994), the 993 (1994-1998) and the 996 (1998-2004). (In Chapter 3 a Porsche 993 Turbo is shown being wool tuft tested on the road.) The 996 figures are for when its retractable rear spoiler is raised – on this data, it appears to have worse front and rear lift coefficients than its predecessor – although they still remain low.

The data is shown in the table below.

STABILITY

Stability can be thought of in terms of the amount of driver correction required to maintain the car on a given path when aerodynamically disturbed. For example, a car being driven along a straight road may be subject to a gusty crosswind. The more stable the car, the less steering correction the driver needs to perform.

However, it's not quite as simple as this, because on-road car stability is also greatly influenced by other factors. For example, some cars 'tramline' – follow longitudinal depressions in the road – and a car tramlining at high speed feels quite unstable. Other cars run very light power steering, and heavier steering at high speed makes a car feel much more stable. (In fact, I once drove down a road while an assistant turned a knob that modified the weight of the steering. As the weight increased, the change in the feel of the car was extraordinary.) The degree of self-aligning torque built into the front steering geometry (eg castor and trail) will also affect perceived and actual stability. I remember road testing a new Mercedes sedan once. The Mercedes, as with many of that era, had a slow steering ratio around straight-ahead. In a cross-wind I found I needed an alarming amount of steering movement to keep the car straight – but it was alarming to me mostly because I was used to cars having much more direct steering that required commensurately small steering wheel movements to provide correction. So, the steering ratio around straight-ahead is also relevant to perceptions of stability.

So while a car may feel unstable at high speed, there are many factors other than aerodynamics that may be contributing to that. And when is it most likely that it is indeed aerodynamic instability, rather than those other factors? If the car's stability decreases markedly as speeds rise, it is more likely to be an aerodynamic problem.

Many years ago, I tested the top speed of two cars I owned in sequence. The first, a 1977 BMW 3.0si, used a body shape that dated back to the late 1960s. I've never seen drag and lift figures for the car, but I have no doubt they were poor. With a modified engine, I think I got an indicated 220km/h (137mph) out of it. That car needed a lot of steering work to stay straight!

The other car was my 1986 Holden Commodore VL Turbo, a car I owned immediately after the BMW. As shown by the amount of change required to make this body shape stable and lower drag (see the 1988 Holden Special Vehicles Group A Commodore in Chapter 2), the standard car was quite poor in both regards. At 210km/h (130mph) – the speedo was probably more accurate than in the BMW – I distinctly remember working out which lane I would try to keep the car in, it was wandering so badly.

However, I did move on to aerodynamically more stable cars – my 1990 Nissan Skyline GT-R did an indicated 260km/h (161mph) on a relatively narrow road with not much steering effort required at all.

In addition to stability at high speeds, another good indication of a car's aerodynamic stability (or

	964	964 Turbo	993	993 Turbo	996 Carrera
Frontal area (m²)	1.79	1.88	1.86	1.93	1.90
C_D	0.32	0.35	0.33	0.34	0.30
C_DA	0.573	0.658	0.614	0.656	0.585
C_{Lf}	-0.01	0.00	0.03	0.01	0.08
C_{Lr}	0.02	0.01	0.07	0.03	0.05

lack thereof) is its behaviour on freeways as you slowly pass other vehicles. For example, when you draw alongside a truck and then gradually pull ahead of it, you will be able to feel the bow wave of the larger vehicle pushing you sideways. I first noticed this in my Gen I Honda Insight, but once you recognise the feeling, you can detect it when driving most cars. The amount that you get pushed sideways depends on both the car you are driving and the aerodynamics of the other vehicle. In a given car, you can feel the different sized bow waves being created by different designs of trucks – slippery trucks have a lesser bow wave. Freeways are also good for assessing cross-wind susceptibility, because changes in exposure are often rapid eg passing through a cutting and then into an exposed area.

Aerodynamic stability is influenced by a number of factors. The first to consider is where the lateral centre of pressure is located (that is, the virtual point through which all side aerodynamic forces act.)

The location of the lateral centre of pressure is dependent on two factors: the shape of the vehicle in profile (more on this in a moment), and effective angle at which the airflow is approaching the vehicle. Considering first the effective angle of the oncoming airflow, imagine a simple vehicle shape, like a bus. The bus is moving forward into a crosswind. If the airflow meets the side of the bus at an acute angle, the bus shape will act as an aerofoil, and so the centre of pressure will be at a point approximately one-quarter to one-third back from the front. However, as the wind speed increases, or the bus speed decreases, the effective airflow direction will eventually be at right angles to the bus, and so the centre of pressure will be halfway along the length of the bus. Therefore, while for convenience we will often describe the lateral centre of pressure for a given vehicle shape as if it is a fixed point, in fact it is not.

The lateral centre of pressure is not necessarily at the same location as the lift/downforce centre of pressure. In most cars, the lateral centre of pressure is located well forward of the centre of gravity. Thus, the application of an aerodynamic side force tends to push the front of the car *away* from the force; eg, a wind coming from the left pushes the car to the right.

If the centre of pressure is moved rearward, so that it is behind the centre of gravity, the side force will tend to push the front of the car *towards* the force. Therefore, too far a rearward movement of the centre of pressure and the car will still need driver correction – it will 'weather-cock' into the wind. However, a smaller amount of movement of the front of the car toward the force is acceptable, because this will help generate the required front tyre slip angle to automatically apply the required correction.

Determining the approximate

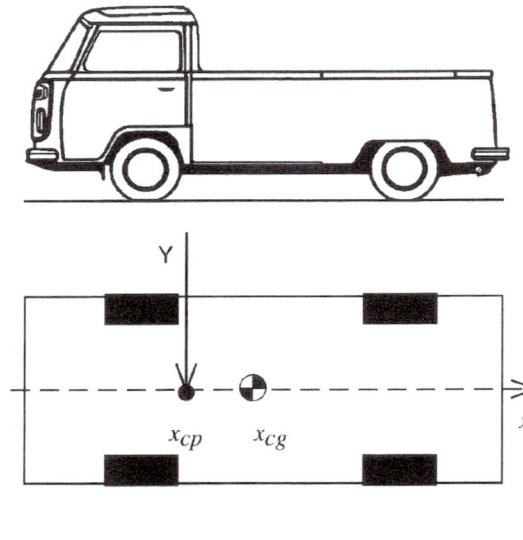

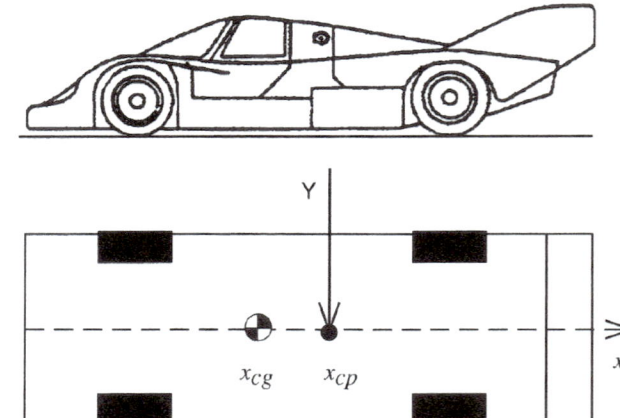

Figure 7-3: Differing relationships between the centre of gravity (CG) and centre of pressure (CP). The Volkswagen Transporter has a CP forward of the CG, giving poor aerodynamic stability. On the racing car, the CP has been moved back by the use of a large rear fin, so that the CP is behind the CG, giving the vehicle better stability. (Courtesy Joseph Katz)

REDUCING LIFT AND IMPROVING STABILITY

Figure 7-4: Donald Campbell's CN7 1960s Land Speed Record Car used a large rear fin, added after a high-speed crash. A rear fin moves the centre of pressure rearward, improving aerodynamic stability.

Figure 7-5: The Mercedes C-111-III research vehicle had 169kW (230hp) at its disposal but could achieve a top speed of no less than 325km/h (202mph). Its C_D was just 0.183. Note the rear fin to give high-speed stability; without it you can see that the lateral centre of pressure would have been well forward. (Courtesy Mercedes)

Figure 7-6: The superb NSU Delphin III speed record motorcycle of 1956. It achieved 305km/h (189mph) on just 56kW (75hp). Note the long, rearward-positioned fin for stability.

centre of pressure and the centre of gravity can be done with just simple tools – see the breakout box on pages 176-177.

FORCES AFFECTING STABILITY

It might seem obvious that the sideways movement of the car is developed by the wind force acting on the windward side, but in fact this pressure is relatively minor. Instead it is the action of *low pressures* on the leeward (downwind) side that creates most of the lateral force.

Because the forward motion of the car is usually relatively fast compared to the speed of a cross-wind, the cross-wind makes its presence felt by generating a yaw (angled) component to the oncoming airflow. For small yaw angles, very low pressures are generated at the leeward A-pillar, while for higher angles of yaw, the flow separates at the A-pillar, and most of the leeward side is in separated flow. Therefore, while we have looked at achieving clean flow separation at the rear of the car with zero yaw airflow (see the previous chapter), to achieve better stability, we need to picture achieving clean separation on the leeward side of the car with an airflow approaching the car at (say) 5° of yaw.

However, there is a fly in the ointment: if the front corners are sufficiently rounded that good flow attachment occurs at zero yaw (thus giving low drag), but are sharp enough that flow separation abruptly occurs once a certain yaw angle is reached, the car may become twitchy in its response. The answer is to have a progressive separation on the leeward side under yaw conditions. However, it's all starting

(continued on page 178)

DETERMINING THE CENTRE OF PRESSURE AND CENTRE OF GRAVITY

The fastest cars in the world go well over 640km/h (400mph). They're the super streamliners that are driven on salt, eg the Bonneville Salt Flats in Utah, USA. The problem with salt is that the surface is similar to a dirt road or packed snow, with the coefficient of friction only a fraction that of pavement. It can perhaps be as high as 0.6 on a really good year, but is typically closer to 0.3 or 0.4. (This can be compared to about 1.0 for ordinary pavement, and up to 3.0 for the dragstrip with a prepped surface.) So how can designers ensure these vehicles go straight? The relationship between the centre of gravity and the centre of pressure is critical.

The easiest way to understand this concept is to think of a dart or an arrow. If you throw a dart with its feathers first, it will turn in the air and hit the board with the steel tip first. A dart wants to go tip-first and is aerodynamically stable in that direction.

To determine if your vehicle is aerodynamically stable, you need to find both the centre of gravity (CG) and the aerodynamic centre of pressure (CP) of your vehicle. We care about the relative positions of the CG and CP because this determines the overall aerodynamic stability of your vehicle. The CG must be in front of the CP for the vehicle to be stable at high speeds. Again, think of that dart. A dart has a heavy nose and feathers in the back. This ensures that the CG is ahead of the CP. You want your vehicle to display the same behaviour as the dart – no matter what happens, you want your vehicle to want to go nose first.

Most folks know (or at least have some idea) what the centre of gravity is. Basically, if you suspended your vehicle from the CG point – wherever it may lie within your vehicle – it would balance perfectly. Once your vehicle was suspended by the CG point, you could reposition it with a light touch of your hand and it would simply hang in that new position.

The centre of aerodynamic pressure (CP) is similar to the CG, but with the CP we are not worried about how gravity will act on your vehicle, but instead we are concerned about the aerodynamic forces exerted on the outer skin of your vehicle by the wind. With the centre of gravity (CG), we find the balance point with respect to the force of gravity, but with the centre of pressure (CP) we will find the balance point with respect to the wind.

While it is possible to determine the centre of pressure

This section is written by Eva Håkansson and Bill Dube', two engineers with vast experience of designing, building and driving vehicles that require good aerodynamic stability at very high speed. Go to http://www.evahakanssonracing.com to see their latest. (Courtesy John Baechtel)

exactly in a wind tunnel or by 3D scanning and use of simulation software, we are going to use a clever (and easy) way of estimating it. While not perfect, it works pretty well for most purposes. It is an old trick used by the model airplane crowd.

First, you must have as accurate a side-view profile of your vehicle as possible. The easiest way to get this is to take a side picture of your vehicle. It is best if you position the camera dead-on perpendicular to centre of the side. Use a telephoto lens setting and take the photo from as far away as possible, with the vehicle still filling most of the frame.

Print the profile photo about 280 x 430mm (11 x 17 inches) and on as heavy a paper as your printer can manage. Smaller and/or lighter tends to be less accurate. Mounting on cardboard or foam board can help, but only if you evenly spray on the adhesive. You can go really big, but it doesn't improve the accuracy of the overall estimate, however. (If you don't have access to a printer, many of the 1-hour print shops offer affordable prints mounted on foam board. This can be a simpler, easier option – and you could get a spare to hang on your wall!)

Carefully cut out the profile of your vehicle with a precision knife or scissors. You can then balance the cut-out on the edge of a ruler to get close to the balance point. Next you will do the final balancing with a push pin, moving the pin slightly forward and backward until you are able to get the exact balance point of your vehicle shape.

You are now pretty darn close to locating the centre of pressure for your vehicle. It is technically the 'centroid of the

REDUCING LIFT AND IMPROVING STABILITY

area,' but for our purposes it is a good enough estimate of the CP.

Now for the centre of gravity.

There are countless ways to find the CG, but this is the standard method that is used on aircraft. It is important that when you are performing these CG measurements, the vehicle is in 100 per cent 'road ready' condition. That is, the driver is in position, there is fuel in the tank, etc.

We begin by choosing some standard fixed 'zero' point. A typical fixed point to choose is the nose of the vehicle. Put your vehicle in a position where it won't roll and mark the point straight down from the nose with tape on your garage floor. Use a plumb bob or a level to get your zero point transferred accurately on the floor. Next, make a line that goes straight out to the left and right of that point, so that it is perfectly parallel with the axles and aligned with the front of your car. This is your zero line (or, formally, your 'datum reference').

Now, measure straight back from your zero line to the centre of the contact patch on each of your tyres. Write down the distance for each of your tyres, keeping track of left-front, right-front, right-rear, left-rear.

Now we determine the weight on each tyre. If you have four identical platform scales, then this is easy. If, like most of us, you have just one scale, then you must make three small platforms that are the same height as your scale. (Typically, you simply use thick wood planks the same thickness as your scale.) You then, somehow, get your vehicle up on the three planks and the scale. Read the scale. Then swap the scale to another tyre. Again, write down the weights and keep track of how much each corner of the car weighed.

We know it would be easier not to bother with the platforms and just move the scale from wheel to wheel, but unless the other wheels remain at the same height as the wheel on the scale, the scale will read incorrectly.

We are now going to calculate the 'moment' for each wheel with respect to our zero line (moment = distance to the zero line x weight).

We make a table like this (values are for a fictitious vehicle):

Tyre position	Distance to zero line (mm)	Weight (kg)	Moment (mm-kg)
Front right	300	412	123,600
Front left	310	418	129,580
Rear right	3000	360	1,080,000
Rear left	3010	365	1,098,650
Total		1555	2,431,830

Now we take the total moment and divide by the total weight: 2,431,830 divided by 1555 = 1564mm.

This is the location of the CG, measured back from our chosen zero line (which happens to be the nose of the vehicle). Carefully measure this spot using your tape measure and transfer it to the side of the vehicle using a little masking tape. The CG should be in front of the CP, or your vehicle is not inherently aerodynamically stable.

(If you have access to a weighbridge, you can measure the weight on the front and rear axles (with you in the car, remember) and then use these weights in the CG calculation.)

If you have discovered a problem with your CG or CP, how do you fix it? The answer is: you can move either the CG forward, or the CP rearward. The choice is dictated by practicality. You may end up moving them both a bit.

The simplest (and probably most common) fix used is to move the CG forward by adding ballast to the front of the vehicle. The farther forward the ballast is added, the more effective it will be. This is why you often see a heavy 12-volt battery or a water cooling tank relocated up in the nose of vehicles at Bonneville. By moving a heavy component from behind the CG to way up in front of the CG, this often doubles the effect of simply adding the weight in ballast. It also does not change the total weight, which is also a good practice.

Removing weight from the rearmost of a vehicle is also very effective. The further back you can remove weight, the better. A thinner rear bumper might be used, for example. Can the rear seats be removed? Spare tyre? Jack? Mother-in-law?

Moving the CP rearward can perhaps be the easiest solution, and also doesn't change the handling balance of the car at lower speeds. If you look at historic photos of streamliner cars at Bonneville, you will notice that the majority of them 'grow' a bigger and bigger fin in the rear over time. This is invariably an effort to add area in the rear of the vehicle (without adding much weight in the rear) to move the CP rearward.

When you are thinking about adding area to the rear of your vehicle, you can test out your ideas by adding a fin to your cut-out photo, and rebalancing the result. You will need to move the CP a bit more than you think, however, because that new fin will add a bit of weight, which will move the CG just a touch.

to get rather complicated, especially for home modification. That said, if you have a car that is very difficult to control in crosswinds, you could trial some changed vertical separation edges at the front and rear of the car, eg on the first and last pillars.

In addition to airflow separating down the sides of the car in yaw conditions, the behaviour of trailing vortices behind the car also alters. SAE paper 2016-01-1626 (*Improvement of Practical Electric Consumption by Drag Reducing under Cross Wind*), written by Nissan engineers, covers what occurs. Both sedan (three box) and hatchback (squareback) vehicles are shown in the paper as having symmetrical trailing vortices for 0° yaw. However, at 4° yaw, the trailing vortices became heavily asymmetric.

These vortices were modelled using CFD; however, testing was also carried out on real cars in wind tunnels. The cars were instrumented across their rear surfaces with pressure sensors. The two different body shapes varied in pressure change behaviour with increased yaw, but in both cases, the overall base surface pressure dropped – thus drag increased.

To see if they could reduce the increase in drag with yaw, Nissan engineers made some changes to the hatchback car. These changes are not described in detail in the paper, but appear to have involved closing-off the wheelarch openings. (Spats over all the wheels would achieve a similar outcome.) The standard car showed flow separation down its leeward side at a yaw angle of 4°, while the modified car retained attached flow down its side, even at the same yaw angle. The modified car reduced the increase in C_D by 80 per cent with 4° yaw. (Incidentally, this side flow separation or attachment can be seen by wool tufting in strong crosswinds, especially at relatively low road speeds.)

Given that the pressure distribution across the rear of the car became asymmetric in yaw conditions, it's likely that this asymmetry also helped 'steer' the car in a cross-wind. Thus, the type of modifications the Nissan engineers performed are likely to help cross-wind stability, as well as reduce drag in those conditions.

But back to the centre of pressure. Given that most cars have a forward location for the lateral centre of pressure, and best stability occurs when the centre of pressure is aft of the centre of gravity, how do we move the centre of pressure backward? As described on the previous two pages, one approach is to add rear fins. Rear fins gained a bad name when they were used for simply styling purposes, notably on US cars of the 1950s. Personally, I love the styling of these magnificent cars – but they were certainly not aerodynamically efficient. However, going back further, fins have been used to provide stability on some significant cars, for example most of the Land Speed Record cars of the 1920s, 1930s and 1950s.

The K-series road research vehicles of the late 1930s were produced at the German research organisation FKFS, and designed by Wunibald Kamm (of truncated Kamm tail fame). They featured two large slotted fins comprising four parts. The patent drawing shows that these fins were either slightly curved in plan view, or were straight and slightly offset from one another, and were angled outward at the rear from the longitudinal axis of the car. Figure 7-7 shows one of the patent drawings.

These cars, covered by Kieselbach (*Stromlinienautos in*

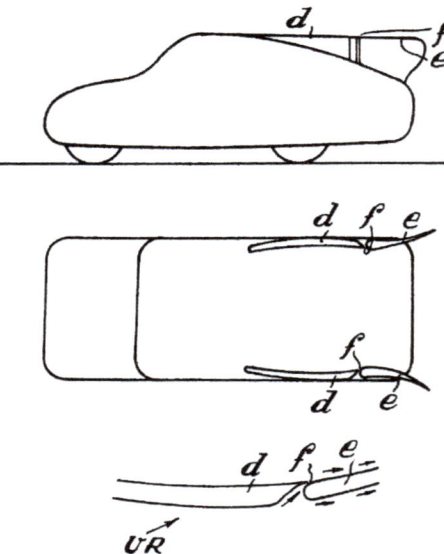

Figure 7-7: From the 1930s German patent Nr 730 027 by Kamm (et al). These fins are fascinating in their shape and design. Translation of some of the accompanying text reads: "Device on motor vehicles with aerodynamic outer shape for maintenance of the direction of travel in crosswind characterised [by the use of] upright stabilising surfaces (fins)." So far, so good! But then the patent goes much further. "[The] apparatus is characterised in that the gap is designed as a nozzle directed obliquely to the vertical longitudinal centre plane." One wonders if this didn't allow the slotted fins to work asymmetrically, depending on the yaw angle of the airflow.

Deutschland – the book is in English and German) fascinate me because of the very high level of research being undertaken. (Arguably, it would take another 30 years before a similar level of proficiency was applied to aerodynamic car design.) In addition to extensive wool-tuft testing carried out on models in wind tunnels, and on the full-size cars on the road, centreline pressures were also measured – perhaps the first time that had ever been done. But

REDUCING LIFT AND IMPROVING STABILITY

back to fins – Kieselbach states that "experiments showed a substantial improvement in cross-wind stability due to these aerodynamic accessories." Notably, in an adjoining picture in the book, the car equipped with fins is shown being tested high in the German Alps – lots of road testing was carried out of these cars. Kamm even patented transparent fins – see Figure 7-8.

The use of fins is also explored in the 1978 SAE paper 650136 (*Effect of Unsymmetrical Incidence of Aerodynamic Forces Acting on Vehicle Models and Similar Bodies*), where testing was carried out on 'car shaped' model bodies in a wind tunnel. The author notes several key points including:

- fins are more effective when used on streamlined bodies than unstreamlined bodies
- a fin that projects above the roof is even more effective, but it can increase body roll
- without exceeding the car's height and length, an average fin area of 13 per cent of the existing side area of the car was effective

I think the use of a fin or fins, especially on fastback cars where they can be easily attached to a hatch and be positioned so that the frontal area of the car is not increased, have major potential for improving stability. (More on fins in a moment.)

The other aspect that influences car stability is lift. The greater the lift, the more unstable the car. Therefore, in addition to improving cornering performance, for best stability the car should not develop lift and, if possible, should develop downforce. But where should that downforce be distributed? References vary as to the appropriate mix of front and rear downforce for a road car.

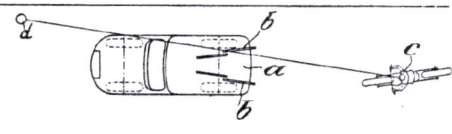

Figure 7-8: From 1930s German patent Nr 724 051 by Wunibald Kamm. Translation of some of the accompanying text reads: "Motor vehicle with aerodynamic outer shape and dorsal fin for stabilisation, characterised in that the fin is made of colourless and transparent material." Note how, as in the previous patent, the fins comprise slotted assemblies. The visual presence of large fins is lessened considerably by making them transparent – easily done these days by using acrylic sheet.

In the development of the 2006 C6 Z06 Chevrolet Corvette (see SAE paper 2005-01-1943), GM engineers used a specific target that correlated lift (the car did not develop downforce) with weight.

The relationship they used was:

Normal force ratio =
$W_f - L_f / ((W_f + W_r) - (L_f + L_r))$

Where:
- W_f = front vehicle weight
- W_r = rear vehicle weight
- L_f = front aerodynamic lift
- L_r = rear aerodynamic lift

They were seeking a figure in the 0.48-0.50 range, and with the Corvette they achieved 0.48 at 300km/h.

The first Audi TT is an interesting example of lift. In 2000, after five deaths had occurred in high-speed road accidents (including one of a former champion rally driver), Audi recalled the car and fitted a rear spoiler. It was suggested that all the accidents had occurred at speeds over 180km/h (112mph) and that rear aerodynamic lift was the culprit. (However, when they modified the cars, Audi also changed the rear suspension and added Electronic Stability Control, so it may have been a combination of factors.) I have not been able to find lift figures for the car (with or without the spoiler) but certainly the shape of the original car looks like it would generate rear lift.

Clearly, rear lift at high speed is likely to provoke lift-off oversteer (the expression referring to a throttle lift, but perhaps it is apt aerodynamically as well!). Some references go further and suggest that, for this reason, more rear than front downforce is beneficial, so giving an increasing understeer bias with speed. (This is the approach taken with the Corvette described above.) However, it seems to me that this is highly dependent on the handling bias the car *already has*. For example, a front-wheel drive without a lot of rear roll stiffness will habitually understeer. In this circumstance, increasing front downforce at a greater rate than increasing rear downforce will tend to get rid of the understeer – the last thing you want with such a car is *more* understeer at speed! However, we're now moving more into the realm of 'handling' rather than stability.

SAE paper 1999-01-0651 takes an interesting approach to assessing the impact of lift on car stability.

SPEEDPRO SERIES

Both straight-line stability and sudden lane-change stability were tested, with professional test drivers giving cars a subjective score out of 10 (10 = excellent). Testing was carried out on six different cars, with many of the cars also tested with different aerodynamic modifications.

The car designated in the paper as 'Car E' is of interest. This was a late-1990s front-wheel drive 'lower-medium' hatchback, that used MacPherson struts at the front and a torsion beam rear suspension. By adding four different types of rear spoiler to the otherwise standard car, five different aerodynamic versions were able to be tested. The following table shows the different front and rear lift coefficients, as tested in the MIRA wind tunnel.

Vehicle	C_{Lf}	C_{Lr}
Base	0.037	0.147
Spoiler 1	0.024	0.053
Spoiler 2	0.022	0.021
Spoiler 3	0.024	0.070
Spoiler 4	0.019	0.038

As can be seen, the front lift dropped by about 40 per cent with any of the rear spoilers fitted. However, the change in rear lift varied substantially with the different spoiler designs: Spoiler 2 reduced rear lift by about 86 per cent, and Spoiler 4 by 74 per cent.

So, could the drivers feel any difference with these aerodynamic changes? In the lane-change test at 125km/h (78mph), all five versions of Car E were rated equally by the drivers, who gave the car a rating of 7.5/10. However, as the testing speed rose, the different aerodynamic versions of the car began to be rated differently. At 160km/h (~100mph) the base version (without a rear spoiler) was rated by the drivers as 5.5/10. However, when the car was fitted with Spoiler 2 or Spoiler 4, the rating was much higher at 7.5/10. At 200km/h (~125mph) Spoiler 4 was rated best (6.5/10) with Spoiler 2 just behind at 6/10. The standard car? At 200km/h, the drivers rated it at 2/10.

This, and similar evidence from the paper looking at the other tested cars, indicates that at high speed, rear lift values have a significant effect on car stability in violent manoeuvres like a sudden lane change. Given that the inertial loads on the car are substantial, and the aerodynamic forces much smaller in comparison, that is remarkable.

This paper also examines a point made earlier in this chapter – what is the best ratio of front to rear lift values? In both straight-line and lane-change stability testing, the greater the front lift compared with rear lift, the higher was the subjective rating for stability. Therefore, $C_{Lf} - C_{Lr}$ is preferably a positive number. To put this another way, it is important for stability that rear lift is low. The paper concludes that the poorest aerodynamic stability occurs in cars with high total lift and a negative (nose down) pitching moment, where lift forces are concentrated at the rear axle.

A more subtle aspect of lift creating instability is covered in SAE paper 2009-01-0004, written by Mazda engineers. They say this of the mechanism by which lift influences vehicle stability:

"It is natural to think that [with aerodynamic lift] the vehicle lifts up and the tyres lose traction at high lift coefficients. But the lift force appears not to be as significant as vehicle weight level, which is over 1t. For example, the lift coefficients of a vehicle at a speed of 200km/h (124mph) with a frontal area of 2.0m² (21.5ft²) [might be] C_{Lf} = 0.1 and C_{Lr} = 0.1, representing lift forces of 38.6kg (85lb) each at its front and rear axles."

In their example, and assuming an equal front:rear weight distribution, the lift force represents only a 7.7 per cent decrease in vertical downward force at each end of the car. Furthermore, they go on to say that, in their experience, changing these lift forces *by as little as 1.9kg (4.2lb)* can greatly improve straight-ahead stability! (Note: F1 aerodynamicist Willem Toet, when reading the draft of this book, commented that he can feel a 1 per cent change in downforce or lift.) So what is going on? The authors of the paper further state that:

"[The] aerodynamic lift coefficient is a time-averaged value measured in a wind tunnel with a fixed vehicle attitude against the ground. But as the moving vehicle is exposed to unsteady aerodynamic forces generated by unsteady flow around the vehicle and unsteady vibration induced by an undulating road, such a vehicle's behaviour is thought to be extremely complex."

They add rather dryly:

"It could be considered that the [measured] aerodynamic lift coefficient may not be capable of fully describing the complex vehicle behaviour during high sped driving."

The engineers undertook testing using two cars, one standard and one modified in an attempt to improve its stability. The modifications made were:

- large curvature A-pillar
- shorter gap between the front wheelarch and the tyre

REDUCING LIFT AND IMPROVING STABILITY

- side skirts
- flat engine undercover
- front tyre deflector
- centre floor undercover
- rear undercover
- rear tyre deflector

The measured wind tunnel data for these two cars is shown in the table below:

	C_D	C_{Lf}	C_{Lr}
Standard car	0.32	0.09	0.11
Modified car	0.30	0.09	0.07

As can be seen, the changes to the modified car reduced drag and rear lift, and in on-road testing, the modified vehicle was judged by drivers to have better stability. However, at 180km/h (112mph), this decrease in C_{Lr} represents a reduction in rear lift of just 13.2kg (29lb).

Testing of both the standard and modified cars showed that, at 180km/h (112mph), the modified car had a rear ride height that was lower than the standard car, with the difference being 10-20mm. Surface pressure measurements were also made on the boot. These two sets of measurements (ride height and surface pressure, both logged at high speed) showed that:

- There was an oscillation of both parameters (ie pressure and ride height) at approximately 2Hz.
- With the standard car, rear ride height rose as surface pressure decreased, but in the modified car, surface pressure increased while ride height increased.

Unfortunately, the paper does not give any technical information on the cars' suspension, so some suppositions are necessary. First, it can be assumed that the 10-20mm change in ride height of the modified car at speed is much greater than you'd expect, with just a 13.2kg (29lb) change in lift – so something else is happening. Second, the oscillation at 2Hz is suspiciously close to the suspension natural frequency you'd expect in a sporty car. (The natural frequency of the suspension is the frequency at which the least energy input will cause the greatest suspension movement – it's the frequency at which the suspension 'prefers' to move, if you like. Without dampers, after being pushed down and released, the suspension would bounce at its natural frequency.)

Further investigation showed that the three-dimensional flow over the boot (trunk) lid varied between the standard and unmodified cars, with some of the cause of this being vortices generated from the A-pillars and then passing over the roof of the car to impinge on the rear.

The paper states:

"In a case where a three-dimensional flow structure with lateral fluctuation exists above the trunk (boot) lid of a vehicle [ie the standard car], the unsteady behaviour of this flow structure induces the unsteady surface static pressure fluctuation that promotes the vertical motion of the vehicle, and destabilises its behaviour. On the other hand, in a case where a weak three-dimensional flow structure with little fluctuation exists above the trunk lid of a vehicle (ie the modified car), the unsteady behaviour of this flow structure induces the unsteady surface static pressure fluctuation that slightly restrains the vehicle's vertical motion, and slightly stabilises its behaviour."

Another paper by Toyota engineers also covers how fluctuations in aerodynamic loads can cause instability. In SAE paper 2015-01-1537 (*Improvement in Vehicle Motion Performance by Suppression of Aerodynamic Load Fluctuations*), the engineers study the behaviour of a hatchback car (from the images in the paper, much more a squareback than a fastback). Their aim was to see if unsteady airflow behaviour could excite the car in roll and yaw. CFD, model wind tunnel testing and track testing were all used.

In addition to the standard car, the testing used modified cars. One car had the rear roof spoiler extended to the upper sides of the car, creating a new vertical face each side. The other car had modified shape rear lights, giving a much cleaner separation line. A further modification to the car with the altered rear lights was the use of single large delta-shaped vortex generators, positioned slightly ahead of the lights on each side of the car. These vortex generators were 140mm (5½in) long and 50mm (2in) high.

Modelling showed that the standard car exhibited rear pressure fluctuations of around 1.5Hz – significant, because as described above, this is in the region of the natural frequency of the suspension (not only in bounce but also in roll and pitch). Simulations showed that the rolling and yawing moments of the modified cars were much reduced over the standard vehicle, with best results achieved by the car with the large vortex generators installed ahead of the modified rear lights. Track testing of a car modified in this way showed that straight-line stability, steering effort around the neutral point, yaw response, and

SPEEDPRO SERIES

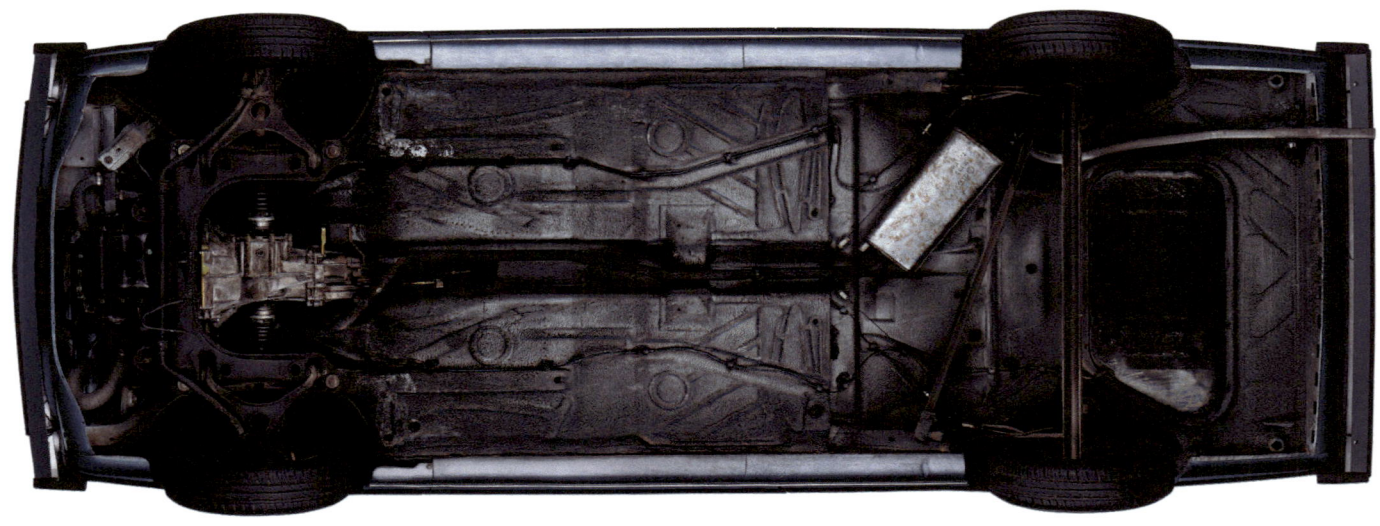

Figure 7-9: How car undersides used to be. This view of a 1972 Audi 80 shows how no attention at all was paid to smoothing undercar airflow. Adding an extensive front undertray to this car would immediately reduce drag and lift. (Courtesy Audi)

linearity were all judged as being improved.

As the paper says with some understatement, "the flow mechanism for the anti-phase fluctuations, together with their frequency selection and scaling relation remain to be clarified." However, it appears that forcing clean separation at the sides of the car may pay dividends in terms of improving stability as well as reducing drag.

So what can we gain from these papers? First, it's a ringing endorsement of the practice of testing vehicles on the road and seeing what is actually happening. Second, the actual mechanism by which aerodynamics affects vehicle stability is likely to be much more complex than an understanding gained only by looking only at steady-state values such as C_{Lf} and C_{Lr}. Third, the interaction between suspension damping, the natural frequency of the suspension and unsteady aerodynamic flows may be quite significant to stability outcomes. Finally, rear lift should be kept low.

UNDERTRAYS AND DIFFUSERS

Despite undercar modifications often being the last thing that aerodynamic modifiers think of doing, changing the flows under the car can be the most effective way of reducing lift. It's also an area of modification where, as mentioned in the previous chapter on using undertrays to reduce drag, aesthetic considerations aren't very important – and so that makes it much easier to construct do-it-yourself modifications.

To reduce lift requires that the undercar airflow is smooth and fast. So, how can we encourage these two behaviours?

First, to retain attached flow, the same type of attention needs to be paid under the car as we do to the upper surfaces. If we were to place suspension arms, springs and dampers on the surface of the bonnet, we'd not expect very good flow past these obstructions – but that's precisely the scenario under the car! As far as possible (refer also to Chapter 6 on this topic), the underside should be smooth. The photographs on these pages of the undersides of various cars show how this outcome can be achieved.

Second, it can be mathematically shown (see Barnard – *Road Vehicle Aerodynamic Design*) that the speed of the air under the car depends on the relationship between the minimum cross-sectional area under the car, and the maximum cross-sectional area, normally achieved at the back of the car by the use of a diffuser. (A diffuser is an upward sloping panel at the rearmost portion of the underside.) In practical terms, the smaller the cross-sectional area under the car and the larger the cross-sectional area at the diffuser, the better.

The cross-sectional area achievable at the diffuser depends on its angle and length, and in road cars these are both severely limited. The maximum upward inclination is influenced by the need to retain attached flow (the boundary layer there will be thickening, making this more difficult), and the diffuser's length limited by things that get in

REDUCING LIFT AND IMPROVING STABILITY

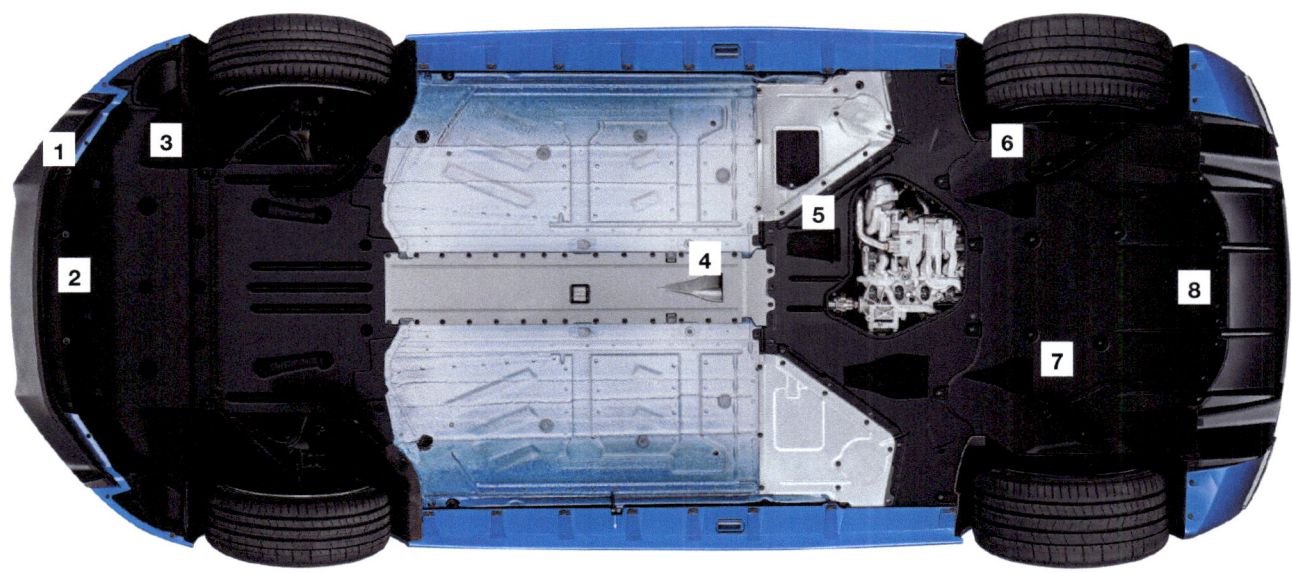

Figure 7-10: The Audi R8 generates downforce – in the case of the V10, 140kg (308lb) at top speed, with 100kg (220lb) at the rear axle. A fixed rear spoiler works together with a large diffuser on the underbody. Two venturi spoilers direct the air into the diffuser at high speed, virtually doubling its effect. Longitudinal fins in the diffuser channel the airflow so that it doesn't rush into the centre. Around the front axle are two small diffusers that send the air through the wheelarches to cool the brakes.
w(1) Front splitter, (2) cooling air outlet, (3) diffuser front, (4) NACA duct for airflow to the engine, (5) duct for airflow to the oil sump/engine compartment, (6) NACA duct for airflow to the transmission/engine compartment, (7) venturi spoiler for increasing the diffuser effect, (8) diffuser fin. (Courtesy Audi)

the way – like bumpers, fuel tanks, and spare wheel wells!

In racing cars, it's been suggested that additionally, it is the vortices created down the outer sides of the diffuser that help generate the low pressures.

A further point to keep in mind is that the steeper the angle of the diffuser, the greater the rearward-facing component of the force it develops – and so the greater the drag. However, even that statement is not always entirely true, as the diffuser may increase base pressure, so, in turn, reducing drag.

SAE paper 2000-01-0354 (*Selecting Automotive Diffusers to Maximise Underbody Downforce*) used a small, simple shape in a wind tunnel to test different aspects of diffusers. The shape was equipped with side plates and a diffuser length that could be varied. This

183

Figure 7-11: Left – in addition to its pop-up rear spoiler, the Lamborghini Aventador uses a nearly flat underside.

Below – the car has nine NACA-style cooling intakes in its underfloor panels. The rear diffuser appears to be about 25 per cent the length of the body. (Courtesy Lamborghini)

research showed that proximity to the ground is important in increasing the pressure reduction achievable by the undertray/diffuser combination. (This is also implied above, where reducing ground clearance decreases the smaller of the two relevant cross-sectional areas.) The same paper suggests an optimal diffuser length that is about 50 per cent the total length of the body – much too long for most normal passenger cars. An upsweep angle of just under 10° was used in this testing.

Katz (*Race Car Aerodynamics*) shows data for a 'generic sedan' that uses a diffuser design more appropriate for road cars, with a length of about 18 per cent of the car's total body length. This data shows that C_{Lr} continues to decrease as the diffuser angle rises from 0 to 11°. However, minimum C_D is obtained with an angle of only 4°. (Note also that this 'generic sedan' has a quite low C_D, dropping to about 0.235 at that 4° diffuser angle.)

In a personal communication, F1 aerodynamicist Willem Toet suggested, that on a road car, the precise angle of the diffuser wasn't of great importance. However, he said that the minimum to maximum cross-sectional area ratio shouldn't exceed 1:6. He also made the point that for an undertray like this to be effective, plenty of air needs to get to it. Note that this isn't an excuse to use an upturned or lifted nose – the lift generated by doing this outweighs the benefits.

A paper written by Audi engineers (SAE 2011-01-0175

REDUCING LIFT AND IMPROVING STABILITY

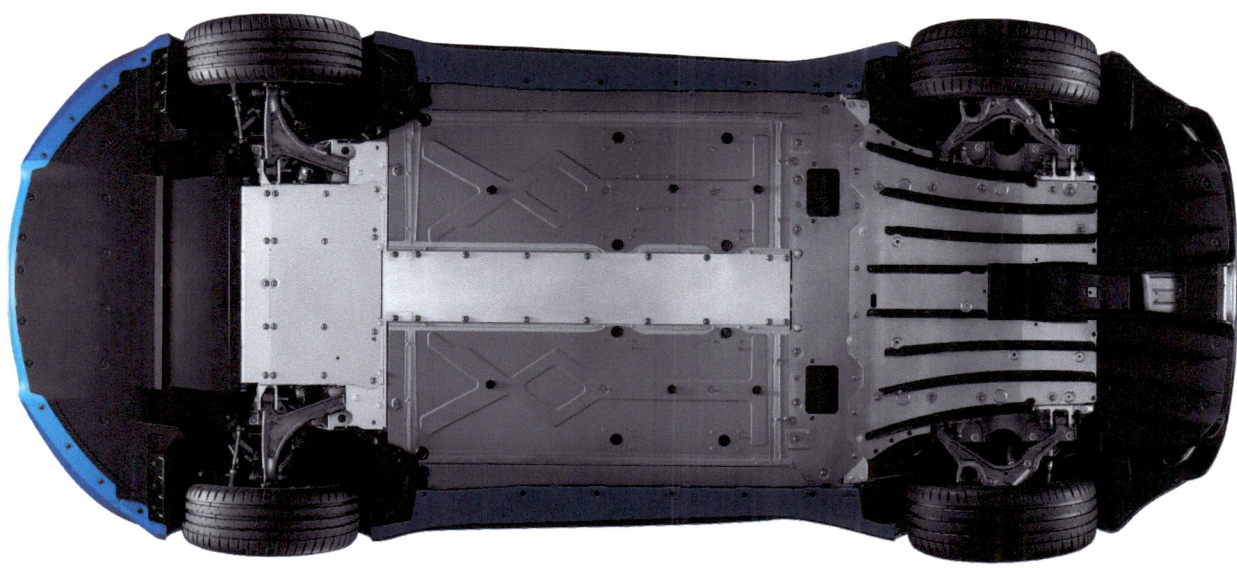

Figure 7-12: Above – Alpine 110 shows an interesting design of rear diffuser. The diffuser is about 30 per cent the length of the body. It uses an extensive number of vanes that direct flow into the two diffusing tunnels located either side of the exhaust outlet. Also note the airflow directors around the front wheels.

Right – this view better shows the rear tunnels of the Alpine 110. Note the flat floor ahead of the diffuser – without that, it won't work! (Courtesy Renault)

– *The New Audi A6/A7 Family – Aerodynamic Developments of Different Body Types on One Platform*) contains some fascinating material on undercar aerodynamic development for reduced drag and lift. (This material in fact crosses over the topic of both this and the previous chapters.)

Audi engineers state:

"A vehicle underbody which yields optimal aerodynamic performance consists of a closed flat plane with a slight diffuser behind the rear axle. With this shape, low drag and lift values can be achieved."

With regard to the A6/A7 cars, they continue:

"A long three-part underbody cover in the area from the front bumper to the gearbox guides the flow efficiently into the underbody. With smooth large-scale covers, a further reduction of the friction resistance between the axles is achieved. Only the area of the transmission tunnel is not covered because for thermal reasons this would have been very complex. Furthermore, the airflow through the engine compartment is inserted into the underbody flow via the transmission tunnel."

The rear diffuser is formed by a further cover, the spare wheel well, and the shaped muffler. A spoiler is fitted to the undertray under the gearbox. This spoiler reduces drag and rear axle lift by improving the way that cooling air exiting the engine bay integrates with the undercar flow.

Testing of the Audi sedan in development showed that the D pillars were forming two longitudinal vortices that caused downwash in the wake of the vehicle. These vortices reattached on the boot lid. The downwash vortices are, in the case of the Audi, compensated for by the upwash created by the undercar diffuser. This approach was effective in the sedan, where drag was reduced and a vehicle C_D of 0.26 was attained. Audi does not quote lift coefficients for this car – other than it met company requirements.

However, the fastback shaped car (the A7) was a new story. Two larger vortices were shed from the D pillars, and the rear of the car, including its flat rear window, was in separated flow. This resulted in a C_{Lr} of 'greater than' 0.100. To reduce this, a moveable rear spoiler was introduced; this operates at speeds above 130km/h (81mph). The spoiler reduces the size of the longitudinal vortices, resulting in less downwash. The vortices therefore attach only partially, reducing the flow speeds on the back window and so increasing pressures. In addition, direct downforce is generated by the airflow being deflected upward by the spoiler. The use of the spoiler caused a decrease in C_{Lr} of -0.130

and a decrease in C_D of -0.012. (Audi says the final C_{Lr} was 'significantly below' 0.10 and the total C_D was 0.28 – still higher than the sedan.)

But what of the A6 Avant – a squareback (wagon) shaped vehicle? For styling reasons the rear window of the Avant was sloped, and so developed a rear flow pattern more akin to the fastback A7 shape than the sedan A6 shape. (In fact, the selected styling approach was a rear window angle of 30°, an angle at which the separation point can jump back and forth between the trailing edge of the hatch and the trailing edge of the roof.) To achieve a true squareback flow pattern (ie the airflow separating at the end of the roof), a roof extension spoiler was added. However, this was found to be ineffective at yaw angles of other than zero, resulting in a steep rise in rear lift in cross-wind conditions. A third rear spoiler configuration was needed, and this resulted in consistent flow separation, even at high yaw angles.

The action of the rear undercar diffuser had been effective with the two previous body shapes – the sedan and fastback. But what about the squareback? With the cleanly separated flows, longitudinal vortices did not form, and so there was no downwash. However, the upwash generated by the rear diffuser was still being developed, creating larger downforce but also increasing drag. Audi couldn't change the whole complexion of the underside for just the squareback model, so needed to take a different approach. The engineers chose to fit spoilers that disrupt the flow through the diffuser, so reducing its effective downforce and thus reducing drag.

Johan Levin and Rikard Rigdal, in a master's thesis in automotive engineering completed at Chalmers University of Technology, Gothenburg, Sweden, carried out modelling on a Saab 9-3 sedan. (*Aerodynamic analysis of drag reduction devices on the underbody of a SAAB 9-3 by using CFD.*) They modelled a rear diffuser that was 1050mm (41in) long, extending from the rear axle line to the rear of the car. Note that this length is 23 per cent of the car's total body length. Diffuser angles varied from 3-14°. An extensive undertray was added to the standard car, and to achieve a diffuser of the required length and angles, the spare wheel well and muffler needed to be substantially modified. Tests were made that included the use of vortex generators to promote better flow attachment (especially on the parts of the diffuser in the wake of the suspension arms), and the use of guidance vanes.

The standard Saab 9-3 had a C_D of 0.288 and the C_L was 0.190. The modelling showed the greatest reduction in C_D with a diffuser angle of 8°. In fact, together with the smoothed underbody, C_D was reduced by 14.5 per cent to 0.249. Lift with the smooth underbody and rear diffuser was also reduced, changing from 0.190 to 0.069 – a substantial 64 per cent decrease. However, looking at the full data, altering the diffuser angle from 5-14° didn't make a major difference to either the drag or lift improvements, implying that it was the presence of the smooth undertray that was making the biggest difference. (Or that any diffuser angle from 5-14° works quite well!) In fact, when the diffuser was tried alone (ie without the undertrays) the C_D *increased* – as the authors say, there was then a gap between the underbody and the diffuser, which acted as an air brake.

Testing vortex generators and guiding vanes on the undertray and in the diffuser made only fractional differences to the drag and lift figures. However, covering the rear wheels resulted in a clear further decrease in C_D, where (together with the undertray and 8° diffuser) a C_D of 0.233 was modelled.

For those assessing drag reduction by fuel consumption changes, the paper has some interesting information. Using Saab computer simulations based on the New European Driving Cycle (NEDC), the change in C_D from 0.288 to 0.233 (a decrease of 19 per cent) resulted in a change in fuel consumption from 7.76 litres/100km (30.31 US mpg) to 7.57 litres/100km (31.07) – an improvement of only 2.5 per cent. Finally, the paper also did extensive modelling on improvements possible on a Saab 9-3 wagon, but achieved much lower gains than on the sedan.

SAE paper 2017-01-2163 (*Numerical Investigation on Aerodynamic Effects of Vanes and Flaps on Automotive Underbody Diffusers*) discusses the results of adding longitudinal strakes (vanes) within the diffuser area. In addition, a wing-shaped flap was trialled at the upper exit of the diffuser. In the paper, a computational simulation of a Formula SAE car is used. In summary the paper found that:

"The addition of a straight vane along the direction of the flow results in an improvement in downforce and the downforce to drag ratio. Best results are observed when the vane is placed at a distance of around a quarter of the width of the diffuser from the outer edge.

"Placing the vanes at an angle such that they face outward, also improves downforce. An angle of 5° to the transverse [I think they mean longitudinal] planes of the car provides the best results at the vane location determined above. A maximum increase [in downforce] of 13% is seen when compared to the base configuration without flaps.

"The addition of a flap at the trailing edge also leads to an increase in downforce (up to 9%) when compared to the base configuration. Placing the vane and flap together at the optimal positions improves the downforce of the diffuser by almost 25% and increases the downforce to drag ratio by nearly 10%."

SAE paper 2008-01-0802 (*Experimental Analysis of the Underbody Pressure Distribution of a Series Vehicle on the Road and in the Wind Tunnel*) looks at the underbody performance of a 2004 BMW 3-series. Rather than concentrating on rear diffusers, the paper examines the performance of the front and centre undertrays. Undercar pressure measurements were made using 124 pressure taps, and testing was conducted on-road as well as in two different wind tunnels.

The undercar panels comprised a front engine undertray, a gearbox undertray and side panels extending back about as far as the pre-axle fuel tank. The exhaust was uncovered for its full length and everything rearward of the fuel tank was uncovered.

The engine undertray was not flat; instead it smoothly curved downward from the underside of the front bumper before flattening under the engine's sump, forming what the authors call an 'S-shaped' undertray. The acceleration of air around this curve developed a significantly low pressure – in fact, the lowest pressure recorded under the car. Other more recent BMW models show a similar design of engine undertray, with some incorporating what may be vortex generators located on the curve (see Figure 7-13).

In practice, if you are after reduced lift, I suggest that you fit:

- A smooth undertray covering as much of the car's underside as possible.
- An engine undertray that incorporates a curve that faces downward.
- A rear diffuser that is 15-20 per cent the length of the car's body and is upturned by 8-10 degrees.
- Be aware that in wagon (squareback) configurations, the diffuser may not be effective in reducing drag.

Figure 7-13: BMW F30 3-series engine undertray, pictured upside-down. Note the downward curve, and what may be vortex generators moulded into its shape.

SPEEDPRO SERIES

Figure 7-14: The airflow through an effective diffuser can be seen here. (Courtesy Mercedes)

HONDA INSIGHT UNDERTRAY AND DIFFUSER

In order to potentially improve drag and reduce lift, I decided to add extensive undertrays to my Gen 1 Honda Insight. In standard form, the slippery car ($C_D = 0.25$) has a small undertray beneath the engine compartment, two smooth undertrays either side of the exhaust between the front and rear wheels, and no rear undertray. The rear of the car is particularly messy, with an exposed torsion beam rear axle, fuel tank and muffler. There is a small deflector fitted in front of the fuel tank, but aerodynamically the underside looks quite poor.

However, the current undertrays are not the first work I have done with the car in this area. In fact, years ago, I trialled a rear-only undertray that was aimed at reducing drag. I made this from old Corflute signs (a stiff but light plastic sheet material) and gaffer tape. Additionally, I made two ramp-shaped fairings to go ahead of the rear wheels. Installing the trial fairings and undertray took about three hours of work.

I then undertook extensive freeway testing, primarily to see if drag was reduced. A 64km (40mi) loop was driven at 105km/h (65mph) for one half of the route and 100km/h (62mph) for the other half. Working with the standard fuel economy display (which reads litres/100km to one decimal place), there was *no change* with or without the rear undertray fitted. However, as described in Chapter 5, sometimes it's possible to get a feel for what is occurring. This I did by estimating

Figure 7-15: A trial rear undertray made from salvaged plastic sheet. It didn't work.

the second (missing) decimal place of the fuel consumption display. I did this in the following way.

For example, say the fuel economy showed 2.9 litres/100km at the end of the freeway test, but as the car exited the freeway and slowed to a stop, this immediately dropped to 2.8 litres/100km. You could therefore surmise that the fuel consumption was very close to the 2.9/2.8 changeover point – say, around 2.86 litres/100km. So, going on both the displayed data (ie to one decimal place) and the data guessed on the basis of how long it took the display to change, once an altered driving style was undertaken (two decimal places), the test results looked like this:

Test	Displayed fuel consumption (litres/100km)	Estimated second decimal place (litres/100km)
Standard	2.9	2.88
Rear undertray	2.9	2.90
Rear undertray and fairings	2.8	2.84

So over the three tests, we're talking an estimated variation of 0.06 litres/100km, or 2 per cent. Even driving carefully over the same route, I'd expect at least this variation in fuel economy – just by everyday driving differences. Therefore, I think that the impact of the rear undertray in reducing drag was negligible. And in reducing lift? I could not pick a difference in any other aspect of car performance – none! So I took the trial undertray off.

It's important to consider the above failure, because lots of aerodynamic modifications on cars *are just like that*. You have an idea,

REDUCING LIFT AND IMPROVING STABILITY

Figure 7-16: I fabricated this front undertray for my Honda Insight. The standard undertrays are shown positioned behind the new one. The new undertray was made from 3mm ABS plastic sheet, stiffened with the aluminium extrusions. Note the high side panels that reduce openings in the wheelarches.

the front bumper to the floor at the base of the firewall, except for the longitudinal opening provided for the exhaust. (Note that the undertray needs to be removed to perform oil and filter changes – there are about 12 bolts to unscrew to do this.)

As described in Chapter 5, testing of downforce was carried out by monitoring ride height sensors. With the front undertray fitted, the *rear* ride height did not change with speed. However, *front* ride height did decrease with speed, indicative of front downforce being developed. At 115km/h (~70mph), I consistently saw a decrease in front ride height of 5mm (just over 3/16in).

Thus encouraged, I decided to again fit a rear undertray. Since the time when the previous design failed, I've fitted a different exhaust which has a smaller muffler that's positioned higher. Therefore, this time, I decided to enclose the muffler. I also decided to incorporate a rear diffuser – however, there were immediately two related problems.

First, and unlike the front undertray, the rear undertray required large spans where there was no existing body support. For example, the longitudinal distance between the rear of the floor, across fuel tank, and then across the rear axle to the spare wheel well, was about 800mm (31½ in). The 3mm ABS didn't have the required stiffness to bridge a gap of this size. Second, if a rear diffuser was to be incorporated into this panel, the required shape would need to be retained, even under aerodynamic forces.

To address both concerns, I made a steel frame from square tube, 15mm x 1.8mm wall thickness. This attached to two existing bolts at the rear of the cabin floor, and then to two new mounts, made in

you trial it, it doesn't work, and then you take it off. Or, you think about why it might not have worked – and pursue another course. In this case, I decided to take the other course – although it took me about nine years to get around to doing it! This time, I decided to improve the underside from the front to the back, and also develop a much better rear undertray design. The results stunned me.

I like to use ABS plastic to form undertrays, and in the case of the Insight's front undertray, I used 3mm ABS, braced internally with two sections of U-shaped aluminium extrusion. The undertray, unlike the short standard one, extended back to the cabin floor. In addition, I made much larger than standard vertical side extensions, to enclose the forward insides of the wheel-wells.

In the transverse-engine car, the exhaust exits nearly centrally at the rear of the engine bay. To provide cooling air to the exhaust, and to prevent the undertray from melting, generous clearance was provided around the exhaust and catalytic converter. I also mounted the centre of the new engine bay undertray a little lower than the standard undertray had been at this point, so resulting in a downward-projecting curve. Elsewhere, while the new undertray is not dead-flat (the ABS conforms to changes in shape where mounting holes are not at exactly the same height), in overall terms it presents a flat surface. This extends from the underside of

SPEEDPRO SERIES

Figure 7-17 (top left): The new rear undertray and diffuser for the Honda. This large (1 x 1.5 metre, or 3 x 5ft) assembly uses a steel square tube frame to support it and maintain the required diffuser geometry. While it is hard to see in the photo, the lower section is angled upward by 10°. The large cut-out is to give clearance to the exhaust pipe. Together with the front undertray, this rear undertray gives measurable and significant downforce, even at 100km/h (~60mph). Handling responded accordingly!

Figure 7-18 (top right): The Honda undertrays, seen from the rear. Note that both the front and rear suspension is at full droop: normally the lower rear spring seats are flush with the tray.

Figure 7-19 (bottom left): The side in-fill panels in the forward part of the front wheelarches can be seen here. Ensure that undertrays are firmly held in place – if they are working well, there will be a downward pull on them.

Figure 7-20 (bottom right): The rear undertray and diffuser can be seen here. Note also the use of the air springing, allowing the ride height to be adjusted to give best drag and downforce outcomes.

the rear of the spare wheel well. An additional advantage of the frame was that it maintained a safe distance between the undertray and the rear muffler.

I initially used the same 3mm ABS as I had used for the front undertray, but I became concerned that the overall weight was getting too high. With the support that the frame was providing, I was able to replace the 3mm ABS with 2mm ABS, so dropping a third of the weight of the sheet material. The ABS sheet is attached to the

REDUCING LIFT AND IMPROVING STABILITY

frame by means of 6mm round-headed bolts that screw into rivnuts inserted in lugs welded to the steel frame.

The diffuser geometry is incorporated into the steel frame – the cladding just wraps around the curve. The diffuser has an angle of 10° to the horizontal and is 420mm (16½in) long. The length was dictated by the car shape (if it were to be any longer, to maintain the angle, the spare wheel well and rear bumper would have had to be modified) and the angle was based on the reference material covered above.

Making the rear undertray without any template to act as a guide (on the front I'd used the small standard undertrays as a template for the front half of the new undertray), and also constructing the steel frame, were both major tasks. On the other hand, it was just a case of taking it slowly and persevering! (Note that I needed to fit only front and rear undertrays – the standard car has excellent trays under the centre section of the body.)

And the results? After the new rear undertray was fitted, the car could immediately be seen to be behaving differently. My 'low ride height' warning (that I've configured to appear on the dash when my air suspension is running too low) was on at almost all speeds over 100km/h (62mph). Furthermore, watching the dashboard-displayed ride heights showed that not only was the front undertray developing more downforce than when the rear tray wasn't fitted, the newly-fitted rear tray/diffuser was now also developing downforce.

Disconnecting the air suspension solenoids (so that active ride height changes couldn't occur) showed the following test results.

Speed (km/h)	Front ride height (mm above bump stops)	Front height decrease over zero speed (mm)	Rear ride height (mm above bump stops)	Rear height decrease over zero speed (mm)
0	56	0	45	0
50	55	1	42	3
110	46	10	32	13
160	38	18	23	22

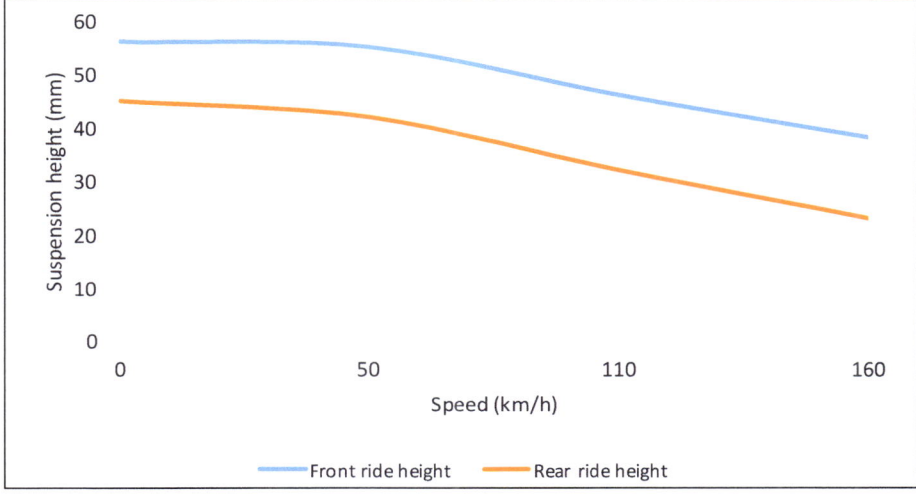

Figure 7-21: The reduction in suspension height with speed, the downforce caused by front and rear undertrays fitted to a Honda Insight. (The rear undertray incorporates a diffuser.) The car uses rising-rate air suspension resulting in a linear change in height with speed, despite downforce increasing rapidly.

In imperial units, that's a ride height drop of about ½in at 68mph and ⅞in at 100mph.

Note that the air springs I am using in the Honda are non-linear – they rise in rate rapidly (ie get much stiffer) with increasing compression. Aerodynamic downforce is also non-linear, so explaining the straight lines shown in Figure 7-21 that plots suspension compression against speed.

By adding weights to the front and rear of the stationary car until the suspension compression was the same as recorded at the desired speed, the actual downforce could be measured. To achieve this, bags of garden potting mix and sand were temporarily placed on the axle lines to create the required suspension deflections. The following table shows the results for 160km/h (100mph).

Speed	Front downforce	Rear downforce
160km/h (100mph)	110kg (240lb)	45kg (100lb)

This is a very strong result that has some important aspects to study. First, you can see that front downforce is, in absolute terms, much greater than the rear. However, the front and rear suspension deflections caused by this downforce are (respectively) 18 and 22mm – quite close. So why is

191

this so? It's because the suspension rates reflect the front:rear weight distribution of the car. The Insight has 60 per cent of its mass on the front wheels, and so has stiffer front suspension.

The other way of looking at this is to show the percentage weight addition of the downforce.

Speed	Increase in effective weight (per cent)	
	Front	Rear
160km/h 100mph	22	13

Or, at 160km/h, the front of the Insight has gone from 510 to 620kg, and the rear from 340kg to 385kg.

When looking at this data, it's important to realise the car doesn't need to be doing 160km/h to be gaining clear downforce. (I measured downforce at 160km/h because the large suspension deflection gives more easily measured results.) Even at 110km/h, the front and rear suspensions are being compressed by 10 and 13mm, respectively.

A further point to note is the way that the fitting of the *rear* undertray changed the behaviour of the *front* undertray. With only the front undertray fitted, the front ride height dropped by 5mm at 110km/h. But with the rear undertray also fitted (and the front undertray physically unchanged), the front dropped by 10mm. With the rising rate springs, that represents a *more than* doubling of downforce at the front with the addition of the rear undertray. Presumably, the much smoother rear flows helps the flow past the front undertray.

With the front and rear downforce now being developed, the car feels substantially different on the road at speed. Because it can develop greater downward tyre forces without having a commensurate increase in inertia, the car clearly corners better. Corners that previously could be taken at a maximum of 100km/h (62mph) can now be taken at 110km/h (68mph) – or more. Furthermore, the ride quality has also improved at speed, as the effective weight of the car has increased.

I am not a racing car driver, but when I used to test new cars for a living, I drove many different cars. I now think that a select small band of those cars developed real downforce: the faster you went, the better they seemed to handle. With the Honda Insight that same – slightly uncanny – feeling now exists as it goes faster.

And drag? As I covered in Chapter 5, it's hard to quantify drag in the same way that you can for downforce (or lift). However, I think that drag is largely unchanged. There is a long, constant-gradient freeway hill that I negotiate frequently (it is about 1.5km – around a mile – long.) Without the undertrays, the Honda would roll down the hill, throttle-off (but with lots of EGR (Exhaust Gas Recirculation), so little engine braking) at the local speed limit – 110km/h (68.3mph). With the undertrays in place and no other changes, the speed is about 109km/h, or 67.7mph. That's very little difference indeed – and it shows that drag hasn't dramatically increased or decreased. (Speed was measured by 10Hz GPS.)

So why do the undertray and diffuser work so well on the Insight? After all, these downforce figures (in both relative and absolute terms) are better than achieved by many sporting cars with extensive professional aerodynamic development. I think the underbody aero works well for a couple of reasons.

First, while I have not done extensive testing of the car in standard form, my memory of watching real-time ride heights shown on the dashboard is that the standard car developed neither significant lift or downforce. Honda engineers were seeking lowest drag for the Insight, and developing either lift or downforce typically has a penalty in induced drag. It's likely that Honda tweaked the car until the C_L was as close to zero as they could. Therefore, the starting point in developing downforce was not to first overcome significant lift!

Second, with my ability to actively adjust ride height, and the non-linear (steeply rising rate) characteristics of the air springs, I run the car quite low. At my normal 'fast road' ride height (as used for the above measurements), the 'knee' of the diffuser is only about 100mm (4in) above the road with the car stationary. At speed, this drops to about 85mm (3⅜ in). This is likely to have increased the effectiveness of the undertray/diffuser – and in fact, when using higher ride heights, the car feels quite different. (But then again, higher ride heights of the air suspension also changes spring rates, and the height of the centre of gravity!)

Either way, the Honda shows how it is possible to achieve significant downforce on a road car by making changes only under the car.

SPOILERS

A spoiler changes the direction of flow of the air. In addition, it can also change the pressure acting on large areas of the car's body.

(Note that here I am *not* using 'spoiler' to mean a simple extension, designed to create clean flow separation. An example of this is a hatchback roof extension, called

REDUCING LIFT AND IMPROVING STABILITY

Figure 7-22: A traditional front spoiler aimed at reducing airflow passing under the car. (Courtesy GM-Holden)

by all and sundry a roof spoiler. However, note that if the roof extension is angled upward, it may in fact act as a spoiler!)

A front spoiler (often perhaps more accurately called an 'air dam') prevents some of the airflow passing under the car. More airflow is therefore directed along the sides and over the top of the car. The extra flow along the sides of the car occurs because of increased horizontal flow along the spoiler (especially near the corners), with this flow able to be seen through wool tufting. The greater amount of air passing over the car can also be seen through wool-tufting, with the stagnation line moving downward with the fitting of the spoiler.

In days of yore, very deep front spoilers were often used. In fact, the Holden Commodore Group A (covered in Chapter 2) had a front spoiler so low that it came with a removable lip. To get the car off the delivery truck, in many cases this lip had to be removed! (And in daily use, most owners left the lip off.) Even the much more recent Chevrolet Volt used a very deep front spoiler (again see Chapter 2).

These large and low front spoilers were effective at reducing both lift and drag. However, their use has declined as more efficient overall outcomes are achievable by using smooth underbodies. This is important to note, as many aerodynamic textbooks and technical papers have lots of data on the benefits of low front spoilers.

And so there were – but primarily on cars with rough underbodies.

Of course, if you need a 'quickie' fix for either front lift or high drag, and the car has an unrefined underside, a front spoiler can still be used. In this case, place it as far forward as possible. Spoilers placed rearward (for example, partly under an engine undertray) trap air that can apply upward pressure. This then generates lift. (Chapter 5 has an example of where I foolishly did just this. It is tempting to do, because the spoiler is then mostly hidden and aesthetics are not as important.)

A forward-facing splitter, angled a little downward (eg by 2°) provides a surface on which the trapped air can provide pressure. However, even if this pressure is relatively high, on a road car the area is also fairly small. Again, gaining a low pressure over (say) a square metre of front undertray is better than getting a higher downward pressure on the (perhaps) 15 per cent of that area

Figure 7-23: About as big a splitter as you could fit to a road car! Jerry Lee measured a pressure of 1145Pa (4.6 inches of water) at 113km/h (70mph) on the upper surface of the splitter. The measured pressure in the middle of the splitter was higher than at the extremities, and Jerry estimates a net downforce of about 23kg (50lb) at this speed. (Courtesy Jerry Lee)

Figure 7-24: This Honda Civic Type R exemplifies the approach being taken to front spoilers in cars developed with attention paid to underside aero. While it includes elements of a front-facing splitter, the difference the spoiler makes to lift and drag is likely to be insignificant. (Courtesy Honda)

that comprises the splitter. Note that if you do fit a front spoiler that incorporates an effective splitter producing real downforce, that force will be applied ahead of the front axle line, and so rear lift will increase.

Canards – small angled planes – are sometimes fitted to the front corners of cars. However, the downforce developed from these appendages must be small unless road speeds are extremely high (or they actively extend, as covered later in this chapter).

And front spoiler design? In addition to incorporating the splitter, ensure that the corners are curved in plan-view so that the airflow can in fact flow around the corners. The spoiler also needs to be as low as possible – leading to the next point.

Perhaps the most important design feature is to make the spoiler and splitter from materials that won't get destroyed the first gutter you nudge or road surface you scrape! Taking a leaf out of car manufacturers' books, flexible rubber works well. It's then a balancing act between using rubber of sufficient stiffness that it will not deflect too far under aerodynamic forces, but will still bend when something unyielding is struck. Note that the front corners of a splitter will be particularly vulnerable on a road car.

But I cannot see myself ever again fitting a large front spoiler to any road car I own; for me, the ease of working under the car and the results that can be gained by the use of an undertray and diffuser are much more attractive. But what about *rear* spoilers? They're a very different story.

Rear spoilers are fitted so that rear lift forces are decreased, or more rarely, downforce is generated. They change the way that rear forces act in two ways. First, a flat plate angled to the airflow that causes air to be directed at any upward angle, will create a downward force. To be able to achieve this, it is best if the spoiler is placed in an area of attached flow, but such is the power of a spoiler, it will still have some effect even if it is working in an area of flow that is only partially attached. Second, and more importantly because of the area over which it acts, a rear spoiler can change

Figure 7-25: This Lexus LFA features front canards. Given the small area of the canard, it's doubtful if much force is developed by these appendages. (Courtesy Lexus)

REDUCING LIFT AND IMPROVING STABILITY

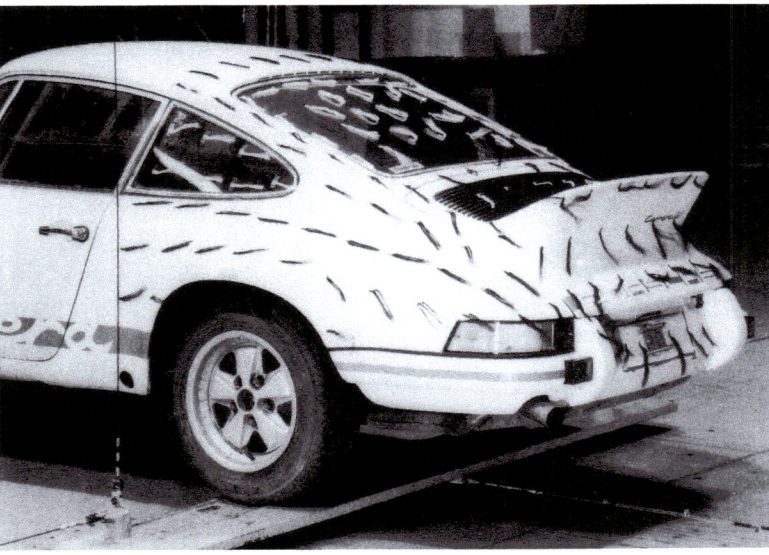

Figure 7-26: Top – a highly successful rear spoiler – and among the first fitted to a production car. It created a stronger negative pressure gradient and caused the flow to separate earlier. The result was a reduction in rear lift from 125kg (276lb) to 31kg (68lb) at maximum speed.
Centre – the Porsche being wind tunnel tested without the rear ducktail spoiler. Note the attached flow over the complete rear of the car (the side behind the rear wheel excepted).
Lower picture – with the spoiler fitted, the flow separates much earlier – it truly does 'spoil the flow.' (All pictures courtesy Porsche)

the pressure that is being applied to the rear surfaces of a car. For example, a spoiler fitted to the trailing edge of a boot lid can cause an increase in pressure across the lid and rear glass. As mentioned, even a small change in pressure acting over a large area can be significant.

One of the classic papers on spoilers (SAE – *770389, The Origin of Drag and Lift Reductions on Automobiles with Front and Rear Spoilers*, published in 1977) shows the results gained with a flat-plate rear bootlid spoiler on a sedan. It was mounted at a 20° angle, tilted back from vertical. Lift decreased almost linearly as the spoiler height was increased from 0 to 100mm (4in). On the other hand, drag dropped to a minimum at 20mm (about ¾in) and then slowly rose as the spoiler height was increased further. Figure 7-27 shows these relationships.

(Incidentally, many rear spoilers – especially on older cars – are 'wing-like' devices that stand proud of the boot lid. Colloquially, many people call these spoilers 'wings.' However, I will define wings as those devices that have attached airflow over both surfaces, with these surfaces shaped to provide a lower pressure on the bottom surface compared with the top surface.)

Another way in which a rear spoiler on a sedan can reduce lift and drag is to simply promote flow reattachment. If the spoiler (especially one that has an upper surface parallel to the ground) is positioned sufficiently high, flow will reattach with a commensurate increase in rear surface pressures. Figure 7-28 shows airflow drawings made of a 1990s car with and without a bodykit. The drawings were

SPEEDPRO SERIES

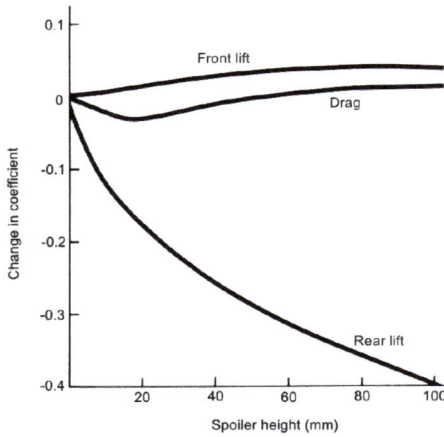

Figure 7-27: Classic patterns of changes in lift and drag with the fitting of a rear spoiler to a three-box sedan. Rear lift decreases as spoiler height increases, and because the spoiler is located behind the rear wheels, front lift rises. Drag falls initially, but then increases with greater spoiler height.
(Data points based on Schenkel)

based on wool-tuft testing. The rear of the 'kitted' car features a large spoiler (that looks a bit like a wing but isn't). The tuft testing showed that flow attachment over the rear window and onto the boot lid was better with the kit fitted. Significantly though, I had a very experienced professional test driver drive the car with and without the kit fitted, and he could detect absolutely no difference in the feel of the car at all. (The testing was done on the track, and included high speeds.)

The difference achievable with a rear spoiler on this sort on a sedan is, however, measurable – at least in the wind tunnel. On the early 2000s Australian VY Holden Commodore sedan, Holden had two slightly different wing-like rear spoilers available. The one fitted to the 'SS' model was a little wider and taller than the one fitted to the 'S' model. You could also buy a version of the car that had no rear spoiler at all – the plain VY. The following table shows the different wind tunnel results:

	C_D	C_L	C_{Lf}	C_{Lr}
No spoiler	0.319	0.113	-0.016	0.128
S Spoiler	0.326	0.027	0.009	0.018
SS Spoiler	0.334	-0.009	0.007	-0.015

Note how the fitting of the rear spoilers increased drag (the higher pressures on the sloping rear window not outweighing the increased wake area), and how increasing rear downward forces caused an increase in front lift, the rear spoilers being positioned behind the rear axle line. It's also interesting to note that the base model car developed front downforce, even though it was only a tiny amount.

SAE papers 950627 and 980394 contain data gathered with differing car configurations, including front and rear spoilers. The car was a 1990 Rover 820Si, a car with a standard C_D of 0.34. (In this figure and the following data, I am using an average of the spread of measurements recorded in the papers. Testing was carried out in different wind tunnels and via coast-down measurements. The coast-down measurements were made with anemometer compensation.)

Firstly, rear boot spoilers: the rear spoiler comprised a simple triangular cross-section lip, with various heights tested. With the different height spoilers fitted to the car with its standard rough underbody, C_D rose progressively from 0.35 to reach 0.37 at 90mm spoiler height. At 100mm spoiler height, the C_L dropped by about

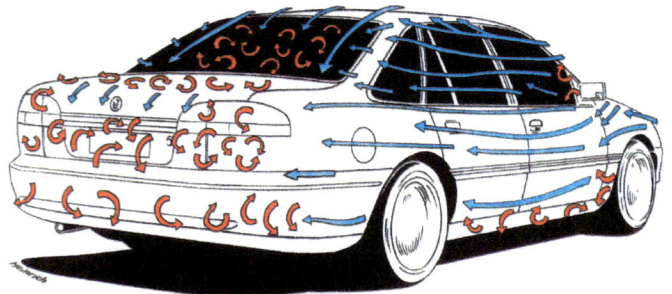 

Figure 7-28: Left – this 1990s Holden Commodore was extensively wool-tufted and then the flows shown in this diagram drawn from the wool tuft photos. Note the separated flow on the rear window and boot lid. Right – a 'bodykit' was then fitted and the car retested. You can see that there is now better flow attachment on the rear window and boot lid, caused by the rear spoiler. (It looks like a wing but isn't.) However, a professional test driver could feel absolutely no difference in the behaviour of the car, even when tested at high speed. (Both drawings courtesy Dave Heinrich)

REDUCING LIFT AND IMPROVING STABILITY

Figure 7-29: This Rover 820si was fitted with front and rear spoilers and then tested. Drag decreased with a large front spoiler but rose with a rear spoiler. The rear spoiler, however, decreased lift. These are typical results on a rough underbody car of this shape and era. (Courtesy Rover)

Figure 7-31: Jerry Lee uses AirTab vortex generators to encourage better flow attachment to his rear spoiler, a large flat plate made from clear plastic. Wool tuft testing revealed better attachment after the vortex generators were fitted. The measured pressure in front of the spoiler was about 820Pa (3.3 inches of water) at 113km/h (70mph). (Courtesy Jerry Lee)

0.15. (Unfortunately, the papers do not state the standard C_L of the car.)

With a smooth floor (the car being modified with extensive underfloor panels, only the centre section of exhaust remaining unpanelled), C_D rose from 0.32 to 0.36 at 100mm spoiler height. That is, the underfloor panelling dropped C_D by 0.03, but the rate of increase of drag with increasing rear spoiler height was steeper.

A simple front spoiler (air dam) was also trialled. As has been described, with a rough underside, increasing spoiler size (ie blocking off more undercar airflow) decreases drag. And so it proved with the Rover, with C_D dropping from 0.35 to about 0.32 at 100mm spoiler depth. But what about a front spoiler with the smooth floor? Frustratingly, the paper doesn't cover this – but given that a deep front spoiler on the rough underside car achieved a C_D of 0.32, and just the same drag coefficient was achieved by the use of underfloor panelling, it's likely that 0.32 is the best drag figure achievable by simple treatments that address underfloor airflow.

If you use the techniques described in Chapter 4, you will be able to directly measure the pressures acting on body surfaces in front of a rear spoiler. For example, in a sedan you can measure the pressures on the boot lid, and if these rise after you have placed a mock-up spoiler on the trailing edge of the lid, then you will have reduced lift. If the pressures on the angled rear window also rise, you may have reduced drag. Given

Figure 7-30: An early use of a rear spoiler. Despite's Saab's aero heritage, you really do have to wonder if the spoiler wasn't positioned in separated flow, and so did little. It would certainly be nice to see that sharp corner at the end of the roofline smoothed into a gentle radius! (Courtesy Saab)

the variations in rear vehicle shape, the ease with which trial spoilers can be mocked-up out of materials like plywood and plastic sheet, and the veracity of the results obtainable by direct pressure measurement, I think that the quickest way to make improvements is to simply get out there and try them!

WINGS

Wings are widely used in motorsport, and there is extensive literature available on the best mounting locations, and best aerofoil shapes to use, in these motorsport applications. However, for a road car, where drag is such an important consideration and speeds are typically much lower, much of this information is of less relevance.

A wing works differently to a spoiler, although there is also one significant similarity. Many rear spoilers – and all wings – develop an upwash of air, so resulting in a downward force. However, in addition, a wing develops a lower pressure on its bottom surface than top surface, providing additional downforce. Figure 7-32 shows the nomenclature used when describing wings.

A lot of data is available for the lift and drag characteristics of different wing profiles, with much of this now available online (eg see http://airfoiltools.com). This data shows, among other factors, how the C_L and C_D vary for different aerofoils with different angles of attack. (Note that you will need to calculate the appropriate Reynold's number first – the cited site has a calculator to allow you to easily do this.) There are also other online calculators that will calculate the actual downforce figures for a selected wing, based on the data you enter on the wing's size, C_L, air density and velocity. (You may need to search under 'wing lift calculator' rather than 'wing downforce calculator.')

When assessing the lift/drag characteristics of a wing, it's important to know that the way in which lift and drag coefficients are calculated for wings differs from the approach taken for cars. As described in Chapter 1, for car coefficients, the projected *frontal area* of the car is used; however, for wings, it is the wing *plan area* that is used. To convert wing data into the more familiar car data, the wing data needs to be multiplied by (wing area divided by the car frontal area). That said, a low C_D wing still has lower drag than one with a higher C_D, and a wing with a higher C_L still has a higher lift than one with a lower C_L. Therefore, you can still directly compare the data for different wing sections.

Note that, a 'high drag' aerofoil of traditional shape is probably not particularly high drag in car terms. For example, a GOE222 profile (one that I'll return to shortly), at Reynolds numbers applicable for car use, is listed as having a C_D of 0.02 at about 10° angle of attack. A wing using this profile, that is 1 metre long and has a chord of 162mm (0.162m), has an area of 0.162m². On a small car, with a frontal area of 1.84m², the calculation to put this C_D into 'car terms' is (0.162/1.84) x 0.02 = 0.0018. To put that another way, a car with a C_D of 0.25, equipped with this wing being used at appropriate angles, would rise to only C_D 0.252. (This takes an additive approach to aerodynamic elements tested separately – always a dangerous thing to do. However, the point remains that 'proper' wings that are not stalled are quite slippery.) In contrast, in a near-stalled condition, the wing's drag is about seven times as high!

On a road car, wings are usually placed at the rear of the car. However, there's no technical reason why a front wing cannot work. In fact, one mounted slightly above the bonnet is likely to work quite well.

Wherever it is mounted, to work most effectively, a wing needs to be mounted in 'clean' air – airflow that is relatively undisturbed by the presence of the rest of the car's body. However, while this may be the case when considering the wing downforce figures in isolation, if the car is also equipped with an undertray and rear diffuser, a rear wing may produce a greater overall

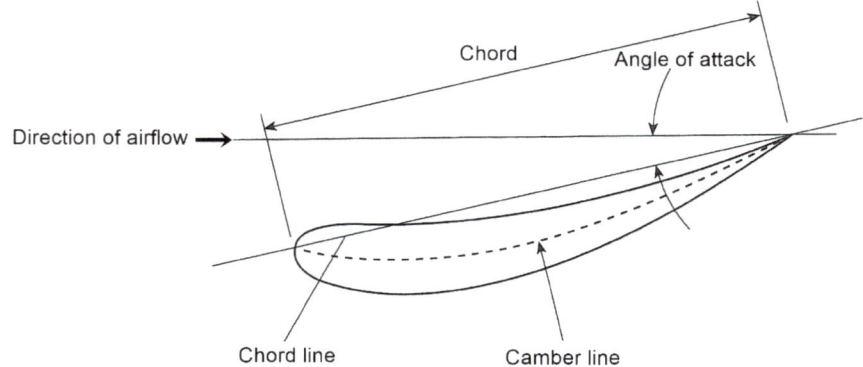

Figure 7-32: Wings are annotated in this way. Note the chord line joins the extreme left and right points, and that the angle of attack is described with respect to the airflow direction. The greater the camber, the greater the 'banana shape' of the wing.

REDUCING LIFT AND IMPROVING STABILITY

car downforce when mounted closer to the body than in free air. This is because the upwash created by the wing increases flow under the car. (As I've said previously, it's not wise to consider aerodynamic changes in isolation!)

There are literally hundreds of different aerofoil profiles available, so how do you go about selecting one for use on a road car? While in an ideal world you might well select the profile solely based on its drag and lift performance, that's no good if you cannot buy or make that wing profile. That said, the following table shows some drag and lift specifications of four profiles appropriate for road car use. The data is for a Reynolds number of about 350,000-600,000, corresponding to a wing chord of 150-200mm (6-8in) at 100km/h (~60mph). The data is derived from http://airfoiltools.com.

Figure 7-33 shows these wing profiles. Note that in the table, the *maximum* C_L is quoted; you may well decide to use a lower angle of attack (so reducing C_L) if you are after the best downforce:drag ratio.

Compared with the data available on the effect of undertrays, diffusers and rear spoilers, the effects of a rear wing on a road car in terms of lift and drag are rarely documented.

The 1970 Plymouth Roadrunner Superbird (detailed in Chapter 2) used a Clark Y aerofoil, 1473mm (58in) in span and 191mm (7½ in) in chord. The wing was mounted 597mm (23½in) above the boot lid. An angle of attack adjustment range of +2 to -10° was provided. One extraordinary aspect of the Chrysler's wing is the height at which it was mounted. Suggested reasons for this vary – some sources say it was done for aerodynamic reasons, and others say the wing was mounted so high simply so that the boot lid could be opened!

As described in the paper on the car (SAE 700036 – *The Aerodynamic Development of the Charger Daytona for Stock Car Competition*), surprisingly, the rear wing doesn't seem to have been all that effective. The rear downforce of the winged car was the same as the previous (unwinged) model at zero yaw, with both cars having a similar negative lift coefficient. However, as yaw increased, the winged car maintained downforce at about the same level, while the unwinged car decreased in downforce. (This data refers to the car configuration for which the engineers chose to present the final data.)

One reason that the wing may not have been as effective as you might think, is that its area is relatively small. The wing's 0.28m² (434in²) is only the equivalent of a square 530 x 530mm (20.8 x 20.8in). To reiterate a point I have already made, if the wing is working primarily as a device to create low pressure on its underside, decreasing the pressure under the car – with the much greater available surface area – is likely to be more effective. Even a spoiler that increases pressure across large areas of the rear of the car's upper body panels may be more effective than a wing – although, as the Superbird shows, not necessarily when the airflow has a yaw component.

The Superbird used a very high-mounted wing, but is such a height necessary? In a 2016 paper entitled "Automobile Aerodynamics Influenced by Airfoil-Shaped Wing" and published in the *International Journal of Automotive Technology*, researchers from the University of Zagreb modelled different wing

Profile	Maximum C_L	C_D at maximum C_L	Angle of attack at maximum C_L (°)
NACA 2412	1.4	0.06	15
Clark Y	1.4	0.03	12
GOE222	1.7	0.02	10
S1223	2.3	0.04	13

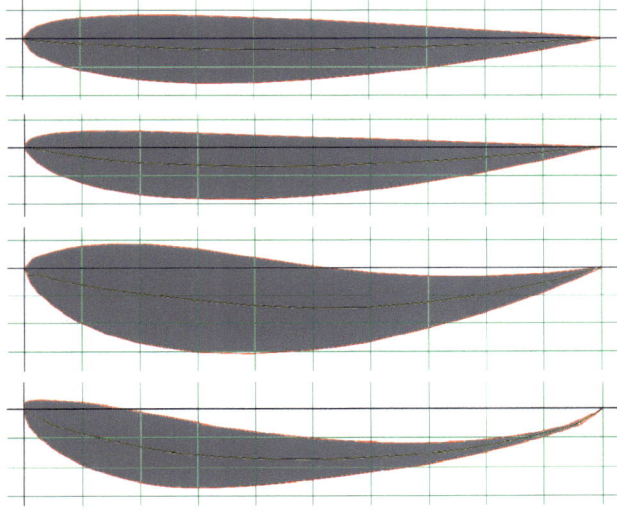

Figure 7-33: The profile of four wings suitable for use on road cars. From top to bottom: NACA 2412, Clark Y, GOE 222, and S1223. (Data points based on www.airfoiltools.com)

SPEEDPRO SERIES

Figure 7-34: Now that's a wing – Plymouth Superbird. Note the size of the supports that act as stabilising fins – an important part of the overall aerodynamic package. See Chapter 2 for more on this car.

heights on the boot of an E38 BMW – a three-box sedan.

In this paper, the height of the wing is expressed as the height above the boot lid divided by the distance the roof is above the boot. That is, a wing positioned at the height of the roof is classified as '100,' and one on the boot lid is '0' – you can think of these figures as 'percentage of the roof height.' Wing heights of 17, 39, 61 and 83 per cent were trialled. The wing had a span of 1800mm (70.9in) and a chord of 150mm (5.9in). The wing profile is not stated in the paper, but it had maximum downforce at an angle of attack of 20° and minimum drag at 7°.

The first point that the researchers describe is that the angle of approaching airflow varies at different wing heights. This is because the streamlines descend toward the rear of the car as the airflow leaves the end of the roof and reattaches on the boot lid. However, the greater the height above the boot lid, the less the downward angle of this airflow. The different wing heights had oncoming airflow angles as shown in the following table:

Height of wing above boot lid (percentage of roof height)	17	39	61	83
Angle of approaching airflow (degrees below horizontal)	12.2	13.4	14.4	15.2

The fact that the flow is not parallel to the ground changes the apparent best wing angles of attack to give greatest downforce and lowest drag.

In other words, wing angles of attack should be set in relation to the angle of incidence of the approaching airflow, and not against a spirit level!

The modelled BMW showed that the most effective height of the rear wing was at 61 per cent of the roof height. The developed downforce increased most steeply as wing height was moved from 0 to 50 per cent height. Note that going higher than 61 per cent decreased downforce only slowly. The lowest drag occurred with a wing height of 39 per cent. Thus, from the perspectives of downforce and drag, the best wing height was around half the vertical distance from the boot lid to the roof. On the modelled BMW, at this height, the angle of attack of the oncoming airflow was about 14° downwards. Thus, when referenced against true horizontal, the best angle of the wing for downforce was not the 'specification' 20°, but in fact only 6°.

Whether wings fitted to most production cars are actually working primarily as wings or as spoilers is contentious. Mitsubishi, in its technical paper *Aerodynamics for Lancer Evolution X*, use the terms 'spoiler' and 'wing' interchangeably. However, according to the CFD

Figure 7-35: Modelling was carried out on an E38 BMW to find the best height for a boot mounted wing. The best wing height for downforce was 61 per cent of the distance from the boot lid to the top of the roof. (Courtesy BMW)

REDUCING LIFT AND IMPROVING STABILITY

Figure 7-36: The rear wing on this Lancer Evo is positioned about half way between the boot lid and the roof height. On a three-box sedan with attached flow onto the trailing edge of the boot, this wing height matches the results found in the research. (Courtesy Mitsubishi)

Figure 7-37: The very large wing on the Lexus LFA is likely to work effectively; smaller wings mounted lower may not achieve much, especially in countries with low speed limits. (Courtesy Lexus)

Figure 7-38: On a three-box sedan, the greater the height above the boot lid, the less the downward angle of the airflow. The optimal angle of attack for a boot-mounted wing therefore varies with height. (Courtesy Mercedes)

Figure 7-39: The Jaguar's XE SV Project 8 uses an adjustable carbon fibre front splitter and carbon fibre rear wing. The car generates 122kg (about 270lb) of downforce – but that is at 300km/h (186mph). (Courtesy Jaguar)

flow diagrams in the paper, there is airflow both sides of the horizontal part of the car's wing (or spoiler!). The earlier model Lancer Evo IV wing had a cross-section shape that appeared to be upside-down, looking more like an aeroplane wing rather than one designed for a car. Given the angle to the horizontal that many aftermarket wings are set at (especially considering the importance of setting this angle against actual airflow direction and not horizontal), many wings are probably functioning in a stalled condition, and so acting as large spoilers rather than wings.

That is not to say that relatively small wings cannot be effective – if you are going fast enough. Jaguar's XE SV Project 8 (at the time of writing, the fastest sedan around Nürburgring Nordschleife), uses an adjustable carbon fibre front splitter and carbon fibre rear wing. The car generates 122kg (about 270lb) of downforce – but that is at 300km/h (186mph).

Honda Insight rear wing and stabilising fins

I fitted a rear wing on my Honda Insight. The wing comprised an aluminium extrusion in the GOE222 profile. I chose to use this wing because:

- As it is a 'proper' aerofoil profile, drag and lift data is readily available for it.
- Compared to many high downforce wings used on racing machines, it has much lower drag and so is more suitable for a road car.
- The extrusion is readily available at a low cost, as it is sold for use in making the blades for Vertical Axis Wind Turbines (VAWT) and Horizontal Axis Wind Turbines (HAWT).

SPEEDPRO SERIES

- The chord of 162mm is a suitable size for the small car.

I bought a 1 metre length of the extrusion from Oz Wind Engineering (www.ozwindengineering.com) – check the website for latest pricing. At the time of writing, the extrusion is great value for money.

The extrusion is designed to be mounted on a 25.4mm (1in) pipe that slides through a matching internal recess in the forward third of the cross-section. Plastic end caps are supplied with the extrusion, with one of these caps equipped with a hole to take the mounting pipe and the other cap not having a hole. (A hole can be easily made in the blank end-cap.) The use of the 'pipe mounting' allows the wing to be easily adjusted for angle of attack. The extrusion weighs 1.3kg per metre (just under 1lb per foot) and has a nominal wall thickness of 1.2mm.

In Chapter 1 we looked at calculating lift and drag. Let's apply that maths to the 162mm chord GOE222 profile. I will assume a short wing length of 800mm (so giving a wing area of 0.13m²) and a speed of 30m/s (108km/h, or 67mph).

The following specs use data from http://airfoiltools.com, and are based on a Reynolds number of about 340,000 (corresponding to a chord of 0.162m and a speed of 30m/s). The GOE 222 has a C_L that rises almost linearly from 0 at -7.5° to 1.75 at 10°. The best lift:drag ratio is at about 5°, where the C_L is about 1.4 and the C_D about 0.02. (Remember, these drag and lift coefficients use wing plan area, not projected frontal area. That's why the numbers look odd in car terms.)

The drag formula is:

Drag force = $0.5 \times \rho \times V^2 \times A \times C_D$

where drag is in Newtons, ρ (rho) is air density in kg/m³, A is area in square metres, V is velocity in metres per second, and C_D is the coefficient of drag.

ρ = 1.2kg/m³
V = 30m/s
A = 0.13m²
C_D = 0.02
Drag force = 0.5 x 1.2 x 30 x 30 x 0.13 x 0.02
Drag force = 1.4 Newtons, or just 143 grams.

Now let's calculate lift.

Figure 7-40: A GOE222 profile sold as an aluminium extrusion for use as wind generator blades. This is a good choice for a low-drag wing with reasonable downforce on a road car. It's cheap, too!

Figure 7-41: Left, GOE22 aluminium wing and plywood rear fins being trialled on a Honda Insight. Testing like this is quick and easy. Downforce was assessed by measuring changes in ride height, sensed by a filtered potentiometer (see Chapter 5). The fins made a radical improvement to straight-line stability.
Right, because of the attached flow on the rear hatch, the wing needed to be set to a positive angle (ie the chord line lower at the rear than the front) to achieve the correct angle of attack to the airflow. This small wing achieved 8kg downforce at 100km/h (about 18lb at 62mph). The fins move the centre of pressure rearward, and so provide better stability. Here the plywood has been replaced with 10mm clear acrylic.

REDUCING LIFT AND IMPROVING STABILITY

ρ = 1.2kg/m³
V = 30m/s
A = 0.13m²
C_L = 1.4
Lift force = 0.5 x ρ x V² x A x C_L
Lift force = 0.5 x 1.2 x 30 x 30 x 0.13 x 1.4
Lift force = 98.3 Newtons, or 10kg downforce.

(We can also check our maths – 10kg is about 70 times the 0.143kg drag, the same ratio shown on the C_L/C_D graph for an angle of attack of 5°.)

So the small GOE222 wing, if mounted in free space at an angle of attack of 5° to the airflow, would provide 10kg of downforce at 108km/h for a drag penalty of just 143 grams force. At 160km/h we'd have about 22kg downforce with about 300 grams drag. But ... all this assumes the airflow on the car at the chosen wing mounting location is at freestream airspeed, that there are no wing mounts creating drag, and that base pressure is not altered. However, the important point is that these calculations give us an *indication* of what we can expect.

I temporarily mounted the wing on the Insight to allow me to do some tests. Earlier in this chapter I described the use of fins to move the centre of pressure rearward. I decided to integrate the wing with two rear fins, with the wing positioned between the fins. (The fins then also acted as wing endplates.) The temporary fins were made from plywood (and the final versions from clear acrylic). The upper edge of the fins horizontally continued the roof line rearward, giving a maximum rear fin height of 270mm (about 10½in) above the rear hatch. The aerofoil was mounted 180mm (7in) above the hatch.

To mount the wing, I used some 25.4mm (1in) aluminium bar. This was cut into two 50mm (2in) lengths, and the ends of the bars drilled and tapped to take 8mm bolts. These threaded spuds were then inserted in each end of the wing extrusion, and glued into place with an anaerobic adhesive – Loctite. (On a longer span wing, you'd use a tube right through the length of the wing. But with only 800mm span, this was not needed.) Two hex socket, stainless steel, button head bolts were used to connect these spuds to the side fins. Loosening the bolts allowed the wing's angle of attack to be easily altered. Testing was carried out with the undertrays and rear diffuser (as described earlier in this chapter) in place.

The testing measured the change in rear ride height at different wing angles of attack, and measured the change on this day, rather than trying to correlate the figures against others taken on a different day. The table below shows the change in rear suspension ride height, compared with slow driving (30km/h). All figures are averages of runs in both directions, and are in millimetres of suspension deflection. Remember when looking at the data, that the diffuser and undertray already cause downforce, so the fact that the rear suspension is being compressed does not indicate that the wing is working. What is wanted is a consistent extra reduction in height (ie at both test speeds) that varies with wing angle.

I started at Position 1 with the wing chord line horizontal, and added another 5 degrees of negative rake for each of the next two positions. For Position 4, I decided to try something radical, and set the wing at about 45° – so acting as a spoiler, not a wing. But looking at the data, none of these positions did very much, especially when it is remembered that (albeit on a different, much cooler day) I'd measured a 13mm drop at the slightly higher speed of 110km/h (that is, from the action of just the undertray and diffuser).

At a bit of a loss, I then went back to my starting point and adjusted the wing in the opposite direction, using a *positive* angle of attack (ie the rear of the wing chord line lower than the front). The first position (Position 6) was about 10° in the 'wrong' direction, and Position 7 added another 5° in this direction.

Speed	Increased suspension deflection from 30km/h (mm)	
	Position 5	Position 6
60	-10	-5
100	-15	-9

Position 5 (highlighted, above) was the first position where there was a clear-cut gain in downforce at both test speeds. At 60km/h, the change in ride height was the greatest recorded in this testing, and furthermore, it was double the average of the previous measurements taken with the wing. At the higher speed of 100km/h, the change in ride height was also the highest recorded, and was 50 per

Speed (km/h)	Position 1	Position 2	Position 3	Position 4	Average
	Increased suspension deflection from 30km/h (mm)				
60	-4	-4.5	-5	-7	-5
100	-9.5	-12	-8	-9	-10

cent more than the average of the previous wing tests. And Position 6? It took me back to where I was with Positions 1-4!

So what was going on? Earlier in this chapter, I mentioned that for a wing to work, its angle of attack has to be set relative to the *direction of actual airflow* over the car, and not to true horizontal. I also mentioned in Chapter 4 that a pitot tube can be used to track this airflow angle. However, in the case of the Honda, I just decided to start testing with the wing set to an angle that would 'obviously' work – except it didn't!

When measured accurately, the 'Position 5' wing angle was +9° to horizontal. The rear hatch angle on the Insight is +14 degrees, and so you can see, if the air is following the hatch angle, the actual wing angle of attack is -5°. Therefore, on this car, because of the attached airflow down the hatch, the wing works *only when set at a positive angle to horizontal*.

You can also think of this in another way. With the wing set horizontally, but with the airflow approaching at 'hatch angle,' the actual angle of attack was already 14°. When I added 5° of negative rake, the actual angle of attack became 19°! You can see that to gain less than 14° angle of attack, I needed to set the wing with a positive rake versus horizontal. (Note that with this wing profile, the same would apply on the BMW sedan cited in the earlier research paper – it too had an airflow angle of 14° at the location where a wing would normally be positioned.)

To avoid major errors of the sort I made above, you can also wool tuft the wing. This will quickly show if the angle of attack is so great that the wing has stalled – in this situation, there will be separated flow on the underside of the wing.

I've covered testing the wing in some detail because not only is the result interesting, it also shows that, first, you should always believe the test data and not your instinct. Second, when doing testing, it's really easy to let hope trump reality. You must mentally step back and say: is this modification *really* doing anything? Finally, wing set-up needs to be done very carefully!

And how much downforce was the wing developing? Looking at the extra suspension deflection created in Position #5 on this day compared with that achieved when the wing was ineffective, and looking at the rear spring rate at this height, I'd suggest about 8kg downforce. As it happens, that matches closely what the maths says should be the downforce with this wing, this angle of attack and this speed – it's good when things stack up! (However, note that as described earlier, a rear wing can cause a diffuser to work more effectively, so it may not be the wing alone that is giving this result.)

And the fins? Testing of the combination of the undertrays and rear diffuser, rear wing and rear fins showed an incredible change in the stability of the car in crosswinds. My test road is a section of freeway often subject to high winds – as evidenced by the wind turbine farms located on nearby hills. This particular road also has a few other aspects that make it a good test route for assessing stability: the road passes in and out of steep cuttings, and in places where it is exposed, the fetch is great. Finally, there are plenty of trucks on the road.

I tested on a day when the wind was blowing across the road, gusting at up to 40km/h (25mph) and averaging about 30km/h (19mph). In these conditions at 110km/h (about 70mph) the car was astonishingly stable; far more so than the standard car and almost unbelievably so for a lightweight car (around 850kg – 1870lb). It was in fact possible to take your hands off the steering wheel: in gusts you could feel the car roll, but it didn't change in direction. (Obviously to achieve this result you also need a car with little bump steer, ie minimal toe change with suspension deflection.)

In summary, wings can be effective on road cars in providing downforce with little drag. However, I suggest that you steer away from motorsport and 'generic' car wings, and pick an aerodynamic profile for which real drag and lift specifications are available. When setting up a wing, always remember that its angle is set with regard to the actual airflow direction, not true horizontal. Measuring ride height is the best way of assessing a wing's developed downforce.

If you wish to improve stability, moving the lateral centre of pressure rearward via the use of one or two fins can be a very effective way of doing so. To provide greater strength and visual cohesion, two fins can be integrated with a rear wing.

ACTIVE AERODYNAMICS

Frequently in this chapter I have described the trade-offs between reducing lift (or gaining downforce) and drag. For example, a rear spoiler sized and positioned to give maximum reduction in lift is very likely to give an increase in drag. A wing angled to give maximum downforce will have a greater drag than when it is angled less radically. The way that these conflicting requirements can be met is to have aerodynamic surfaces that change, depending on the requirements at that time.

The most common example of this is a pop-up rear spoiler. It stays flush with the bodywork at

REDUCING LIFT AND IMPROVING STABILITY

> ## DOWNFORCE IN A ROAD CAR
>
> The notion that 'downforce = good' is of course widespread. For example, where downforce aids are not banned, cars in all motorsport categories use them. But there is surprisingly little material published about how downforce changes the behaviour of *road* cars.
>
> The single most important aspect about downforce to understand is that *downforce increases the downward load or force of the car without changing its inertia*.
>
> Let's think about this in more detail. Let's say that you have a lightweight, small car, and you place 30kg (66lb) in the boot. If you are a sensitive driver, you will immediately feel the presence of that extra weight. For example, when you corner hard and lift off, you'll feel that the tail is more likely to come out with the change in weight distribution. If you go quickly over a hump, you'll feel that the back of the car rises more off the crest – and then comes down with a bigger thump! (Inertia is the property of an object to keep moving as it was – so over the hump, the back of the car wants to keep on going upward at the original angle, and then when it starts coming down, it wants to keep on coming down!)
>
> Now let's move that 30kg (66lb) weight to the middle of the car. The car's weight distribution probably hasn't changed, and to a large extent, neither has the car's grip. The tyres are pushing down more firmly (and that increases grip) but when you corner, you have more weight to accelerate laterally. The upshot is that one aspect pretty well balances the other – and so there's no gain.
>
> But let's now take out that weight and instead add downforce – and we'll say equal amounts of downforce front and back. Now, when we go over that hump, the vehicle lifts off less – the inertia hasn't changed, but the push downwards has increased. When we go around a corner, the tyres are gripping better – and this time, there's no more weight to accelerate laterally and so cornering speeds can be higher. When we brake at high speed, there's more tyre grip, but no more weight to slow.
>
> And the advantages don't stop there. I think that in a road car, ride quality is very important. And downforce helps here too. With downforce, the effective weight of the car has increased, and so too has spring deflection. This lowers the natural frequency of the suspension (in effect, the suspension becomes softer) and so a given size of upward bump doesn't vertically accelerate the body so much. Instead, it just compresses the springs to a greater extent. (If you're interested in suspension, you can also consider this – with downforce, the sprung/unsprung weight ratio has increased! That simple statement warrants some careful thinking …)
>
> Of course, all is not rosy. If you are adding significant amounts of downforce, you will need to consider the amount of suspension travel you have left at high speed. Also, unless you are very careful to minimise it, downforce will result in increased drag. More subtly, you also end up with a 'two-mode' car – spring and damper settings that work best at high speed will likely no longer be suitable for low speed.
>
> And one final thing – be careful. To be honest, I was quite stunned at the improvement in cornering prowess of the Honda Insight resulting from the added front and rear downforce. But the Insight is hardly a paragon of grip, so the starting point was – in absolute terms – fairly low. I think if instead you began with a car that was already highly competent in handling, with added downforce you might find yourself going *extremely* fast when exploring the new limits. As I said, be careful!

speeds of up to (say) 120km/h (75mph) but above that, rises into position. However, in addition, front spoilers can be designed that lower only at higher speed (so giving better clearance when negotiating driveways and poor roads), and as we've already seen in the previous chapter, radiator blinds can open and close to change cooling airflow.

Hindered – as so much transport development has been – by stupid rules imposed in racing, moveable aerodynamic surfaces have not been given the attention that they deserve. However, this is likely to change – patent applications for moveable aerodynamic surfaces are now being lodged by many mainstream car companies. It's also a particularly fertile area for people modifying cars for themselves. Microprocessor-based control systems are now incredibly cheap, and geared DC motor drives, servos and even pneumatic actuators are now often priced at only one-tenth of what they commanded a decade ago.

Here are some of the potential active aerodynamic components:

- Active height rear spoiler
- Active angle change rear wing
- Active height front spoiler
- Active angle change rear diffuser
- Active deployed rear fin
- Active cooling control of airflow (in or out)
- Active rear rudders

In addition, active ride height change (eg through the use of air suspension) can alter the effect of undercar aero.

In an MSc thesis written by J Atkinson in 2014 (*Research into the Potential of Variable Aerodynamic Properties to Modify Ground Vehicle Behaviour*) at Coventry University, UK, an additional interesting variety of active aerodynamic surfaces are shown. These include:

- An active rear wing that is used to provide downforce, but can also be deployed when aerodynamic braking is needed.
- An active rear wing that has side plates that extend from the wing to the car body, thus increasing exposed side area and so moving the centre of pressure rearward for better stability.
- Dual rear fins on the boot lid of a sedan that lie flat against the surface when not deployed, but can pivot to a vertical position as needed, including during the active countering of yaw.
- A single rear fin whose angle to the longitudinal and vertical axes can be altered as required to give optimum approach angles to the apparent wind (similar to the active rudders mentioned above).

The 2012 Pagani Huayra can actively vary the front ride height, and adjust four control flaps, one on each corner of the car. The flaps are controlled by a dedicated ECU that is fed information from the ABS and engine control ECUs that communicate information about the car's speed, yaw rate, lateral acceleration, steering angle and throttle position.

In a 2015 paper published in the *Archive of Mechanical Engineering* (*Numerical Analysis of Aerodynamic Characteristics of a High-Speed car with Moveable Bodywork Elements*) Janson and Piechna look at the effectiveness of Huayra-style moveable flaps front and back. The authors say:

"Based on the results of calculations presented in the paper, it seems that tailgate flaps [ie those located at the rear of the car] are quite effective, but front flaps appear to be beneficial for improving the cooling system, but do not […] generate aerodynamic downforce.

"One of interesting and important results of recent investigation is finding that there exist configurations of flap positions that don't generate the downforce as expected. Depending on whether the rotation axis [ie hinge] is located at the front or rear edge of the flap, there are two types of aerodynamic characteristics. When the rotation axis is arranged at the front edge of the flap, the changes in lift force are smooth, and proportional to the opening angle of the flap. When the rotation axis is disposed at the rear flap edge, there is a delay in generating the lift force. It only increases beyond a certain opening angle. There is still further dependence on the position of flaps. Flaps on the back of the body generate more downforce and the changes are smoother. Flaps placed closer to the centre of the vehicle produce less downforce and need to open at least to 30° to start generation of the aerodynamic downforce.

"The best solution is to place flaps behind the vehicle rear edge. They give the highest values of force in relation to the flap surface, and the force changes are smooth. In certain driving conditions, they allow for a slight reduction of aerodynamic drag. The flaps placed on the car body, in most cases, in addition to the expected downforce, generate lifting forces on other surfaces. The flaps located behind [ie at the rear extremity of] the body do not produce such negative effects."

The 1989 Mitsubishi HSR-II concept car had an intriguing number of moveable aerodynamic surfaces. As SAE paper 910597 (*Numerical Simulation of Road Vehicle Aerodynamics and Effect of Aerodynamic Devices*) describes, these comprised:

Figure 7-42: The 2012 Pagani Huayra can actively adjust four control flaps, two at each end of the car. The flaps are controlled by a dedicated ECU that is fed information about the car's speed, yaw rate, lateral acceleration, steering angle and throttle position. (Courtesy Pagani)

REDUCING LIFT AND IMPROVING STABILITY

Figure 7-43: The Porsche Boxster active rear spoiler is a good example of integrating an active element into rear bodywork. (Courtesy Porsche)

- A front spoiler that moved up and down.
- Two front canard wings that extended outward from either side, positioned ahead of the front wheels. In their undeployed position, they were flush with the bodywork. The two canards could be deployed individually.
- Two horizontal rear flaps, hinged on their forward edges. They were able to rise and were deployed individually.
- A rear diffuser that moved up and down.

The surfaces were deployed as shown in the table below.

In Condition 1 the paper states that the aerodynamic package allows:

"… the balance of lift between the right and left sides [to be] changed, in addition to a great reduction in lift. This force increases cornering power during high-speed cornering, and increases the vertical loads on the inside wheels by restraining rolling of the vehicle."

- Condition 1 (cornering) increased the C_D by 0.1 and decreased the C_L by about the same amount.
- Condition 2 (acceleration) resulted in about a 0.05 decrease in C_L with only a tiny increase in C_D.
- Condition 3 (braking) made radical changes to both drag and lift, with C_D increasing by 0.2 and C_L decreasing by 0.4!

Information on the HSR-II is scarce, but it is suggested that the car in optimal low-drag configuration had a C_D of 0.2. If that is the case, in active aerodynamics braking mode, the drag of the car doubled!

In SAE paper 2017-01-1592 (*Effects of Active Aerodynamic Wings on Handling Performance of High-Speed Vehicles*) an interesting idea is introduced: that of splitting front and rear wings into two halves and independently altering their angles of attack. In addition to collectively altering the angle of attack to provide appropriate high-speed downforce and low-speed drag reduction, the system also allows differential operation of the wings in cornering. For example, by using a larger angle of attack on the inner wing half, greater downforce can be applied to the inner wheels than the outer wheels, so resisting body roll.

In order to reduce drag, especially when there are yaw angles present in the approaching airstream, SAE paper 2017-01-7000 (*Adaptive Base-Flaps Under Variable Cross-Wind*) discusses the use of active vertical flaps placed either side of the base area. Testing was carried out on a simple shape in a wind tunnel. The paper states:

"Optimal flap positions are those that balance a low lateral force on the vehicle with a base pressure increase through boat-tailing, while also minimizing the drag force on the flap. This entails both ruddering the flaps in an asymmetric manner to minimize the lateral force and boat-tailing them inward to generate a narrower wake.

"The requirements for minimum drag mean that, for non-zero yaw angles, the optimal positioning of the flaps is necessarily non-symmetric, and therefore cannot be achieved with a static system. […] It has been shown that a system that adapts to changing yaw angles is able to perform up to 70% better than an optimized static system using surfaces of the same length."

This paper concentrated on reducing drag. However, the development of aerodynamic devices at the back of the car that provide clean separation edges and minimal wake size in a low-drag position, but change position to provide braking and/or downforce configurations as needed, should be straightforward for many car shapes.

The 991 Porsche 911 Turbo uses active aerodynamics comprising an inflatable, three-section front spoiler, and a moveable rear wing capable of two erect heights and also able to be adjusted in its angle of attack. Three different aerodynamic modes are used. In normal mode, as when the car is first started, both the front spoiler and rear wing are in retracted positions.

Condition		Front spoiler	Canards	Rear flaps	Rear diffuser
1	High-speed cornering	Down	Out (inner side)	Up (inner side) 20°	Down
2	Acceleration	Down	In	Down	Up
3	Braking	Down	Out (both sides)	Up (both sides) 30°	Down

SPEEDPRO SERIES

Figure 7-44: Left, the Porsche 911 Turbo S spoiler in its low position … and right, in the first of its two raised positions. The second position extends the spoiler further and angles it by up to 15°. At the same time, an inflatable front spoiler lip also deploys. (Both photos courtesy Porsche)

At the next level of performance, called by Porsche 'Speed' mode, the outer sections of the front spoiler are extended and the rear wing rises 25mm (1in) to its first position. At the final level of performance, called by Porsche 'Performance' mode, the entire front spoiler is extended and the rear wing rises by 75mm (3in) and is tilted forward by 7°. (Incidentally, when the sunroof is open, the vertical extension of the rear spoiler, and its attack angle, are further increased, presumably because the airflow reaching the rear wing is disrupted by the open roof.)

The following table shows the aerodynamic performance of the car in these different modes.

	Start mode	Speed mode	Performance mode
C_D	0.31	0.31	0.34
C_{Lf}	0.04	0.02	-0.05
C_{Lr}	-0.01	-0.03	-0.10
$C_{Lf} - C_{Lr}$	0.05	0.05	0.05

Note how the difference between front and rear lift does not change in the different modes. But what about when the system is actively changing? When changing from Performance to Speed modes, the front spoiler is retracted prior to the rear wing being lowered. This keeps the $C_{Lf} - C_{Lr}$ close to 0.05, even during the mode changeover. Without taking this staged approach, Porsche data shows that the $C_{Lf} - C_{Lr}$ figure would have been negative for a few seconds during the mode transition, causing a pitch change.

When developing active aerodynamic surfaces, there are some important aspects to keep in mind. Active devices can be subjected to high aerodynamic loads. If you have a 1500 x 150mm (59 x 6in) panel that is acting as an airbrake, a force of nearly 18kg (40lb) can be developed on it at 130km/h (81mph). (And at 300km/h this would rise to 98kg, 216lb). Therefore, both the aerodynamic surface and its hinge and link mechanisms need to be sturdy.

Irrespective of how you trigger it, the control system will need to have sufficient hysteresis. Hysteresis in this case refers to the difference in speed between deployed and non-deployed states. For example, if you chose to deploy a spoiler at 100km/h (62mph), you would not want it retracting at only 95km/h (59mph). With such little hysteresis, as your speed varied (eg from corner to corner on a winding road), the aerodynamic behaviour of the car would keep altering. Instead, a better approach would be to retract the spoiler only when speed has dropped to perhaps 40km/h (25mph).

Another design challenge is the powering-mechanism. Any approach that uses electric actuators will also need to have position sensing. This is so that the control system knows the position of the device. For example, a simple DC-motor, geared-down appropriately (eg a car electric window motor) can be used to power a moveable spoiler. Such a motor is reversible just by swapping the polarity of the connections – easily done via a suitable relay. However, how will the system know when to stop the motor? It can do this only if it has position sensing. This is most easily done by a pot (eg a suspension position pot of the sort we used in Chapter 5). On the other hand, model servos have position sensing built in. In either case, a programmable microcontroller board will need to be used as the brain of the system. Note that eLabtronics (www.elabtronics.com) is able to produce low-cost custom controllers with these capabilities. For more complex systems, I recommend

REDUCING LIFT AND IMPROVING STABILITY

Adaptronic (www.adaptronic.com.au). The Adaptronic devices are fully programmable and, as they are usually employed as engine management controllers, have a large number of inputs and outputs. Furthermore, they can be tuned 'live,' and have gauges that can be viewed on the laptop screen.

If all of that sounds too difficult, consider using pneumatics. Miniature pneumatic cylinders are available that can be powered by a small compressor and/or air tank. Unlike an electric actuator, the air pressure can be maintained on the cylinder, even when the aerodynamic device is in the fully-deployed position. (In fact, the pressure can be used to hold the aero device in that position.) This means that position sensing is not needed. Return to the non-deployed position can be by compressed air release and a spring, or you can use a two-way cylinder and switch pressure to the other port. Two-way pneumatic valves for this purpose are readily available. In some cases, eg an air brake, aerodynamic air pressure alone may be enough to return the device to its non-deployed position.

The main input for activating and deactivating an aerodynamic device is likely to be road speed. Road speed is already sensed by car electronic systems (eg the dashboard, engine ECU, ABS ECU, etc), and this signal can be tapped into. For example, a high input impedance frequency switch module with adjustable hysteresis can be used. Some aerodynamic devices can be triggered more simply – eg an air brake via the brakelight switch – although to avoid frightening people in stop-go traffic, you may wish to place a speed switch in series!

Finally, don't forget that moveable aerodynamic devices can be made completely manual. If you are embarking on a long-distance drive, configure the surfaces for lowest drag. On the other hand, when you have decided that you're going to go for a blast on a challenging, winding road, configure the devices for maximum downforce.

Figure 7-45: Top, rear spoiler/air brake in non-deployed position. Note how it incorporates a duck tail ('kick-up') even in this mode, reducing lift and giving clean separation.

Centre, the device in the first of two deployed positions. Pressure has been increased on the surfaces ahead of the spoiler, but note how the wake has substantially increased in size.

Bottom, the panel is fully raised, acting now as an air brake. (All photos courtesy Mercedes)

Chapter 8
Improving airflow through heat exchangers

- ♦ **Pressure differentials**
- ♦ **Improving underbonnet intercooler airflow**
- ♦ **Testing a diesel intercooler**
- ♦ **Designing and installing an alternator cooling duct**
- ♦ **Turning vanes**
- ♦ **Bonnet (hood) vents and ducts**

All modern car designs have spent thousands of hours in the wind tunnel, while the engineers refined and altered, tested and assessed. But, contrary to popular belief, many of those hours were not used to create a body shape with great drag and lift figures. Instead, they were spending the time optimising the cooling system airflows – making sure that plenty of air reached the radiator(s) and could then leave without obstruction.

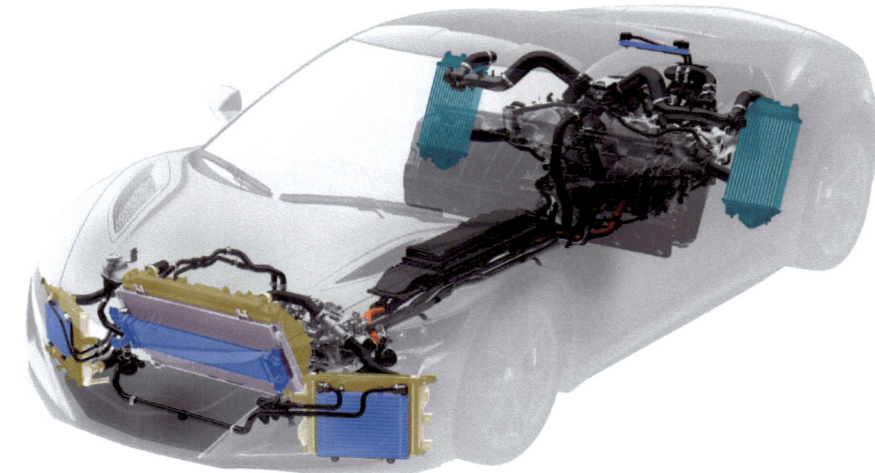

Figure 8-1: This Honda NSX uses no less than ten different heat exchangers. Openings at the front of the vehicle supply cooling airflow to the front engine radiators, twin motor unit cooler, condenser, transmission gear cooler and hybrid Power Distribution Unit. Air flowing over the roof and down the rear hatch glass feeds the transmission clutch cooler. (Courtesy Honda)

The average modified car is very different. Perhaps there's a randomly designed bodykit in place, and there might also be added bonnet (hood) vents. But there's invariably not a lot of rhyme or reason to it all – sometimes directional vents are put in backwards, while some scoops

IMPROVING AIRFLOW THROUGH HEAT EXCHANGERS

intended to let air *in* are almost certainly letting air *out*. Aftermarket undertrays designed to improve heat exchanger flow? Well, while there's plenty of discussion online about them, fitting and then testing their effectiveness on road cars is almost unknown.

In this chapter I want to take a detailed look at designing and fitting a front undertray, front spoiler, and bonnet vent. These were aimed at improving the airflow through the front heat exchangers – radiator, transmission oil cooler, air conditioner condenser, intercooler and power steering cooler. (Here I'll concentrate on intercooler flows, but the others apply just as much.) The whole exercise proved to be a learning experience in more ways than one – like finding out that at some speeds, the guinea pig car's underbonnet intercooler had in fact, *zero* outside airflow through it ... Surely not? – but that's the truth.

After we have looked at the successful modifications that were performed on that car, we will also look at some other examples of modification (some not so successful), and then summarise the whole approach.

PRESSURE DIFFERENTIALS

Air will only flow if there is a pressure differential. This is an important point to grasp – air doesn't pass through the radiator just because the car is moving forward. Instead, there needs to be a higher pressure in front of the radiator, and a lower pressure behind it – that is, a *difference* in pressure.

To better understand this, imagine that the engine bay is completely sealed, top and bottom. The car with the sealed engine bay moves forward, and air initially flows in through the radiator. However, without any escape route, the engine bay soon 'fills up' with air, until the pressures either side of the radiator become equal. In this situation, no more air will flow through the radiator.

Obviously, engine bays are *not* sealed top and bottom – so what happens when there are the normal

Figure 8-3: The amount of air that will flow through the under-bonnet intercooler is governed by the pressure developed in the bonnet scoop, minus the pressure developed within the engine bay on the underside of the intercooler. (Courtesy Subaru)

openings beneath the engine? These openings allow some air to flow out of the engine bay. The pressure build-up under the bonnet is therefore reduced (although it may be still higher than ambient). So, to develop a large pressure differential across the radiator, the greater the 'escape routes' we can provide for underbonnet air, the better.

You can now see how important the underbonnet pressure is in influencing airflow through front-mounted heat exchangers. So what determines this pressure? As we've seen, the effectiveness of the air 'escape routes' is important. But any design feature that bypasses the heat exchangers and adds air to the engine bay will also influence the underbonnet pressure build-up. For example, if you've installed a bonnet scoop, it's quite likely that there's an additional pressure build-up under the bonnet which will result in decreased flows through the radiator, intercooler, oil cooler and/or air-conditioning condenser.

If you are trying to improve the flow through front-mounted (or scoop-fed) heat exchangers, the first measurements that you should make are of the pressures occurring when

Figure 8-2: The Audi R8 e-tron uses good heat exchanger airflow management, with the front air entrance in a high-pressure zone, and the exit in a low-pressure zone. (Courtesy Audi)

SPEEDPRO SERIES

the car is moving. (See Chapter 4 for how to make these measurements.)

IMPROVING UNDERBONNET INTERCOOLER AIRFLOW

The car I will use as the main example in this section is a Nissan Maxima V6 Turbo. The car was originally not intercooled, but I added a small intercooler located horizontally in the front-left corner of the engine bay. (It replaced the battery that was moved to the other end of the car.) I then added a bonnet scoop to feed air to the intercooler. I initially fitted a small scoop, but the result was poor – and so I then fitted a much larger scoop. However, the results were still not up to scratch, so I started investigating the aerodynamic flows through the intercooler.

The first step in measuring how well the scoop was working, was to measure the pressure of the air in the engine bay. (Remember, in this car, this is the pressure that the exit air from the intercooler is pushing into.) I soon found that with the car moving, the pressure rise in the engine bay was considerable.

For example, at 80km/h (50mph), there was no less than 100 Pascals (Pa) (0.4 inches of water) pressure build-up in the engine bay. Given that I had previously measured a pressure in the intercooler scoop of 100Pa (0.4 inches of water) at the same speed, it looked very much like the air movement through the intercooler was zero. (Remember, if the pressure either side of a heat exchanger is the same, no airflow will occur through it.) Therefore, the first measurements I made indicated that the Maxima's underbonnet intercooler was getting *no outside airflow through it*, even at high road speeds, and despite using a large forward-facing scoop.

I then made more measurements of the pressure under the bonnet of the Maxima. Two measuring locations were picked:

- against the underside of the bonnet, in the middle of the panel towards the rear of the engine bay
- near to the intercooler air outlet, which is towards the front-left of the engine bay

The road speed for all the measurements was 80km/h (50mph) and the electric radiator fans were not operating when the measurements were taken. (Interestingly, the underbonnet pressure measurably rose when the radiator fans were working – they push more air into the space.)

The first measurements were taken with the standard undertray configuration. As the table on the right shows, and as already found in my first measurements, the pressure build-up in the engine bay at 80km/h (50mph) was 100Pa (0.4 inches of water).

Figure 8-4: This Nissan Maxima V6 turbo proved to be a fascinating exercise in improving heat exchanger flow. The small bonnet scoop leads to an intercooler that was added to the car. The result was poor and so a much larger scoop was fitted, but the results were still not up to scratch. Then I started investigating the aerodynamic flows through the intercooler …

Figure 8-5: The Maxima's small intercooler (arrowed), mounted in the front-left corner of the engine bay. Flow through this core was massively improved by reducing engine bay pressures.

212

IMPROVING AIRFLOW THROUGH HEAT EXCHANGERS

STARTING POINTS

The desire to make some aerodynamic changes to the front of the Maxima also came about because I'd found that any changes to the undertray had a noticeable effect on intercooler efficiency, as measured by an intake air temperature gauge.

In one case I'd had the car wheel aligned. The mechanic, noticing how the standard undertray drooped down at the rear, had tied it upwards with a cable-tie. The intake air temperature immediately rose, indicating that the outside airflow through the intercooler was poorer. In another case, I'd been driving the car with the undertray completely removed. This time, the performance of the intercooler seemed to be even poorer.

It therefore seemed that some tweaking of the undertray had the potential to dramatically improve intercooler (and perhaps also radiator) efficiencies.

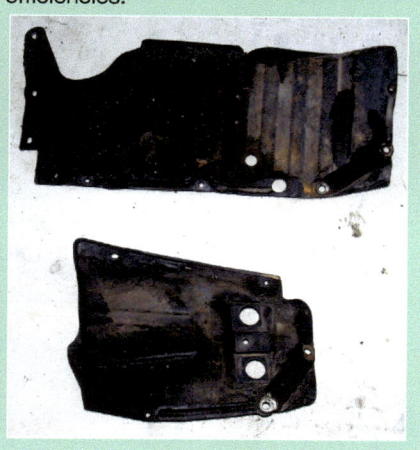

	Standard undertrays
Rear of bonnet	100Pa (0.4 inches of water)
Near intercooler	100Pa (0.4 inches of water)

The short factory undertrays were then removed, leaving an opening that allowed the road to be seen by peering down between the radiator and the engine. The table below shows the measurements that then resulted.

	Standard undertrays	No undertrays
Rear of bonnet	100Pa (0.4 inches of water)	100Pa (0.4 inches of water)
Near intercooler	100Pa (0.4 inches of water)	125Pa (0.5 inches of water)

As can be seen, the pressure at the rear of the engine bay stayed the same at 100Pa (0.4 inches of water) but the pressure at the front of the engine bay (which is where the intercooler was mounted) rose to 125Pa (0.5 inches of water). This explained why the radiator and intercooler didn't work as well with the small standard undertrays removed – the pressure in the engine bay was higher with these undertrays off (see box on the left).

A test undertray was then fabricated from some plastic sheet that had once backed an advertising hoarding. The front of the trial undertray was held in place with cable ties connecting to the lower edge of the bumper, and at the rear, by a wire that connected it to the anti-roll bar. The undertray was allowed to droop down at the rear, creating a gap that reached about 100mm (4in) in the middle.

Figure 8-6: A trial undertray, made from plastic material that was previously used as an advertising sign. Always trial aerodynamic changes, using cheap material that can be readily put into effect as a mock-up for testing. In this case, fitting the trial undertray decreased the measured pressure in the engine bay, something that will improve flow through the intercooler.

The measurements taken with this undertray are shown in the table below. As can be seen, the pressures found towards the front of the engine bay are markedly altered by the presence and shape of the front undertray. Over having no undertray at all, the trial plastic undertray reduced the pressure in the front part of the engine bay by 40 per cent. Significantly, it was also much better than the standard undertrays, reducing the pressure build-up over them by 25 per cent.

	Standard undertrays	No undertrays	Plastic undertray
Rear of bonnet	100Pa (0.4 inches of water)	100Pa (0.4 inches of water)	100Pa (0.4 inches of water)
Near intercooler	100Pa (0.4 inches of water)	124Pa (0.5 inches of water)	75Pa (0.3 inches of water)

When looking at these measurements, remember that the lower the pressure we can achieve in the engine bay, the better the likely airflow through the heat exchangers.

Next a spoiler lip (air dam) was added to the undertray. The trial lip was formed from foam rubber, was 50 x 50mm (2 x 2in), and attached to

213

the undertray 150mm (6in) back from the leading edge. The measured results are shown in the table below and Figure 8-7.

	Standard undertrays	No undertrays	Plastic undertray	Plastic undertray with spoiler
Rear of bonnet	100Pa (0.4 inches of water)	100Pa (0.4 inches of water)	100Pa (0.4 inches of water)	75Pa (0.3 inches of water)
Near intercooler	100Pa (0.4 inches of water)	125Pa (0.5 inches of water)	75Pa (0.3 inches of water)	75Pa (0.3 inches of water)

Figure 8-8: A trial spoiler added to the undertray. Using foam rubber for the spoiler allowed the shape and location of the spoiler to be easily changed for testing. The spoiler reduced the underbonnet pressure still further.

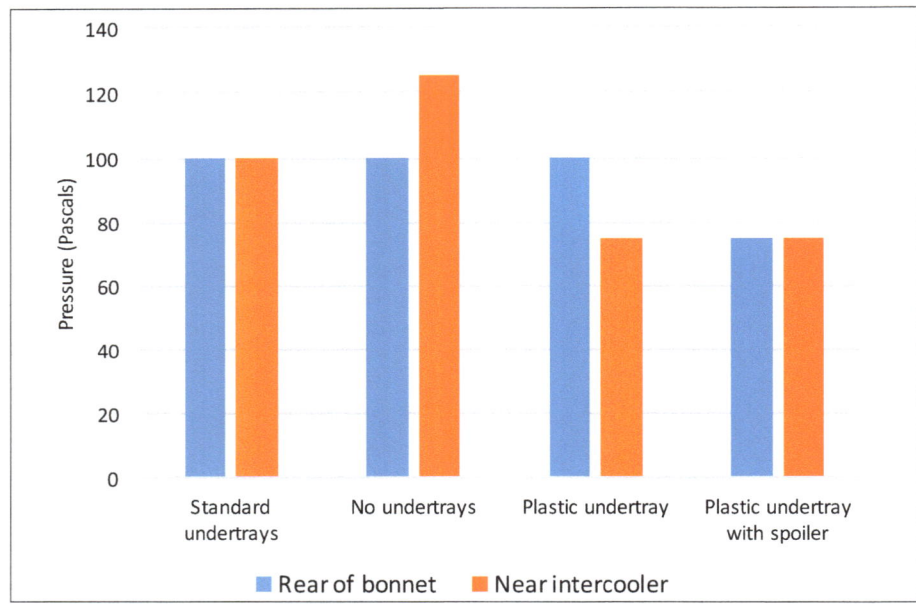

Figure 8-7: The pressures measured in the engine bay of the Maxima with different undertray configurations. When seeking to improve flow through heat exchangers that exhaust into the engine bay, the lower the pressure, the better. Best results in these initial tests were achieved with a trial plastic undertray fitted with a spoiler.

So compared with standard, the new trial undertray and spoiler lip decreased the front and rear pressure engine bay build-ups by 25 per cent. Over having no undertray at all, the new combination dropped the front and rear underbonnet pressures by 25 per cent and 40 per cent, respectively. This major pressure change (thus improving airflow through all the front-mounted heat exchangers) is especially important to consider if your car has no undertray at all.

Intercooler measurements

The measurements taken so far were referenced against cabin pressure, measured with the windows open. I then decided to do some more measurements that focused on pressure *differentials*. These were carried out by running two sensing tubes from the measuring instrument to either side of the intercooler and then measuring the actual pressure difference across the core. The table below, and Figure 8-9, shows the results.

	No undertrays	Plastic undertray	Plastic undertray with spoiler
Pressure differential across intercooler	-25Pa (-0.1 inches of water)	0	+25Pa (+0.1 inches of water)

IMPROVING AIRFLOW THROUGH HEAT EXCHANGERS

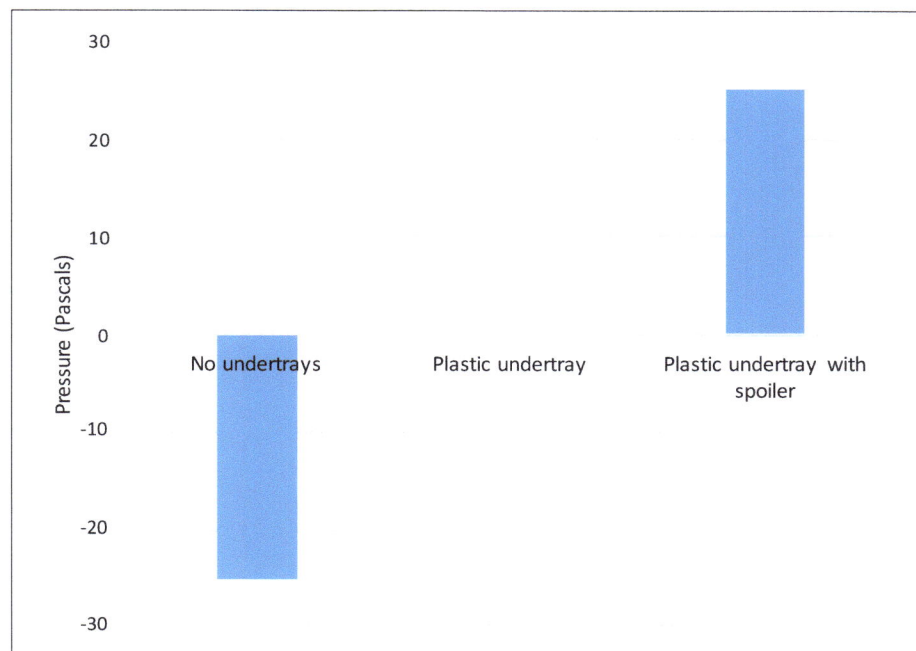

Figure 8-9: The pressure differential, measured across the underbonnet intercooler in various undertray configurations. This graph shows that with no undertrays fitted, the pressure was actually higher on the underside of the intercooler than the side connected to the feed scoop! With just the plastic mock-up undertray fitted, there was zero pressure differential, and with the plastic undertray and spoiler fitted, the pressure on the upper side of the core was greater than on the lower side, meaning that flow would occur in the correct direction.

Note: a positive sign is used where the pressure on the front face of the intercooler is higher than the pressure in the engine bay.

These measurements were an eye-opener! Without any undertrays in place, the air from under the bonnet is likely to have been flowing *out* through the bonnet scoop! That is, there was a measured higher pressure under the intercooler than on top, even at 80km/h (50mph). No wonder the intake air temperatures were high in this configuration – the air was coming in through the radiator, being heated, then passing through the intercooler and out of the large forward-facing bonnet scoop! I had had a pre-heater happening, not an intercooler ...

With my mock-up plastic undertray in place, the pressure at 80km/h (50mph) was the same top and bottom of the intercooler. That is, *no airflow through the core would have been occurring*. And with the trial plastic undertray and the foam rubber spoiler, I had a small positive differential of +25Pa (+0.1 inches of water), that is, the pressure on top of the intercooler was a little higher than underneath.

From these measurements it could be seen that a new undertray would give better intercooler flow, but the exact gain was heavily dependent on the precise complexion of the undertray (eg whether or not it used an additional spoiler). Therefore, rather than continuing with trial undertray combinations, I decided to develop and test the real one.

The new undertray

Undertrays of this sort can be made from a variety of materials, including fibreglass, ABS plastic, polycarbonate (Lexan) or marine plywood. I chose ABS – a tough plastic that can be bent when heated and is easily drilled, sanded and cut with woodworking tools.

I won't take you through every design iteration made with the ABS, but after hours of constructing and testing, constructing and testing, I struck gold. With a short undertray configured flush with the lower bumper at the front, angled slightly downwards and open at the rear, and with the sides sealed-off with angled folded panels, the measured pressure differential across the intercooler rose to 75Pa (0.3 inches of water) at 80km/h (50mph). And did the intercooler then work more effectively? It certainly did! Over the worst-case scenario (ie, no undertrays), the intake cruise air temperature on a 30°C (86°F) day dropped from 65°C (149°F) to 47°C (117°F) – yes, the intercooler was now working.

So could the pressure difference across the intercooler be improved even further with the addition of a spoiler? The answer was 'no'. Despite trialling lip spoilers in different positions and of different heights on the undertray, no significant further gain in pressure across the intercooler could be made with this version of the undertray.

And what about underbonnet pressures – the starting point of the measuring process? Towards the front of the engine bay, the underbonnet pressure was reduced by 75 per cent, from 100Pa (0.4 inches of water) down to 25Pa (0.1

SPEEDPRO SERIES

Figure 8-10: The final undertray was made from ABS plastic sheet. This material can be heated and bent – note the folded-up side flaps. This undertray worked very well at reducing engine bay pressure – but interestingly, adding a spoiler to this final version achieved no further benefit.

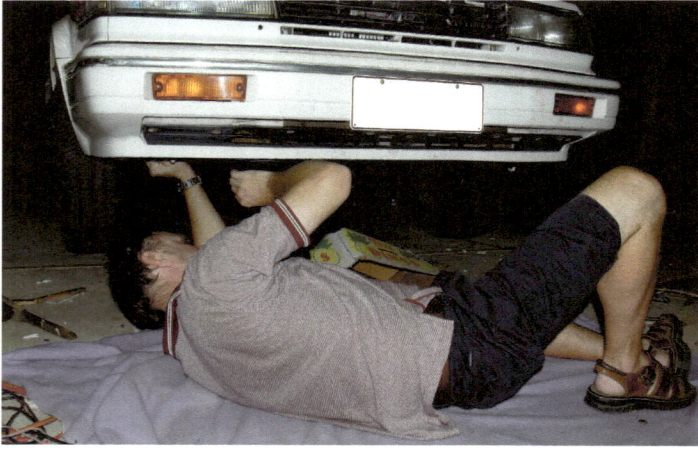

Figure 8-11: The (younger) author fitting the undertray. You don't need elaborate facilities to do work like this.

inches of water), both at 80km/h (50mph). However, the final version of the undertray does not reduce pressures towards the rear of the engine bay at all. To a large extent, getting a big pressure decrease at the front of the engine bay is far more important on this car than at the rear – the intercooler and radiator are both mounted near the front of the car, and so have their exit air exhausting at this position.

(Of course, a car with a top-mount intercooler mounted over – or behind – the engine woud be a completely different story!)

Next, I decided to see what could be achieved with some bonnet vents.

Siting vents

Bonnet vents are openings designed to exhaust air – to promote the flow of air out from the engine bay. As covered earlier, air movement will occur only when there is a pressure differential. So, to cause air to flow out from under the bonnet, what is needed is an underbonnet pressure that is greater than the pressure on top of the bonnet – *at the location where the vent is*. As described in Chapters 1 and 4, most cars have a low pressure at the leading edge of the bonnet and a high pressure at the base of the windscreen. So on this basis, you wouldn't site a bonnet outlet vent close to the windscreen – in fact that's usually where the cabin ventilation inlet ducts are, because they are taking advantage of that high pressure.

Measurement was made of the pressures on the surface of the Maxima's bonnet. At 80km/h (50mph) road speed, these ranged from -125Pa (-0.5 inches of water) at the leading edge of the bonnet to +150Pa (+0.6 inches of water) at the base of the windscreen. Figure 8-12 shows these measurements. (It's interesting to compare these measurements with those taken on the two Mercedes in Chapter 4. Note that the Mercedes tests were done at 70km/h (43mph), not the 80km/h (50mph) used for the Maxima.)

Earlier, I measured the maximum underbonnet pressure at 80km/h (50mph) as +100Pa (+0.4 inches of water) – and here, near to the windscreen, the surface pressure was +150Pa (+0.6 inches of water). In other words, place a bonnet vent at this location, and air will be flowing in from outside, through the vent and into the engine bay! (And the same applies if you were to prop the rear of the bonnet up with spacers, a suggestion I often see proposed to improve radiator efficiency.) Clearly, the further forward that the vents were to be placed, the lower the available outside pressures.

IMPROVING AIRFLOW THROUGH HEAT EXCHANGERS

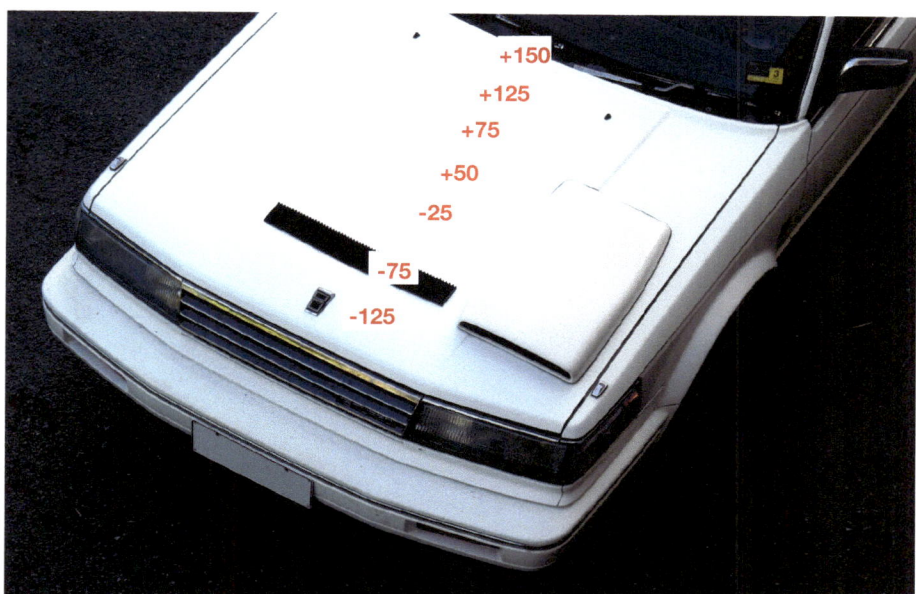

Figure 8-12: The surface pressure measurements on the top of the bonnet, measured at 80km/h (50mph). Note that the image shows the car with the exit grille installed, but it was not present when these measurements were made.

But the outside pressures are literally only half the story. What about the underbonnet pressures? As mentioned above, with the new-design undertray in place, the maximum pressures under the bonnet (again at 80km/h (50mph)) was 100Pa (0.4 inches of water). This was recorded across the rear half of the engine bay. At the front of the engine bay, the underbonnet pressure was 15Pa (0.1 inches of water).

So what were the alternatives for vent siting? The greatest difference between the underbonnet and over-bonnet pressures was at the very leading edge of the bonnet, where the pressure in the engine bay was +150Pa (+0.6 inches of water) higher than the body surface pressure. However, it's very hard to site a vent here and moving backwards a little to the front third of the bonnet still gave 100Pa (0.4 inches of water) pressure difference.

Vent selection

For reasons of aerodynamic drag, it is normal to direct the air out of vents as parallel to the surface as possible. This causes least turbulence. But of course bonnet vent selection also depends on issues like price, durability and aesthetics – bonnet vents are much more visible than undertrays! Taking into account all of these factors, I purchased from a boating supplies shop some stainless steel louvred vents. The vents were bought prior to making the above pressure measurements (not a good idea!) and I had thought they would be installed north-south, ie, with their long axis parallel with the car. However, after making the pressure measurements, I realised that this would put a considerable portion of each vent in a less than ideal area of pressure differential – it would be as if each vent was only half as big as it really was.

I then reconsidered. What was really required was a vent that could be mounted east-west, ie, across the bonnet. This would be best sited as far forward as possible. In fact, the shape of the vent wouldn't be all that different to a ventilation inlet grille at the base of the windscreen – except it would be sited a long way forward, and would be exiting air (rather than drawing it in). A trip to a car wrecking yard and a long walk around the yard found the ideal vents – the ventilation inlet vents on a Holden VL Commodore, vents which are normally positioned in the trailing edge of the bonnet.

The chosen vents come in two pieces. If required – as was the case here – they could be shortened by chopping off a section at each end. (The vent assembly is slightly curved in plan, so this kept their symmetry.) A hacksaw was used to do this, then an electric jigsaw fitted with a metal-cutting blade was used to cut out

Figure 8-13: The selected bonnet vents were salvaged from a wrecking yard – they are vents usually used as cabin ventilation inlets.

SPEEDPRO SERIES

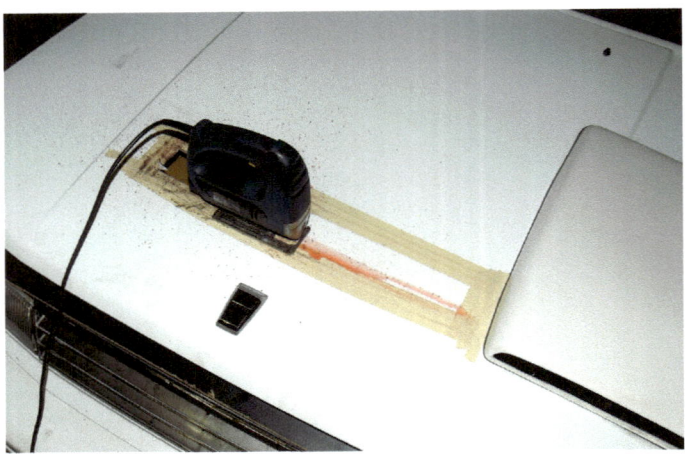

Figure 8-14: The hole for the vents was cut with a jigsaw fitted with a metal-cutting blade. Note the masking tape to protect the paint from the saw's shoe, and the automatic transmission fluid being used to lubricate the blade.

Figure 8-16: The finished intercooler airflow system – vents, large scoop, and (not visible) the new undertray.

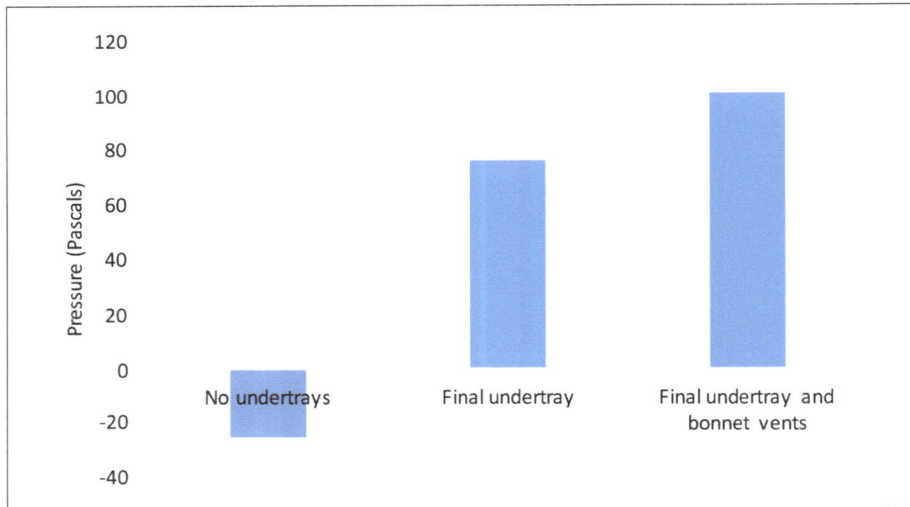

Figure 8-15: Compared with having no undertrays, the final modifications increased the pressure difference across the intercooler core by a factor of five – from -25Pa to +100Pa (-0.1 to +0.4 inches of water). The test speed was 80km/h (50mph).

the long, slightly curved hole in the bonnet in which they sat.

Results

The previous best pressure differential I'd achieved across the underbonnet intercooler with the new undertray installed was 75Pa (0.3 inches of water) at 80km/h (50mph). But with the new vents installed, this rose to 100Pa (0.4 inches of water). Therefore, over having no undertray at all, the fitting of the new undertray and bonnet vent lifted the pressure difference across the intercooler from -25Pa (-0.1 inches of water) to +100Pa (+0.4 inches of water). This is shown in Figure 8-15. Or to put it another way, the pressure difference across the intercooler was increased by a factor of five! That is an excellent result.

The step-by-step modification of the Maxima to greatly improve its intercooler flow was achieved at low cost, and relatively easily. Note also that if the appearance of the car needed to remain unchanged, much of the final benefit could have been achieved by confining the changes to those carried out under the car.

TESTING A PEUGEOT 405 DIESEL INTERCOOLER

When I owned a Peugeot 405 SRDT turbo diesel, I decided to measure the likely outside airflow through the intercooler. This car used a duct integrated into the underbonnet sound insulator to feed the intercooler. The intercooler was positioned towards the rear of the engine bay. The location from which the duct picks up air wasn't all that clear, but it appeared to gather air from the very small gap between the bonnet and its locking platform. However, as indicated above, it's the pressure *differential* that matters, so even if the feed duct was poor, perhaps the engine bay was optimised to create a low pressure

IMPROVING AIRFLOW THROUGH HEAT EXCHANGERS

Figure 8-17: This Peugeot 405 diesel picks up air for its underbonnet intercooler from the tiny gap between the bonnet and its locking panel. Measurements showed that the pressure differential across the intercooler was just 0 to 25Pa (0-0.1 inches of water) – very poor.

below the intercooler? Some measurement would find out.

The pressure probes were placed in the middle of the top and bottom faces of the intercooler. At a test speed of 100km/h (62mph), there was a pressure on the top surface of the intercooler of +100 to +125Pa (+0.4 to +0.5 inches of water). (The fluctuation was caused by wind gusts and the presence of other vehicles.) Under the intercooler, at the same speed and in the same conditions, there was a pressure of +100Pa (+0.4 inches of water). (The under-bonnet pressure fluctuated less as wind gusts and the presence of other vehicles had less impact.)

Those figures mean the pressure differential across the intercooler was just 0 to 25Pa (0 to 0.1 inches of water). Compared, for example, with the Nissan Maxima described above, where modification achieved a final differential pressure four times as great as this *at a lower speed*, the pressure differential across the Peugeot intercooler was very small.

That's especially concerning in a diesel turbo that is on boost a lot of the time – and so where you want high intercooler efficiency, even at low speeds. As another comparison, the pressure difference across the Peugeot's radiator/air conditioner condenser at 100km/h (62mph) was a constant 60Pa (0.25 inches of water).

To put this another way: at speed, the airflow through the underbonnet intercooler was terrible. Therefore, installing a larger intercooler core would have achieved little – in this car, the first step in improving intercooling efficiency would be the creation of a greater pressure on top of the core (eg by an external bonnet scoop) or the reduction in pressure under the core (eg by experimenting with different shaped undertrays, or by fitting bonnet vents).

I think the important lesson here is that before upgrading an intercooler core, you need first to assess the airflow that is likely to be passing through it. A core with very little pressure difference across it will not work effectively as a heat exchanger.

TRYING TO IMPROVE FLOW THROUGH AN AUDI S4 INTERCOOLER

When I owned an Audi S4 I experimented – unsuccessfully – with improving the flow through the side-mounted intercooler. The 5-cylinder turbo used a small intercooler placed to one side of the engine, in the forward portion of the wheelarch. It was fed air by a dedicated forward-facing duct that was sealed to the intercooler core. However, the rear of the core appeared to have no exit at all – the plastic lining inside the wheelarch completely covered the rear of the intercooler!

The pressure in front of the intercooler was measured at +350Pa (+1.4 inches of water) at 120km/h

Figure 8-18: The Audi S4 used an intercooler positioned ahead of the right-hand front wheel. The air exit for the intercooler was poor, resulting in a major pressure build-up behind the core.
(Courtesy Audi)

SPEEDPRO SERIES

(75mph). When the probe was placed behind the intercooler (ie in the space between the intercooler core and the wheelarch lining), a pressure of +250Pa (+1 inch of water) was measured at the same speed. Therefore, the pressure differential across the intercooler was 100Pa (0.4 inches of water). (Remember that this measurement is at much higher speed than the measurements we've looked at so far.)

To try to lower the pressure build-up behind the intercooler, a rectangular hole about 125 x 50mm (5 x 2½in) was cut in the wheelarch plastic lining directly behind the intercooler core. A stainless steel, louvred grille was then pop riveted over the opening. (A grille needs to be used, otherwise the core could be damaged by stones or choked by mud.)

Unfortunately, after this had been done, the measured pressure behind the intercooler remained exactly the same at +250Pa (+1 inch of water) at 120km/h (75mph)! There were a number of possible reasons for this:

- The grille being used didn't have sufficient flow capacity.
- The pressure inside the wheelarch was higher than atmospheric.

I then measured the pressure inside the wheelarch – something that I should have done before any of the modifications were undertaken! But this measurement showed zero pressure build-up, so the reason that a positive pressure could still be measured after the intercooler was because the flow capacity of the grille was insufficient. In retrospect, this makes sense – the grille openings were collectively far smaller than the intercooler feed opening at the front of the car.

Figure 8-19: Extensive louvres used behind a heat exchanger mounted ahead of a wheel-housing, XE Jaguar.

So, another reminder that the exit path for the air leaving a heat exchanger is very important to the overall flow that is achieved.

HONDA INSIGHT – ALTERNATOR COOLING DUCT

This chapter is on improving the flow through heat exchangers, but I'd like to cover another modification which, while it doesn't relate to heat exchangers, is very similar in approach.

I installed an alternator on my Honda Insight. (The hybrid car doesn't normally have an alternator as it uses a DC:DC converter on the high voltage battery. However, I am running the car as a non-hybrid, thus the need for the alternator.)

The alternator was mounted at the rear of the engine bay, close to the turbo that I have also installed. However, in this position, I found that the alternator got very hot, and in fact I had one fail due, I believe, to overheating.

The easiest way of addressing this was to fit a small cooling duct. But would the cooling duct actually flow any air? To determine if it would, I measured pressure difference between the location of the proposed mouth of the duct and the alternator location. This measurement found that even at 60km/h (37mph), there was pressure difference of 50Pa (0.2 inches of water). At 100km/h, the pressure difference was 150Pa (0.6 inches of water).

I installed a duct, using a small diameter ex- vacuum cleaner hose and forming a bell-mouth entrance to the duct made from PVC pipe. (To make such a bell-mouth, heat the end of the PVC pipe with a heat gun and then force it over an inverted, suitably sized china bowl. It will then form a nicely-shaped flare.)

Measurement showed a clear reduction in the alternator temperature, and at the time of writing, the alternator has not suffered failure. The same approach

Figure 8-20: An alternator cooling duct. The measured pressure differential between the area where the duct picks up air and the alternator is 50Pa (0.2 inches of water) at 60km/h (37mph). At higher speeds, the differential rises rapidly. Because of these pressure differentials, even a small duct like this flows well.

IMPROVING AIRFLOW THROUGH HEAT EXCHANGERS

can be taken with any cooling ducts, including brake ducts.

TURNING VANES

In some cars, the airflow is required to undertake a sharp bend to reach the heat exchanger. For example, if a bonnet scoop is being used to feed air to an intercooler located in the engine bay, the airflow will have to undertake a right-angle turn to pass from the scoop entrance through the core. Corners of this type can create turbulence and so reduce the actual flow.

The description above matches the intercooler feed system of the Nissan Maxima covered earlier in this chapter, and so I decided to fit turning vanes to improve flow through its intercooler core. But firstly, what actually are turning vanes?

Turning vanes have been used in industry for many years. All wind tunnels, for example, have turning vanes at the corners of the air recirculation ducting. Turning vanes are also used in industrial air-conditioning systems – especially those where space constraints means that right-angled mitre-type joins are used. Some cars use turning vanes in their engine air intake ducts.

Turning vanes are simply curved inserts placed in the bend to help the air get around the corner. Their presence reduces turbulence and so the pressure drop normally experienced is lower. Their need becomes greater as the radius of the corner becomes tighter, and also if there is a change in cross-sectional area that occurs at the same time.

The textbooks discuss in detail the optimal design of turning vanes – including points such as the radius, shape and thickness, angle of attack and how much they should extend upstream and downstream. All that's well and good – but in practice, building the vanes to this degree of

Figure 8-21: Turning vanes used in a large wind tunnel. (Courtesy NASA)

precision is difficult. However, a few rules can be followed.

If multiple vanes are used, more should be positioned towards the inner radius of the bend (which makes sense since that's where you'd expect greatest flow separation). Turning vanes should also be shaped to suit the different centreline radii of the 'corner,' much like lane markings on a freeway where the inside lanes have a tighter radius.

The first step on the Maxima was to measure the existing pressure on the upper surface of the intercooler. This testing showed that showed that at 100km/h (~60mph), a pressure of 125-150Pa (+0.5 to +0.6 inches of water) was present. This was with the scoop connected to the intercooler via a simple foam rubber seal. In this configuration, the post-turbo inlet air temperature in cruise on a 26°C (about 80°F) day was measured at 55°-60°C (130°-140°F).

I decided to make two turning vanes – one positioned at the back of the opening to the intercooler and the other positioned about midway in the opening. 90mm (3½in) PVC plastic stormwater pipe was used to make the vanes. The pipe was first cut longitudinally and then the ends of the vanes were flattened by being

Figure 8-22: Turning vanes used in the intake of a Mercedes V6 engine. (Courtesy Mercedes)

softened with a heat gun and then pressed between two surfaces. A lot of trial and error fitting was needed, and a number of vane iterations were made (no problem since the pipe is so cheap!). The vanes were installed inside the scoop, with the top of the rearmost vane glued to the inside roof of the scoop. (This section of vane was made longer so that it could project flush with the roof.) If you wanted to trial some vanes prior to making final versions, you could

Figure 8-23: Turning vanes made for an underbonnet intercooler fed by a scoop. These vanes were made from PVC pipe cut, heated and flattened appropriately. The measured pressure on the upper surface of the intercooler core increased after they were fitted.

SPEEDPRO SERIES

use cardboard or thin plastic sheet held in place with adhesive tape.

Testing with the added turning vanes showed that the pressure on the upper surface of the intercooler had increased. At 100km/h (62mph) it rose to 150-200Pa (0.6-0.8 inches of water), an average improvement of 27 per cent. The greater pressure *range* noticed at 100km/h (60mph) with the vanes in place is also interesting; watching the gauge, it appeared that the effect of any minor gusts of wind was much better transmitted as pressure build-ups to the intercooler. Talking these two factors together suggests that with the vanes fitted, there were fewer aerodynamic losses in the intake to the intercooler. The intake air temperature in cruise also dropped, decreasing to 50-55°C (122°-131°F), an improvement of about 5°C (10°F).

MORE ON BONNET VENTS AND DUCTS

Note that when selecting engine bay vent and duct locations, some other considerations also apply.

Firstly, vent locations that are orientated north/south and are close to the edge of the vehicle may not show a strong pressure differential at zero yaw conditions (ie with the airflow parallel with the long axis of the car). However, in non-zero yaw conditions (ie there is a crosswind component) there may be quite a strong pressure differential. This is because, as airflow wraps around the curve between the side of the car and the bonnet, a low pressure is generated. Vents in this location will also tend to disperse exit air away from the cabin ventilation entrance duct at the base of the windscreen.

Secondly, if the duct is designed to allow air to enter (rather than exit) the engine bay, the normal need for a higher pressure on the surface of the bonnet (or a projecting scoop) can be overcome by the use of a NACA duct. These ducts, which are sunk below the surface of the bodywork, develop an opposing pair of vortices which produce a strong flow of close-to-body air into the engine bay. Testing carried out by Opel showed that when a NACA duct was installed in even an area of negative pressure, (eg behind the leading edge of the bonnet), cool air from outside was still forced into the engine bay.

THE OVERALL APPROACH

When assessing the flow through heat exchangers, the first step is to measure the pressure differential occurring across the heat exchanger at speed. If the pressure differential is small, little flow will occur. If the pressure differential shows that the pressure on the rear of the heat exchanger is higher than that on the front, reverse flow will occur.

If the pressure on the front face of the heat exchanger is well below the pressure recorded at the stagnation area of the car, examine the way in which the outside air reaches the heat exchanger. Is all air entering the grille (or duct) forced to flow through the heat exchanger, or are there 'escape routes' where the air can bypass the core? Foam rubber strips can be easily used to seal paths that allow air to sneak around the cores. (Spray-paint the foam black and it will be quite unobtrusive.) Is the air pressure on the front face of the heat exchanger distributed evenly across its surface? If it isn't, you might be able to use turning vanes or revise (or install) ducting leading to the core.

If the pressure recorded on the rear face of the core is overly high, examine the air exits. If the air exit is into the engine bay (very common), what can be done to reduce engine bay pressure? As covered above, good approaches include the use of undertrays, spoilers or bonnet vents. Exit vents need to be located where there is a low surface pressure and a high engine bay pressure – normally, as far forward on the bonnet as possible.

If the air exit from the heat exchanger is not into the engine bay, can it be channelled to a low-pressure area? Consider the fact that many buses with rear-mounted engines draw air in from the side of the vehicle (which is not a high-pressure area) but then exhaust it into the wake (which is a low-pressure area), therefore still achieving the required pressure differential across the radiators.

The following table covers some common scenarios.

Figure 8-24: Large foam rubber strips (arrowed) that have been added to prevent the airflow from bypassing the heat exchangers when the bumper bar cover is back on the car. Using foam is good because the shape doesn't have to be a perfect match, as it will deform as required when the bumper is put back in place.

IMPROVING AIRFLOW THROUGH HEAT EXCHANGERS

In front of the core		Behind the core		Flow
High pressure	+	Atmospheric pressure	=	Good Flow
High pressure		Below atmospheric pressure		Excellent flow
Atmospheric pressure		Below atmospheric pressure		Good flow
High pressure		High pressure		Poor flow
Atmospheric pressure		Atmospheric pressure		Poor flow

It's also very important to not make judgements just by looking at the car. Few would have looked at the Nissan Maxima's huge forward-facing scoop and suggested that at speed, air was in fact flowing *out* of the scoop, not into it! The small diameter alternator cooling duct on the Honda Insight looks to the eye like it would be quite ineffective, but pressure measurement showed it would work – and it has.

Finally, note that increasing the airflow through heat exchangers may worsen the drag coefficient of the car. Therefore, if you are also chasing low drag, you may need to consider active devices such as shutters that partially or completely close off the heat exchangers when they are not required to work at full capacity (see Chapter 6).

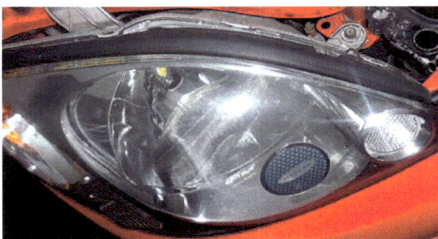

Figure 8-25: Sealing front-facing openings will force more air through heat exchangers, as well as typically lowering drag. Left, above, is the standard headlight. The middle photo shows it after a rubber sealing strip has been applied. On the right, the finished result is neat, unobtrusive and inexpensive. (All pics courtesy Pascal Dunning)

Figure 8-26: This vent on a Cadillac ATS V coupé is sited to take advantage of the low surface pressures found in this area of the car. (Courtesy GM)

Chapter 9
Engine intakes

- **Benefits**
- **Measuring pressures**
- **Measuring engine intakes**
- **Building a new engine intake**

In the previous chapter we discussed techniques that can be used to improve the flow through heat exchangers like radiators and intercoolers. Similar techniques can be used to site engine air intakes at the best position to take advantage of ram airflow and, in many cases, also provide the engine with cool intake air.

BENEFITS

Almost all technical aerodynamics books make the point that ram air pressure isn't very great. For example, calculations show that at a speed of 130km/h (81mph), a maximum of 800Pa (3.2 inches of water, or 0.11 psi) is available. This

Figure 9-1: A good example of a standard high-performance intake. Air enters the main grille (arrowed) and a high pressure is created in front of the radiator. The smaller grille (directly above the arrow) feeds this air to a large duct that flows to the intake air filter. The duct's upper surface is sealed when the bonnet is closed. (Courtesy Ford)

represents an increase over normal atmospheric pressure of only about

ENGINE INTAKES

0.7 per cent. You would therefore think that the advantages of siting an engine intake in an area of high aerodynamic pressure are small.

However, four points need to be made.

First, we're all aware of the flow restriction that occurs across air filters, and how engine power suffers as the flow becomes more restricted through the use of a dirty filter. Well, the pressure drop across a dirty filter is also often very small – for example, only a few inches of water. And yet we are able to be feel and measure a power drop caused by this! If we can use a well-sited intake that takes advantage of ram airflow, it's possible to compensate for this air filter pressure drop – that is, to actually have a positive pressure in the airbox following the filter at part-throttle, and to have zero pressure drop immediately after the filter at full-throttle.

The second point is that manufacturers strive to locate engine intakes in areas of high aerodynamic pressure. Many high-performance motorcycles have internally streamlined ducts that join front-facing ducts with the airbox; and many high-performance cars have equally elaborate ducts that (sometimes expensively) pick up air from a high-pressure location. If it achieved so little, you may wonder why manufacturers would bother.

Third, in over 30 years of modifying cars, I have often been able to feel (and measure) the difference achieved by using a new intake system prior to the airbox. Those changes may well be a combination of things: less pressure drop caused by better designed ducting; less pressure drop caused by taking advantage of aerodynamic pressure to offset flow losses (eg across the filter); and the inhalation of cooler air that is more dense, and so can increase engine power output.

Finally, because modifying intake systems can be done so cheaply, in terms of cost per unit of power, intake modification usually gives you the best bang for your buck of any power-producing modifications you can make.

To be clear, I am not suggesting that siting an engine intake duct in an area of a high aerodynamic pressure will suddenly turn your car into a fire-breathing powerhouse, but I do suggest that it is worthwhile taking advantage of this pressure to help offset inevitable losses in the intake.

MEASURING PRESSURES

Measuring air pressures on a moving car to find the best location for an engine intake is done in a similar way to pressure testing covered in Chapters 4 and 8. Because the pressures are (relatively) large and absolute accuracy is not needed, the use of a reference reservoir for one side of the Magnehelic gauge is not required. (But of course, if you want to be as accurate as possible, you could use one.) Place one end of a hose at the location you're interested in, run the other end of the hose back to the positive port on the gauge, and then drive the car and read the gauge.

I carried out testing on a 2003 Lexus RX330. A Magnehelic 0-1 inch of water gauge was used and a road speed of 60km/h (37mph) worked well. The following table shows the measured pressures, with the data presented in order from the highest measured pressure to the lowest.

Probe location	Measured pressure	
	Pascals	Inches of water
Middle of front numberplate	+137	+ 0.55
Front tow-hook blanking plate	+100	+ 0.40
Leading edge of front undertray	+75	+ 0.30
Middle of headlight	+50	+ 0.20
Middle of Lexus badge in grille	+50	+ 0.20
Base of windscreen	+25	+ 0.10
Below front foglight on bumper	+12 to +25	+ 0.05 to + 0.1
Front wheelarch	-25	- 0.10
Outer edge of headlight	-112	- 0.45
Top of windscreen	-150	- 0.60

Figure 9-2: Pressures were measured across the front of this Lexus RX330. Taking such an approach allows you to find a good location for an engine intake duct. (Courtesy Lexus)

SPEEDPRO SERIES

It can be seen that the numberplate is in the stagnation zone, with a much higher pressure than the other measurements. However, it's rather hard to site an engine intake duct in the middle of the numberplate – but there's also relatively high positive pressure at the leading edge of the front undertray, a much easier place to locate the mouth of the duct.

But what about at the base of the windscreen? Since time immemorial, the intake to the cabin ventilation system has been sited here, because this is generally a high-pressure area. In modified cars it's also common to place the mouth of a cold air intake through into this plenum volume. But these measurements show that on the RX330, the available pressure – while still being positive – is much lower than can be obtained across the front of the car. I've also seen air intakes placed through to the wheelarch area, but on the RX330, the pressure in the wheelarch is actually *lower* than atmospheric – something I have found to be the case in many car measurements.

Once the location has been found that looks good for an engine intake, testing can be carried out at a variety of speeds. This step is needed in case the pressure distribution across the front of the car changes as other aerodynamic effects occur. In the case of the RX330, the 0-1 inch of water gauge was too sensitive to measure the pressure build-up at a variety of speeds, so a 0-150 inches of water Magnehelic gauge was used instead. The chosen measuring point was near to the tow-hook blanking plate on the front bumper. The measurement showed that, as expected, pressure rose exponentially with speed – from 100Pa at 60km/h, to 750Pa at 120km/h, to 2500Pa at 150km/h. Figure 9-3 shows these measurements.

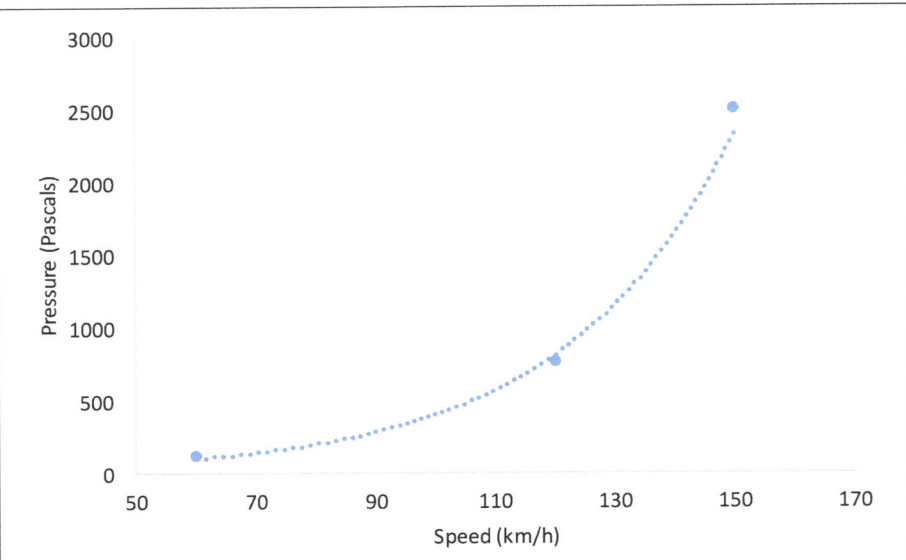

Figure 9-3: Rise in measured pressure with speed on the front-facing tow-hook blanking plate on the front bumper, Lexus RX330.

MEASURING ENGINE INTAKES

The above section looked at the different pressures that can be quickly and easily measured across the front of a car. But what about going to the next step, and making a new intake? Well, there's actually something we can do before that. When working on a car, it makes sense to first measure how well the intake system is working in standard form. And how do we do this? Again, with a trusty Magnehelic gauge.

The intake system of a typical car comprises an intake snorkel that leads to an airbox, inside which is positioned the filter. If we drill a small hole in the airbox on the engine side of the filter, and then screw in a small, self-tapping plastic fitting (eg a connector sold for small diameter plastic irrigation piping), we can use plastic tube to connect

Figure 9-4: A tapping made in the airbox to allow pressure drop measurement to occur. The greater the flow restriction, the lower the pressure recorded at this point.

ENGINE INTAKES

this fitting to the low-pressure port on a Magnehelic gauge. (When testing is finished, just unscrew the plastic fitting, and put a dab of black silicone over the hole – it will then be nearly invisible.)

With the gauge connected to the airbox in this way, consider what happens if we load the engine. If there is no flow restriction between the outside air and the engine side of the airbox, the gauge will stay on zero. However, any restriction will cause a pressure drop, and will be seen on the gauge. The more restrictive the intake, the greater will be the measured pressure drop.

(This is just like the vacuum reading you can make with a gauge connected to the intake manifold at idle. In that case, the pressure drop is caused by the nearly-closed throttle body. Open the throttle and the amount of vacuum decreases.)

Over the years, I have measured the intake systems of many standard cars. Measured at full-throttle and maximum engine speed in second gear, those readings have varied from 1000Pa (4 inches of water) to an astonishing 7700Pa (31 inches of water). Incidentally, that last measurement was the first I ever did in this way, and in that case, I was using a simple vertical water manometer I'd built. With a measured pressure drop this large, I could barely fit the manometer in the cabin of the car. Just as well the pressure drop wasn't any higher, or I'd have needed a car with a sunroof!

You can see that the measurements described here are likely to require a higher scale gauge than the Magnehelic gauge we used to measure low-speed aerodynamic pressures. So, to do these measurements, you can either buy another gauge specifically for intake system measurements (and it will also be useful for high speed aerodynamic measurement), buy a digital manometer, or make a simple water manometer as described in Chapter 4.

The first step is to measure the performance of the standard intake. If it is already very good (eg less than 2500Pa – 10 inches of water pressure drop at full power), consider carefully whether it is worthwhile making changes. My experience is that the law of diminishing returns applies quite strongly here, and with such a good system to start with, you may put in a lot of effort and achieve relatively little. (But that hasn't stopped me doing just that!). On the other hand, if the pressure drop of the standard system is two or even three times that figure, you can be confident of achieving gains.

Let's start by looking at a system and then making improvements. Note that in this particular case, the mouth of the intake was already located in an area of high aerodynamic pressure, and so was not moved. However, I still wish to cover the modification, because it shows how you can go about making a new intake duct.

BUILDING A NEW ENGINE INTAKE

The car in question was a 2008 Skoda Roomster, fitted with a 1.9-litre diesel engine. Most diesel engines don't regulate airflow with a throttle, and so typically have a much higher average airflow than an equivalently-sized petrol engine.

Figure 9-5: The standard intake system of the Skoda Roomster (arrowed).

Figure 9-6: The mouth of the intake system of the Skoda is already located in an area of high aerodynamic pressure, so no changes in this aspect were needed.

SPEEDPRO SERIES

In fact, off-boost, the diesel always flows the airflow equivalent of a same-size naturally-aspirated petrol engine *operating at full-throttle!* And, with a turbo diesel being on boost for so much of the time, the actual airflows can easily be more than 50 per cent higher than indicated by the swept volume. In short, for maximum power and minimum fuel consumption, allowing a diesel to freely breathe is not just important, but because of the huge volumes of air going in and out of the engine, is also much harder to achieve than in a petrol engine.

In standard form, the Roomster's intake was a mix of good and bad. Starting at the beginning of the intake system, air was drawn-in through an opening in the panel above the radiator. (As stated, the intake was sited at a position of high aerodynamic pressure.) The rectangular opening was also very large – about 150 x 50mm, or 75cm^2 (11.6in^2). Collecting air from this large opening was a rectangular plastic moulding that had an inlet area of about the same cross-sectional area as the panel hole – again, excellent. However, for some odd reason, the exit to this plastic moulding was, in relative terms, tiny – just 15cm^2 (2.3in^2)! That's right, the intake necked-down by no less than 80 per cent! Connecting to this tiny inlet was a flexible tube that carried the air to the airbox. This started with a cross-sectional area of about 30cm^2 (4½ in^2) and joined the airbox with about 42cm^2 (6½ in^2) cross-sectional area. However, the actual opening in the airbox wall was much larger again – about 50cm^2 (7¾in^2).

So where did all that leave us? In short, the opening in the panel above the radiator (ie the mouth of the system) was very generously sized, and the opening in the wall of the airbox was also good. But between those extremes, all sorts of restrictions to flow were present.

And what did on-road measurement show? The testing was done at full throttle in second

Figure 9-8: This tiny entrance to the intake duct was able to be more than doubled in size.

gear, with the data measured at 500rpm intervals. The pressure-drop peaked at just under 2500Pa (14 inches of water) at 3500rpm. The full measurements are shown in Figure 9-7. Note that, at maximum speed in second gear (80km/h, 50mph), the measured pressure at the intake mouth's location was 175Pa (0.8 inches of water). Without this positive aerodynamic pressure, the pressure measured in the airbox would have been even lower!

The first step in improving the intake was to cut off the rear of the first section of the intake duct, so that the necking-down no longer occurred. To make the new intake duct, I used 80mm (just over 3in) PVC plastic pipe and fittings. This approach has a number of advantages: different fittings are readily available (eg pre-formed bends) and the inner surface of the pipe has a smooth surface. The pipe can also be heated and then shaped and formed. You would not use this pipe in close proximity to an exhaust manifold or turbo, but situated in cool areas of the engine bay (and with intake air constantly passing through it) the pipe will survive without problems.

The first task proved to be the most difficult – making a

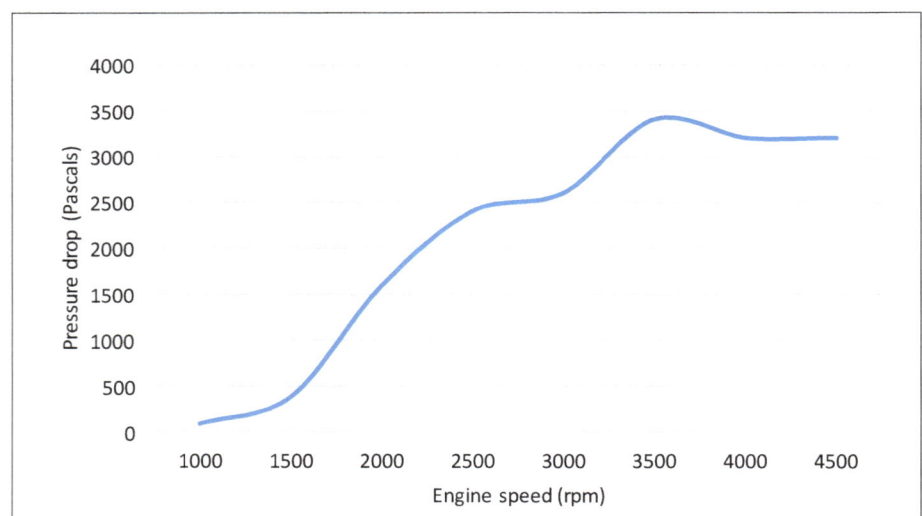

Figure 9-7: The pressure drop of the standard intake system on a Skoda Roomster diesel. Note that this was with the intake already picking up air from an aerodynamic high pressure area; without this, the pressure drop in the airbox would have been even greater!

ENGINE INTAKES

Figure 9-9: The new plastic pipe being shaped to match the enlarged duct entrance piece. PVC pipe can be easily bent and stretched once heated.

Figure 9-10: The opening to the airbox was also enlarged.

Figure 9-11: Part of the new intake duct. Note the depressions that have been made in the pipe to give clearance to other parts in the engine bay.

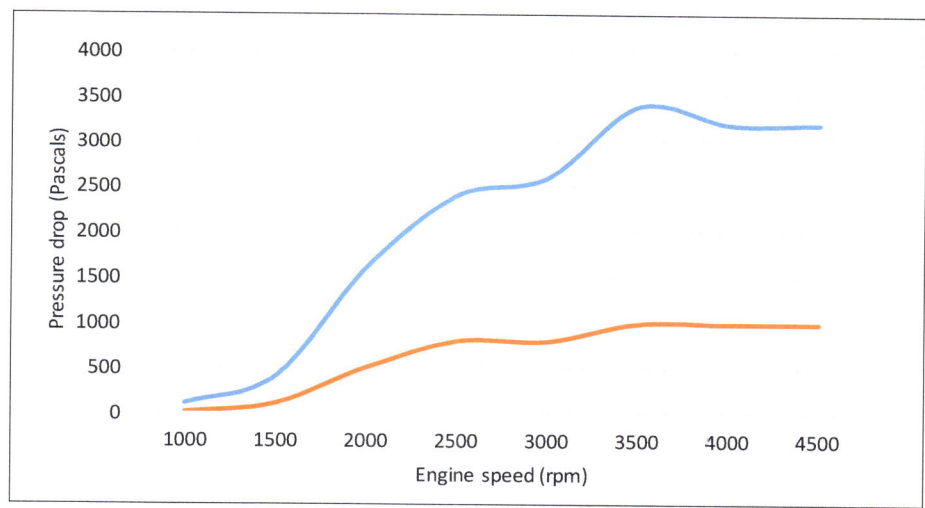

Figure 9-12: The pressure drops of the standard versus modified systems. The restriction to intake flow has been greatly reduced.

transition piece that would go from the rectangular mouth duct to the round 80mm diameter tube. Unfortunately, no 'gutter type' adaptors available off the shelf had the right shape or smooth internal contours. In the past I've made lots of adaptors like this by heating the pipe (using a heat gun) until it is soft and then moulding it into shape, either by gloved hands or with hand tools. However, even that proved difficult in this case, and so I made a wooden block that replicated the dimensions I needed the tube to take. I then cut the tube off at a slight angle (so increasing the open area), heated the end of the tube and forced the shaped wooden block into the mouth. This approach formed the correctly-shaped opening, but I found it very difficult to mould the piece without wrinkles occurring in the walls. However, after lots of trial and error, I developed a technique that got rid of the wrinkles. I found a wrinkle-free shape could be formed if the pipe was heated and stretched after it had been shaped. I used my engine crane to do the stretching, but the required tension isn't that great – so a rope over a pulley could probably also be used.

Next I tackled the airbox. The standard intake was cut off with a hacksaw and the resulting hole was enlarged. A standard 90mm (3½in) right-angle plumbing adaptor could then be placed through the hole in the airbox. To seal the gap between the bend and the airbox, I used a flexible neoprene collar. The use of the neoprene collar on the bend allowed it to be rotated when the duct was being fitted, but still seal when the duct was in place. (The bend needed to be rotated because the duct had to be assembled *in situ* – it couldn't be fitted into place if glued together outside of the engine bay.)

With each end of the system done, the connecting pipe could then be shaped. This required various indents to provide clearances; these were achieved by heating the pipe until it was soft, and then using gloved hands to wriggle it into the required position – the wriggling caused engine bay items to dent the heat-softened pipe and so give the required clearances. One side of the pipe was done at a time in this way.

The duct complete, pressure drop testing was again carried out. Figure 9-12 shows the measured

SPEEDPRO SERIES

Figure 9-13: Far left, the new intake duct in place for testing. Unpainted it stands out, but when it has been painted, as on the right, it's nearly invisible. Measured intake pressure drop decreased by as much as 69 per cent.

results, with the blue line showing the standard pressure drop, and the brown line showing the pressure drop achieved with the new intake. As can be seen, there is a massive reduction in the restriction. Whereas in standard form, the airflow restriction peaked at just under 3200Pa (14 inches of water), with the modified duct, the maximum was just 1000Pa (4 inches of water)! This was a decrease by as much as 69 per cent. You can also see that an improvement has been achieved throughout the rev range – although as the amount of airflow increases, so does the percentage improvement.

This sort of duct, fabricated from PVC pipe and using a mix of bends, custom shapes and standard parts, shows an effective approach to making a new pre-airbox intake system.

EXAMPLE NEW INTAKE – HONDA INSIGHT

Above I described measuring the pressure drops through standard intakes, and said that the best standard car I have ever measured had a pressure drop of only 1000Pa (4 inches of water). That car was my Honda Insight. So why bother trying to improve it?

Well, I was initially prompted to make the measurement when someone on a Web discussion group said that the Insight's performance could be noticeably improved by removing the airbox intake snorkel. This implies, of course, that the snorkel is restrictive to flow. The Honda Insight is an efficient car, so it seemed very unlikely that Honda's engineers had made such a mistake.

To see if this were the case, I whipped out a Magnehelic gauge and went for an around-the-block test drive. And when I saw that 4 inches of water figure, I realised that pulling off the snorkel would achieve nothing. (Why? Well, even if the pressure drop decreased to zero, the engine would be breathing hot intake air from directly behind the radiator – not good for performance. In this case, the potential gain of reducing the restriction would not be enough to make up for the hot air.)

Time passed, then one day I thought of an improvement: why not connect the end of the snorkel to a location of high aerodynamic pressure? (The mouth of the standard intake is positioned within the engine bay.) Inspection showed that if a plastic blanking plate was removed and a right-angled tube added to the original snorkel, the mouth of the tube would be pressurised by the forward motion of the car. However, would this upset the very finely optimised intake? I added a short length of snorkel salvaged from a Nissan Skyline and then used foam rubber to fill the gap around the opening. Foam rubber was also used to seal the gap between the original and new snorkels. It wasn't pretty, but it would be an interesting test.

The test I had in mind was fuel economy – would the open-road cruise economy improve? However,

ENGINE INTAKES

Figure 9-14: The standard intake of the Honda Insight is located within the engine bay. A very efficient intake system, it has a maximum measured pressure drop of only 1000 Pascals (4 inches of water).

ascertaining this was very difficult – when you get extremely good open-road fuel economy, you need to make a big improvement to see a measurable change! However, my *feeling* was that, yes, with the trial snorkel fitted, there was a slight improvement.

The next step was to make a proper intake. After trialling some different approaches, I settled on truck radiator hose that had the right shape bends. This original end of the Honda's intake snorkel was cut off, and the end heated and shaped until it was round (it's usually an odd semi-rectangular shape). A flared bell-mouth was made from the middle of a plastic cake dish, and foam rubber used to seal the gaps so that all of the air entering the grille had to either pass through the radiator or go into the

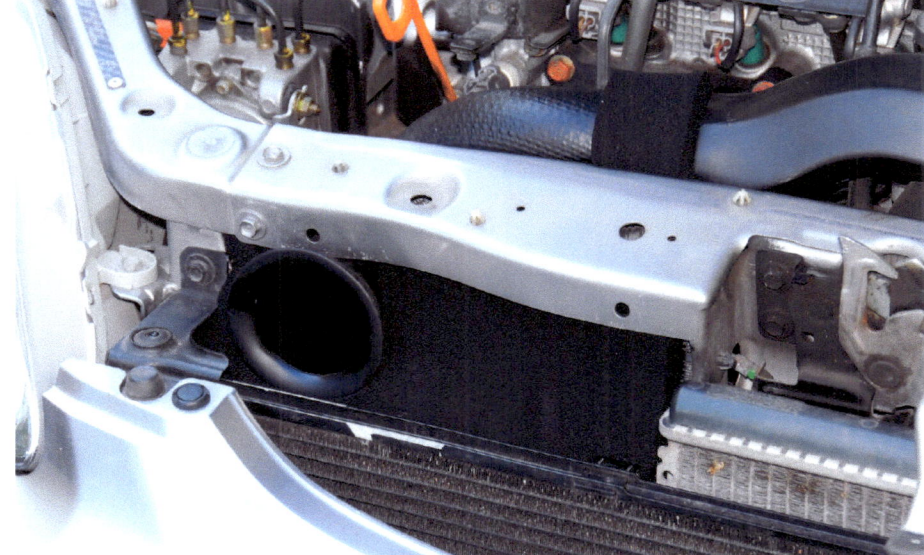

Figure 9-15: The new Honda intake picks up air from the high-pressure area in front of the radiator. In cruise conditions, the new intake system poses less than zero restriction. The flared bellmouth was made from the centre of a plastic cake dish; a subwoofer port could also be used.

231

SPEEDPRO SERIES

engine through the snorkel. Contact adhesive was used to hold together the bits.

And the test results? After the new snorkel had been installed, the maximum recorded pressure drop of the complete intake system – snorkel, airbox and filter – halved to just 500Pa (2 inches of water)! Even better, in any constant throttle cruise over 40km/h (25mph), there was, in fact, a positive pressure on the throttle side of the airfilter. This pressure was typically about 125Pa (0.5 inches of water). To put this another way, in cruise conditions, the modified intake system was posing *less than zero* restriction, and even at maximum flow, the throttle was seeing 99.5 per cent of atmospheric pressure.

Downsides? Or, why didn't Honda do it this way? In very dusty conditions, the filter would need to be changed more often, and there was a just-audible induction note that could be heard at full throttle with the windows down.

OTHER EXAMPLES

Over the years I have modified many car intake systems. The photo on the right shows an Audi S4, for which I made a large duct that connected the airbox to the standard air-feed duct for the oil cooler. A positive pressure of 750Pa (3 inches of water) was measured in this duct at 100km/h (62mph). That, and a larger diameter intake, was enough to reduce the measured pressure drop on the duct side of the engine air filter from a maximum of 2200Pa (9 inches of water) to zero at full load in second gear.

On the left can be seen the modified intake to a Ford Falcon six-cylinder. The arrow points to the added forward-facing intake that feeds air out through the semi-circular opening seen directly above. Here it can in turn feed the over-radiator duct that connects to the airbox (the two being connected when the bonnet is down). Note that the mouth of the duct doesn't need to face forward; anywhere in front of the radiator will have high pressure.

www.velocebooks.com / www.veloce.co.uk
Details of all current books • New book news • Special offers • Gift vouchers • Forum

Chapter 10
Reducing aerodynamic noise

- ♦ **Causes of aerodynamic noise**
- ♦ **Testing and measurement**
- ♦ **Decreasing noise production**
- ♦ **Decreasing noise transmission**
- ♦ **Damping versus isolating**
- ♦ **Modifications to reduce noise**

Figure 10-1: Sophisticated noise measurement in a wind tunnel using an external parabolic microphone in addition to an in-car dummy equipped with ear microphones. (Courtesy Mercedes)

One area where cars have made enormous advances is in the reduction of in-cabin noise. Current cars are typically very quiet, with tyre, engine and aerodynamic noise all well-suppressed. But what if you have an older car, and you'd like to quieten it? In that case, you'll need to ascertain the sources of aerodynamic noise and then make changes that can reduce it.

CAUSES OF AERODYNAMIC NOISE

As described in Chapter 1, most of the attached flow on a car comprises a turbulent boundary layer. This turbulence excites panels – it causes them to vibrate. The second cause of aerodynamic noise

233

occurs where the flow separates and then reattaches, with these pressure fluctuations creating noise. This type of noise can be associated with external rear vision mirrors, and the A-pillars. (In addition, noise-generating vortices are often formed at the A and D pillars.) Lastly, cavities formed by body openings (eg at the doors) can both generate noise and allow it to be transmitted to the interior.

Once I was lucky enough to sit in a car positioned in a wind tunnel, as the huge fan was brought up to speed. I remember two aspects most clearly: the fact that the dashboard vents started to flow air, and the deep throbbing roar from the rear of the car as the airflow separated, and the wake and shedding vortices excited the rear panels. And that was in a Lexus, a brand known for its very quiet cars!

SAE paper 900315 (*Automobile Aerodynamic Noise*) takes an interesting approach to describing noise generation. It looks at typical automotive aerodynamic velocities and compares them to the scale lengths that apply. For example, when considering the whole car, a typical length might be 5 metres and a typical speed, 30m/s. The frequencies associated with large scale flow separation are therefore about 6Hz – too deep for us to hear, but associated with vibration. However, if we consider separation and turbulence on a smaller scale associated with body details ranging from 10-100mm in size, the frequencies are in the range from 300 to 3000Hz (from a deep note to a higher squeal).

TESTING AND MEASUREMENT

The simplest way of assessing the contribution made by aerodynamic noise to the interior noise level in the car is to make changes and then see their effect. For example, drive along at 100km/h (~60mph) and place the car in neutral. The engine noise will decrease, allowing you to assess the contribution of other noise sources. Tyre noise will not change, so do this test on the smoothest road surface you can find. When doing this test, you can often detect aerodynamic noise that was previously contributing to overall noise levels, but which was not readily identifiable.

Taping the joints and gaps in the exterior body (eg around the doors, bonnet and boot) and then testing the car will show the influence that pressure variations in these cavities is having on noise production and transmission. (The tape stops only airflow; it doesn't have enough thickness to reduce noise transmission, so any change is caused by alterations to oscillating air pressures.) Placing a temporary undertray under the car (eg made from plastic sheet) will show you how much of a contribution the underfloor turbulence is making to noise levels.

In SAE 2009-01-0180 (*Scaling Laws in Automotive Aeroacoustics*) the authors tested noise levels while making various changes to the vehicle in a wind tunnel. Taping all external body gaps caused a significant decrease in noise at frequencies over about 300Hz; using a large block of foam at the front of the car to prevent undercar airflow decreased noise levels (again by a significant amount) in the 90-300Hz range; and then taping thick cardboard over the glass (spaced from the glass by 50mm, and so replicating a form of double glazing) decreased noise at frequencies over 300Hz. The result of all the changes was a decrease in noise levels over the baseline by 10db(A). While obviously you cannot drive around with a large foam block stopping under-car airflow, nor with cardboard taped over the windows, the pattern in the measurements is clear. Deeper noise frequencies, at least with a rough underbody, are associated with airflow beneath the car.

To ascertain where noise is coming from, a length of flexible plastic tube can be used as a stethoscope. With someone else driving the car, place one end of the tube at your ear and move the free end around inside the cabin. For example, noise penetrating through door seals, noise from the A-pillars, and noise from rear vision mirrors passing through the window glass can all be identified in this way.

Measuring noise levels is complex and difficult, and I am not sure that it's any better than simply using your ears. This is the case for a few reasons. First, noise is a perceptual variable: a particularly annoying noise to you might be my gentle rustle. Second, because it depends on the frequencies in question, and in turn the weighting curve that is used, quite noticeable changes in noise levels may be unreadable on a normal sound level meter. (The 'A' weighting curve is meant to replicate the sensitivity of the ear to different frequencies, but there is a lot of debate as to whether it does so – especially with louder sounds.) I've used sound level meters to measure everything from exhaust noise to interior cabin levels – and in every case, the measurements have never matched my perceptions.

For measuring the spectrum of the noise (ie the frequency distribution), again your ears are probably your best tool. That said, apps exist for smartphones (eg the 'Vibration' app for iPhones) that can perform a spectrum analysis

REDUCING AERODYNAMIC NOISE

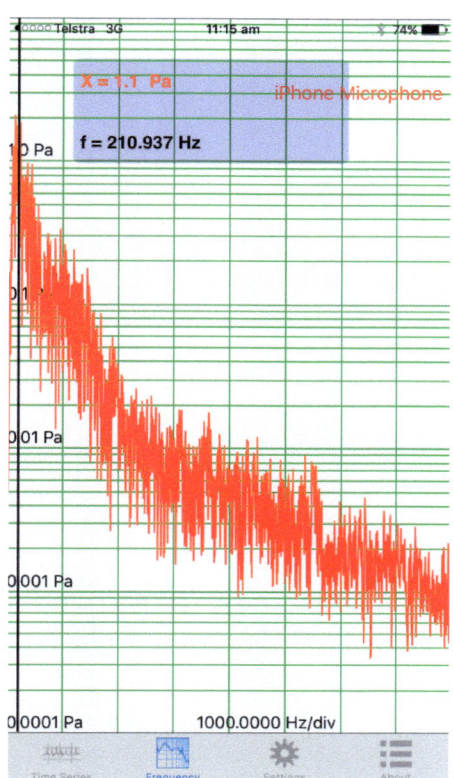

Figure 10-2: The iPhone 'Vibration' app can perform a spectrum analysis on in-cabin sound, showing you the frequencies that are loudest. However, your ears can perform this function, too!

of noise. I did some testing of this app in my fairly aerodynamically quiet E500 Mercedes (see Figure 10-2), and this showed that deep frequencies dominated. However, that was probably the road noise from tyres, I think. Slightly lowering a window changed the frequency distribution – but again the instrument's analysis didn't tell me anything that my ears weren't already telling me!

DECREASING NOISE PRODUCTION

As we've seen, much aerodynamic noise is created through pressure variations on panels. The panel vibrates through being excited by vortices or pressure variations occurring at high frequency. Therefore, if we can reduce panel vibration, noise levels will decrease.

Panel vibration can be reduced by adding damping material, or by increasing panel weight – or a combination of both approaches. Damping material reduces the propensity of the panel to ring – when tapped, it will sound a dull thump rather than a 'boing'. Increasing the panel mass changes the resonant frequency of the panel – that's the sound frequency at which the panel will most readily vibrate.

Both effects can be achieved by the addition of heavy, viscous sheet materials like bituminous compounds, butyl and mass-loaded vinyl. Note that this material must be effectively stuck to the metal panel across all its area, and must not come loose. If the attachment to the panel is not sufficient, the panel will be able to vibrate without flexing the damping material – so nothing will have changed. Large panels with externally attached turbulent boundary flow (eg the roof) can be treated in this way.

Heavy bituminous spray-on compounds are also available. These are easier to apply where access is tight and/or the panel is shaped in a way that makes adhering sheet material difficult. When using sound deadening sprays, build up the thickness by using multiple layers, rather than attempting to spray it on all at once. Mask areas from overspray, and give it plenty of time (ie days) to harden. Commercial paint-on sound deadening coatings are available – these are used in trains and ships. I haven't used this material in a car, but if you have a car stripped to a bare shell, I'd look at using professional coatings of this sort.

Figure 10-3: Heavy mass-loaded vinyl like this is stuck to vibrating panels to change their resonant frequency. To be effective, the material must strongly adhere to the panel over its full surface.

Flow separation and reattachment – for example, under the car – can be addressed using the techniques that we've covered in the Chapter 6 on reducing drag. For example, a smooth undertray will reduce undercar noise created by the turbulence. On a three-box sedan, achieving flow reattachment on the rear window and boot lid will reduce wake noise.

Exterior mirrors can present unique problems. For example, SAE 2007-01-1549 (*Laminar Flow Whistle on a Vehicle Side Mirror*) describes a phenomenon where the laminar boundary layer separation point oscillates rapidly, causing high-pitched whistles. Quite subtle changes in mirror shape and housing texture were used to prevent these whistles occurring. Modifications to the original design included shape changes of only a few millimetres, graining the plastic of the housing (or using sandpaper or trip wires on the surface), and filling small depressions. Whistles of this sort are unlikely to be found in production car mirrors, but may occur if you are adapting a mirror from a completely different type of vehicle, eg a motorcycle. Aerials, that can shed vortices at a periodic frequency

SPEEDPRO SERIES

and so also cause a whistle, can be treated by changing their cross-sectional shape – experimentation is the key here.

DECREASING NOISE TRANSMISSION

There are two ways in which noise can be transmitted to the inside of the car's cabin.

The first is air-borne noise. This type of noise is transmitted to the interior of the cabin by air movement. For example, a hole in the engine firewall will allow the entry of airborne noise. If you are stuck in traffic in a city and wind down the window, the noise level inside the cabin will increase. The change in noise level is caused by the transmission of airborne noise. Note that the window has to be down only a tiny amount for the noise level to dramatically increase – airborne noise travels through even a small opening.

The second type of noise transmission is that transmitted through the car's metalwork – structure-borne noise. It's rather like when you put your ear on a railway line. You can hear the train coming long before you can hear it through the air – the noise is being transmitted along the steel railway line. In current cars, structure-borne noise is more common than airborne noise.

Airborne noise that is transmitted through holes and gaps is best addressed by closing those openings. For example, the openings for hoses and cables that pass through the firewall should be completely sealed. Openings in the floor of the boot or hatch space (for example where there are missing grommets or rubber blanks) should be resealed. All door and boot seals should be in good condition. If door seals leak and yet they are in good condition, add extra seals. (Most current cars that are aerodynamically very quiet have two or even three sets of seals on the doors, for example.) An old trick in cars where the door window frames suck out at high speed is to lower the window and then carefully bend the doorframe inwards a little. When shut, the door will then push more firmly on the upper seals.

One approach to reducing structure-borne noise transmission is to let the structure vibrate as it did previously, but to separate the cabin space from that panel vibration. In other words, to place an effective noise barrier between the panel and the cabin. For example, if the floor panels are being excited by undercar airflow, and you don't wish to fit a large undertray, you can place a noise barrier on the floor. A good noise barrier comprises a sandwich of mass-filled vinyl between two layers of foam 'rubber'. The heavy vinyl blocks the sound wave transmission but can still vibrate within the foam rubber layers

Figure 10-4: Any opening into the cabin can be the source of air-borne noise. Whenever you pass cables through into the interior, use a grommet that seals well, and if there are any holes, fill them with blanking plugs.

DAMPING VERSUS ISOLATING

As described above, there are two different ways to reduce noise. However, since both approaches use layered material that is placed on panels, they're often confused.

Damping materials reduces noise *at the source*. They decrease the ability of the panel to vibrate when excited. On the other hand, isolation materials prevent the noise *being transmitted*. That is, the panel is still vibrating but you can no longer hear it.

The most frequent mistake that people make is to use damping layers when they should be using isolating materials. For example, using damping barrier layers on the inside of door trims is to waste the material – the door trim is unlikely to be creating sound through high frequency vibration. In this case, you would attach a noise barrier to the inside of the door trim.

REDUCING AERODYNAMIC NOISE

– the barrier is 'decoupled' from the adjoining surfaces by the foam so that vibration is not transmitted.

Figure 10-5: This is an isolating type of noise barrier. The heavy vinyl layer (arrowed) blocks the sound wave transmission but can still vibrate within the foam layers.

MODIFICATIONS TO REDUCE NOISE

Modifications to specifically reduce aerodynamic noise can vary from being quick and simple to being major and complex! Covered here are two modifications at the 'simpler' end of the range.

Prius windscreen wipers

When not being used, the best position for windscreen wipers is below the trailing edge of the bonnet – that way, they're completely shielded from the airflow so are quiet, and also don't contribute to aerodynamic drag. However, taking this design approach requires a separate 'parking' position for the wipers, which adds cost to the mechanism. The result is that many cars use a design that leaves their wipers exposed to the airflow all the time, creating noise and drag.

In some cases, the shape of the plenum area ahead of the windscreen, or the trailing edge of the bonnet, is used to direct air over the wipers, leaving them in a separation bubble. In other words, the airflow separates from the surface of the car just enough to pass over the wipers, reattaching on the windscreen a little above the wiper line. This results in much reduced aerodynamic noise from the wipers.

If your car (a) doesn't run wipers with a separate below-the-bonnet park position, and (b) doesn't appear to do anything to direct airflow over the wipers, then you might want to add a deflector that will reduce aerodynamic noise.

In this case, the car was a 1999 Toyota Prius. Despite having an excellent (for the time) C_D of 0.29, little attention seemed to have been paid to the wipers. On this car they didn't park beneath the level of the bonnet, and, at speed, aerodynamic noise could be clearly heard from their arms.

To see what the air was doing around this part of the car, the area around the wipers was wool tufted and the car was driven at 60km/h (37mph). This testing showed that there was attached flow up the bonnet, and then the full flow of air meeting the wipers – the attachment line on the windscreen was actually behind the wipers!

A 'quick and dirty' prototype wind deflector was made from an old plastic sign and duct tape. This fitted under the trailing edge of the bonnet (hood) and directed air up and over the wipers. The trial deflector was left on the car for a week while several hundred kilometres of freeway driving was undertaken. Two effects were noticeable: wind noise from the windscreen wipers could no longer be heard, and the cabin ventilation system tended to breathe hotter air. The latter point is important: most cars pick up their cabin ventilation air from directly in front of the windscreen and so any change in the aerodynamic pressure at this point can affect ventilation.

The final deflector was constructed from 4mm ABS sheet, bought from a plastics wholesaler. This plastic is tough but can be bent (if heated first, the bend is

Figure 10-6: Wool tufting of the area around the wipers of this NHW10 Toyota Prius showed that they were exposed to the main airflow.

237

SPEEDPRO SERIES

Figure 10-7: The degree to which the wipers were exposed to the airflow can be seen in this wool tuft shot. Note the attached flow below the black line on the windscreen.

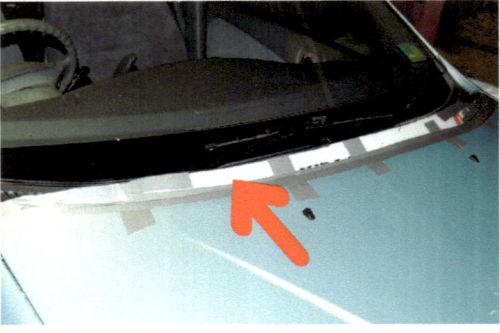

Figure 10-8: A trial deflector was made from plastic and tape. It worked in reducing wind noise from the wipers, so a final design was then made from ABS plastic.

Figure 10-9: The final deflector in place. Unless you looked for it, you wouldn't see it.

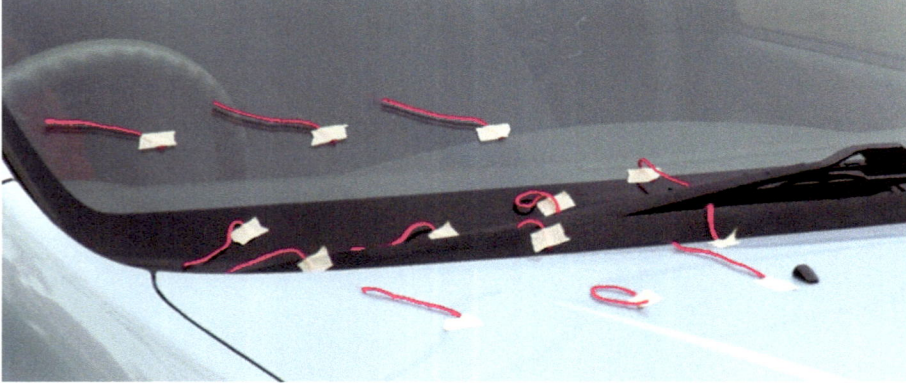

Figure 10-10: The airflow after the deflector was installed. Note how the flow attachment is now well above the black windscreen line, so the wipers are now in a small separation bubble. Wiper noise disappeared!

retained), cut and filed. ABS is much more resilient than acrylic, for example.

To angle the airflow upwards, the deflector needed to be heated and bent along its length. This was achieved a small section at a time, using a heat gun and few pieces of particle board to create a clamp and a lever. It's tricky to get the bend even all the way along, but ABS can be reheated and rebent as often as you like. Practice on a scrap bit first! The deflector was held in place with double-sided tape. Unless you knew what you were looking for, you couldn't even see it.

Wool tuft testing was again undertaken after the deflector had been fitted. This showed that the flow reattachment was higher up the windscreen – with the wipers now positioned in a small separation bubble.

The deflector worked very well. Aerodynamic noise from the wipers was no longer audible, while the final version of the deflector (which was a bit shorter than the white plastic prototype) didn't change the behaviour of the cabin ventilation system.

Note: since this modification was made, a product has been released that would achieve the same outcome much more easily. The product is a rubberised strip that is often fitted to cars as a small lip-type rear spoiler. It comes with pressure-sensitive adhesive on one surface, and so can be quickly cut to length and then stuck in position.

Volkswagen Beetle A-pillars
We met Gerrelt Molhoek in Chapter 6, where he fitted a rear spoiler to the top of the rear window of his 1973 Volkswagen Beetle to separate the flow. Well, Gerrelt's Beetle runs an Alfa Romeo flat-four engine, and he drives his Dutch

REDUCING AERODYNAMIC NOISE

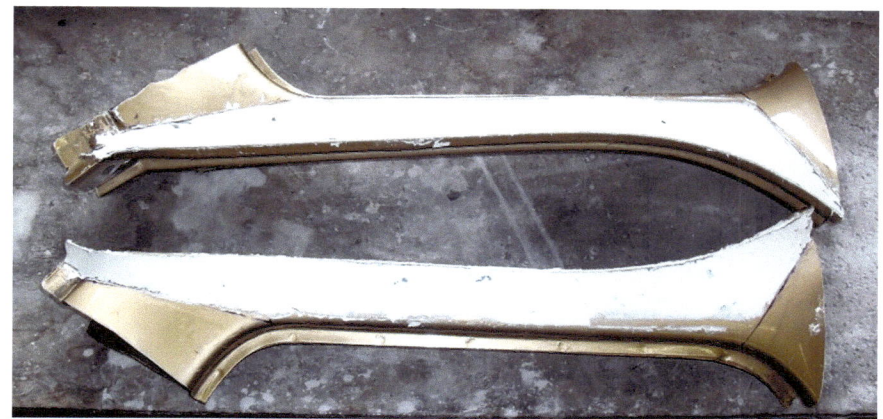

Figure 10-11: To reduce wind noise from the gutters on the A-pillars of his 1973 Volkswagen Beetle, Gerrelt Molhoek decided to make some vacuum-formed plastic covers. He obtained some A-pillars off another Beetle and then filled the gutters with fibreglass resin and body filler. These became the male moulds for vacuum-forming the thin plastic.
(Courtesy Gerrelt Molhoek)

highways at 120km/h (75mph) – at which speed aerodynamic noise from the A-pillars was intrusive. Looking at the pillars, he realised that the rain gutters stood very high – could they be smoothed to reduce noise? Early Porsche 911s use the same type of pillar design, and in those cars, plastic mouldings are available to cover the exposed rain gutters. People who fit these covers report a reduction in noise.

To test whether such an approach would be effective, Gerrelt covered the pillars with tape and went for a drive. With the tape in place, he noticed that the engine was now louder than the wind noise – and so the wind noise must have been reduced. However, plastic covers aren't commercially available for Beetles, so they needed to be made.

Gerrelt decided to vacuum-form some new covers in thin plastic. Not to do things by halves, he bought a front window frame from another Beetle, and then cut out the A-pillars. He then filled the rain gutters on the newly-acquired A-pillars with fibreglass resin, and then used body filler to smooth the upper surface. Using these pillars as the male moulds, he then heated ABS plastic sheet and, using a vacuum cleaner to provide the low pressure, vacuum-formed the covers to shape. (If you are not familiar with home vacuum forming, look it up – it's ideal for making small three-dimensional parts.)

The covers were held in place by the windscreen rubber along the length of the pillar, and some double-sided tape at the top and bottom. Wind noise has been reduced, and rain can still flow along the gutter through the top and bottom openings in the covers.

Figure 10-12: Here are the vacuum-formed plastic covers in place. Rain can still flow along the gutter through the openings top and bottom in the cover. Wind noise is reduced. (Courtesy Gerrelt Molhoek)

References

The following publications are referred to in this book. Note that while researching this book, I read many more publications than those cited, but these are the ones that I think are most useful to you.

BOOKS

(Note: for books with multiple editions, the edition that was used for this book is indicated.)

Allen, J. *Aerodynamics – the science of air in motion*, 2nd edition, Granada Publishing, 1982.

Barnard, RJ. *Road Vehicle Aerodynamic Design*, 3rd edition, MechAero, 2009.

Frere, P. *Porsche 911 Story*, Haynes Publishing, 7th Edition, 2002.

Hucho, WH (Ed). *Aerodynamics of Road Vehicles*, Butterworths, 1987.

Katz, J. *Race Car Aerodynamics – Designing for Speed*, Robert Bentley Publishers, 1995.

Kieselbach, RJF. *Stromlinienautos in Deutschland*, Kohlhammer, 1982.

Scibor-Rylski, AJ. *Road Vehicle Aerodynamics*, Pentech Press, 1975.

Tamai, G. *The Leading Edge – Aerodynamic Design of Ultra-streamlined Land Vehicles*, Bentley Publishers, 1999.

PAPERS

Atkinson, J. "Research into the Potential of Variable Aerodynamic Properties to Modify Ground Vehicle Behaviour." Coventry University, 2014.

Bannister, M. "Drag and Dirt Deposition Mechanisms of External Rear View Mirrors and Techniques Used for Optimisation." SAE Technical Paper 2000-01-0486, 2000.

Barnard, RH, Bullen, PR, Qiao, J. "Fixed and Variable Cooling Outlet Geometries for the Minimisation of Associated Drag." *Procs 5th MIRA International Vehicle Aerodynamics Conference, Warwick, UK*, 13-15 October 2004.

Barth, R. "Effect of Unsymmetrical Wind Incidence on Aerodynamic Forces Acting on Vehicle Models and Similar Bodies." SAE Technical Paper 650136, 1965.

Beauvais, F, Tignor, S, and Turner, T. "Problems of Ground Simulation in Automotive Aerodynamics." SAE Technical Paper 680121, 1968.

Buckley, F. "ABCD – An Improved Coast Down Test and Analysis Method." SAE Technical Paper 950626, 1995.

Buljac, A, Džijan, I, Korade, I, et al. "Automobile aerodynamics influenced by airfoil-shaped rear wing" *International Journal of Automotive Technology*, 2016.

Buscariolo, F, Magazoni, F, Maruyama, F, Alves, J, et al. "Aerodynamic Enablers Review for Automotive Applications." SAE Technical Paper 2016-36-0203, 2016.

Cai, J, Kapoor, S, Sikder, T, and He, Y. "Effects of Active Aerodynamic Wings on Handling Performance of High-Speed Vehicles." SAE Technical Paper 2017-01-1592, 2017

Chaligné S, Turner R, Gaylard A. (2018) "The Aerodynamics Development of the New Land Rover Discovery." In: Wiedemann J (eds) *Progress in Vehicle Aerodynamics and Thermal Management, FKFS 2017*. Springer, Cham.

Cooper, K, Syms, J, and Sovran, G. "Selecting Automotive Diffusers to Maximise Underbody Downforce." SAE Technical Paper 2000-01-0354, 2000.

D'Hooge, A, Palin, R, Johnson, S, Duncan, B, et al. "The Aerodynamic

REFERENCES

Development of the Tesla Model S – Part 2: Wheel Design Optimization." SAE Technical Paper 2012-01-0178, 2012.

Emmelmann, H, Berneburg, H, and Schulze, J. "The Aerodynamic Development of the Opel Calibra." SAE Technical Paper 900317, 1990.

Evrard, A, et al. "Comparative effects of vortex generators on Ahmed's squareback and minivan car models." *Journal of Automobile Engineering, I Mech E*, 2017.

Froling, T and Juechter, T. "2006 Chevrolet Corvette C6 Z06 Aerodynamic Development." SAE Technical Paper 2005-01-1943, 2005.

Garcia de la Cruz, J, Brackston, R, and Morrison, J. "Adaptive Base-Flaps Under Variable Cross-Wind." SAE Technical Paper 2017-01-7000, 2017.

George, A. "Automobile Aerodynamic Noise." SAE Technical Paper 900315, 1990.

Howell, J and Le Good, G. "The Influence of Aerodynamic Lift on High Speed Stability." SAE Technical Paper 1999-01-0651, 1999.

Janson, T, and Piechna, J. "Numerical Analysis of Aerodynamic Characteristics of a High-Speed car with Moveable Bodywork Elements." *Archive of Mechanical Engineering*, Volume 62, Issue 4, Pages 451–476, ISSN (Online) 2300-1895, 2015.

Kataoka, T, China, H, Nakagawa, K, Yanagimoto, K, et al. "Numerical Simulation of Road Vehicle Aerodynamics and Effect of Aerodynamic Devices." SAE Technical Paper 910597, 1991.

Kataoka, S, et al. "Aerodynamics for Lancer Evolution X." *Mitsubishi Motors Technical Review*, No 20, 2008.

Kawakami, M, Murata, O, and Maeda, K. "Improvement in Vehicle Motion Performance by Suppression of Aerodynamic Load Fluctuations." *SAE Int J Passeng Cars – Mech Syst* 8(1):205-216, SAE Technical Paper 2015-01-1537, 2015.

Kawamata, H, Kuroda, S, Tanaka, S, and Oshima, M. "Improvement of Practical Electric Consumption by Drag Reducing under Cross Wind." SAE Technical Paper 2016-01-1626, 2016.

Koike, M, Nagayoshi, T, and Hamamoto, N. "Research on Aerodynamic Drag Reduction by Vortex Generators." *Mitsubishi Motors Technical Review*, 2004, No 16.

Landström, C, Walker, T, Christoffersen, L, and Löfdahl, L. "Influences of Different Front and Rear Wheel Designs on Aerodynamic Drag of a Sedan Type Passenger Car." SAE Technical Paper 2011-01-0165, 2011.

Larose, G, Belluz, L, Whittal, I, Belzile, M, et al. "Evaluation of the Aerodynamics of Drag Reduction Technologies for Light-duty Vehicles: a Comprehensive Wind Tunnel Study." *SAE Int J Passeng Cars – Mech Syst* 9(2):772-784, 2016.

Le Good, G, Howell, J, Passmore, M, and Garry, K. "On-Road Aerodynamic Drag Measurements Compared with Wind Tunnel Data." SAE Technical Paper 950627, 1995.

Le Good, G, Howell, J, Passmore, M, and Cogotti, A. "A Comparison of On-Road Aerodynamic Drag Measurements with Wind Tunnel Data from Pininfarina and MIRA." SAE Technical Paper 980394, 1998.

Levin, J, and Rigdal R. "Aerodynamic analysis of drag reduction devices on the underbody of a SAAB 9-3 by using CFD." Chalmers University of Technology, Gothenburg, Sweden, 2011.

Littlewood, R, Passmore, M, and Wood, D. "An Investigation into the Wake Structure of Square Back Vehicles and the Effect of Structure Modification on Resultant Vehicle Forces." *SAE Int J Engines* 4(2):2629-2637, SAE Technical Paper 2011-37-0015, 2011.

Leuschen J, Cooper KR. (2009) "Summary of Full-Scale Wind Tunnel Tests of Aerodynamic Drag-Reducing Devices for Tractor-Trailers" In: Browand F, McCallen R, Ross J (eds) *The Aerodynamics of Heavy Vehicles II: Trucks, Buses, and Trains Lecture Notes in Applied and Computational Mechanics*, vol 41 Springer, Berlin, Heidelberg.

Lew, C, Gopalaswamy, N, Shock, R, Duncan, B, et al. "Aerodynamic Simulation of a Standalone Rotating Treaded Tire." SAE Technical Paper 2017-01-1551, 2017.

Lounsberry, T, Gleason, M, and Puskarz, M. "Laminar Flow Whistle on a Vehicle Side Mirror." SAE Technical Paper 2007-01-1549, 2007.

Marcell, R and Romberg, G. "The Aerodynamic Development of the Charger Daytona for Stock Car Competition." SAE Technical Paper 700036, 1970.

Mayer, W and Wiedemann, J. "The Influence of Rotating Wheels on Total Road Load." SAE Technical Paper 2007-01-1047, 2007.

Mayer, W and Wickern, G. "The New Audi A6/A7 Family – Aerodynamic Development of Different Body Types on One Platform." *SAE Int J Passeng Cars – Mech Syst* 4(1):197-206, SAE Technical paper 2011-01-0175, 2011.

Meder, J. "Aerodynamic Shapeshifter – The New Porsche 911 Turbo." FKFS –

SPEEDPRO SERIES

9th Aerodynamic Conference, 2013.

Nakamura, D, Onishi, Y, and Takehara, Y. "Flow Field Analysis in the Development of the 2013 Model Year Accord Hybrid." SAE Technical Paper 2015-01-1534, 2015, https://doi.org/10.4271/2015-01-1534.

Ng, E, Watkins, S, Johnson, P, and Mole, L. "Use of a Pressure-Based Technique for Evaluating the Aerodynamics of Vehicle Cooling Systems." SAE Technical Paper 2002-01-0712, 2002.

Okada, Y, Nouzawa, T, Nakamura, T, and Okamoto, S. "Flow Structures above the Trunk Deck of Sedan-Type Vehicles and Their Influence on High-Speed Vehicle Stability 1st Report: On-Road and Wind-Tunnel Studies on Unsteady Flow Characteristics that Stabilize Vehicle Behavior." *SAE Int J Passeng Cars – Mech Syst* 2(1):138-156, SAE Technical Paper 2009-01-0004, 2009.

Palaskar, P. "Effect of Side Taper on Aerodynamics Drag of a Simple Body Shape with Diffuser and without Diffuser." SAE Technical Paper 2016-01-1621, 2016.

Park, H. "A Rear-View Side Mirror with Exterior Lens to Improve Field of View and Aerodynamics of Automobiles." SAE Technical Paper 2017-01-1358, 2017, https://doi.org/10.4271/2017-01-1358.

Passmore, M and Le Good, G. "A Detailed Drag Study Using the Coastdown Method." SAE Technical Paper 940420, 1994.

Pitman, J and Gaylard, A. "An experimental investigation into the flow mechanisms around an SUV in open and closed cooling air conditions." *11th FKFS Conference: Progress in Vehicle Aerodynamics and Thermal Management*, 2017.

Sapnaras, D and Dimitriou, I. "Experimental Analysis of the Underbody Pressure Distribution of a Series Vehicle on the Road and in the Wind Tunnel." SAE Technical Paper 2008-01-0802, 2008.

Santer, R and Gleason, M. "The Aerodynamic Development of the Probe IV Advanced Concept Vehicle." SAE Technical Paper 831000, 1983.

Schenkel, F. "The Origins of Drag and Lift Reductions on Automobiles with Front and Rear Spoilers." SAE Technical Paper 770389, 1977.

Sebben, S. "Numerical Simulations of a Car Underbody: Effect of Front-Wheel Deflectors." SAE Technical Paper 2004-01-1307, 2004

Shigarkanthi, V, Damodaran, V, Soundararaju, D, and Kanniah, K. "Application of Design of Experiments and Physics Based Approach in the Development of Aero Shutter Control Algorithm." SAE Technical Paper 2011-01-0155, 2011.

Simmonds, N, Pitman, J, Tsoutsanis, P, Jenkins, K, et al. "Complete Body Aerodynamic Study of three Vehicles." SAE Technical Paper 2017-01-1529, 2017, https://doi.org/10.4271/2017-01-1529.

Tortosa, N and Karbon, K. "Aerodynamic Development of the 2011 Chevrolet Volt." *SAE Int J Passeng Cars – Mech Syst* 4(1):166-171, 2011.

Unni, T. "Numerical Investigation on Aerodynamic Effects of Vanes and Flaps on Automotive Underbody Diffusers." SAE Technical Paper 2017-01-2163, 2017.

Vdovin, A. "Investigation of Aerodynamic Resistance of Rotating Wheels on Passenger Cars" Chalmers University of Technology, Gothenburg, Sweden, 2013.

Wang, F, Yin, Z, Yan, S, Zhan, J, et al. "Validation of Aerodynamic Simulation and Wind Tunnel Test of the New Buick Excelle GT." *SAE Int J Passeng Cars – Mech Syst* 10(1):195-202, 2017, SAE Technical Paper 2017-01-1512.

Wickern, G and Brennberger, M. "Scaling Laws in Automotive Aeroacoustics." SAE Technical Paper 2009-01-0180, 2009.

MAGAZINE

Sherman, D. "Drag Queens: Aerodynamics Compared." *Car and Driver*, June 2014

Veloce SpeedPro books –

 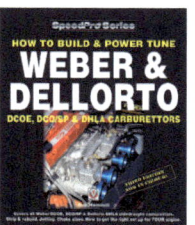 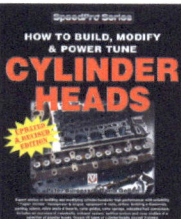  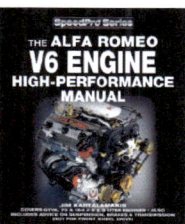 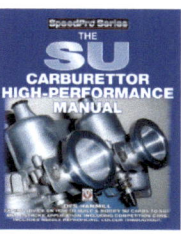 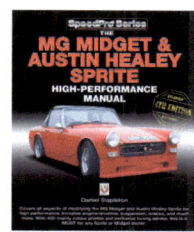

978-1-903706-59-6   978-1-903706-75-6   978-1-903706-76-3   978-1-903706-99-2   978-1-845840-21-1   978-1-787111-68-4   978-1-787110-01-4

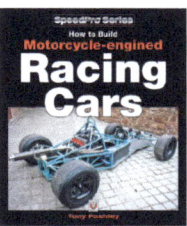  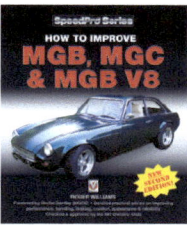 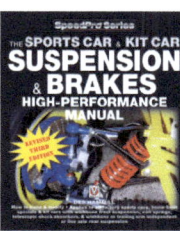 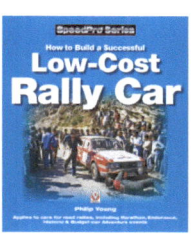 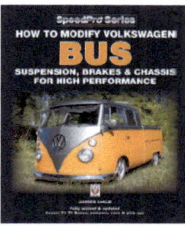 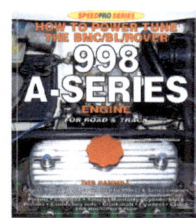

978-1-787111-69-1   978-1-787111-73-8   978-1-845841-87-4   978-1-845842-07-9   978-1-845842-08-6   978-1-845842-62-8   978-1-901295-26-9

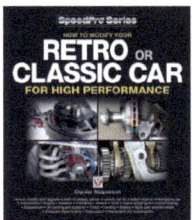 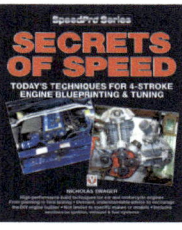 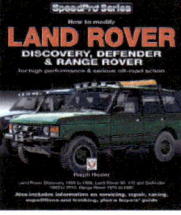    

978-1-845842-89-5   978-1-845842-97-0   978-1-845843-15-1   978-1-845843-55-7   978-1-845844-33-2   978-1-845844-38-7   978-1-787113-34-3

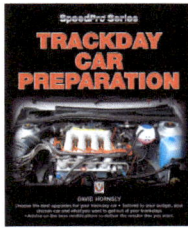 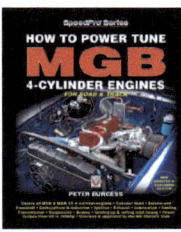  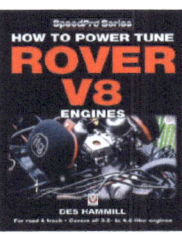   

978-1-845844-83-7   978-1-787113-41-1   978-1-845848-33-0   978-1-787111-76-9   978-1-845848-69-9   978-1-845849-60-3   978-1-787110-91-5

 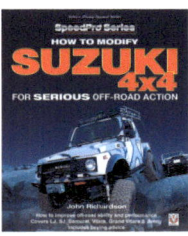 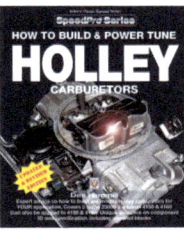 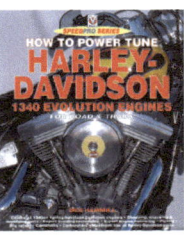 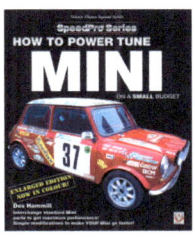 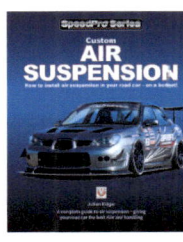 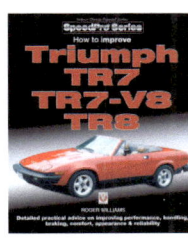

978-1-845840-19-8   978-1-787110-92-2   978-1-787110-47-2   978-1-903706-94-7   978-1-787110-87-8   978-1-787111-79-0   978-1-787110-88-5

  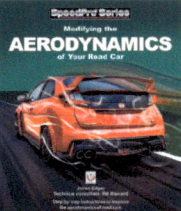

978-1-903706-78-7   978-1-787113-18-3   978-1-787112-83-4

– more on the way!

MORE FROM VELOCE ...

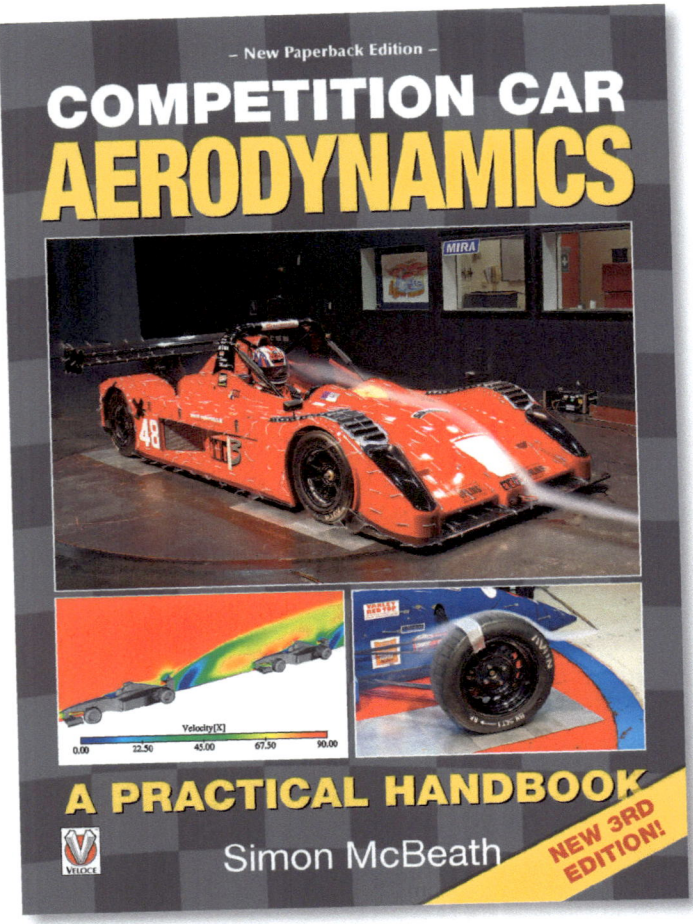

From historical background to state of the art techniques, and with chapters covering airdams, splitters, spoilers, wings, underbodies and myriad miscellaneous devices, *Competition Car Aerodynamics* also features in-depth case studies from across the motorsport spectrum to help develop a comprehensive understanding of the subject.

ISBN: 978-1-787111-02-8
Paperback • 23.3x16.9cm • 320 pages

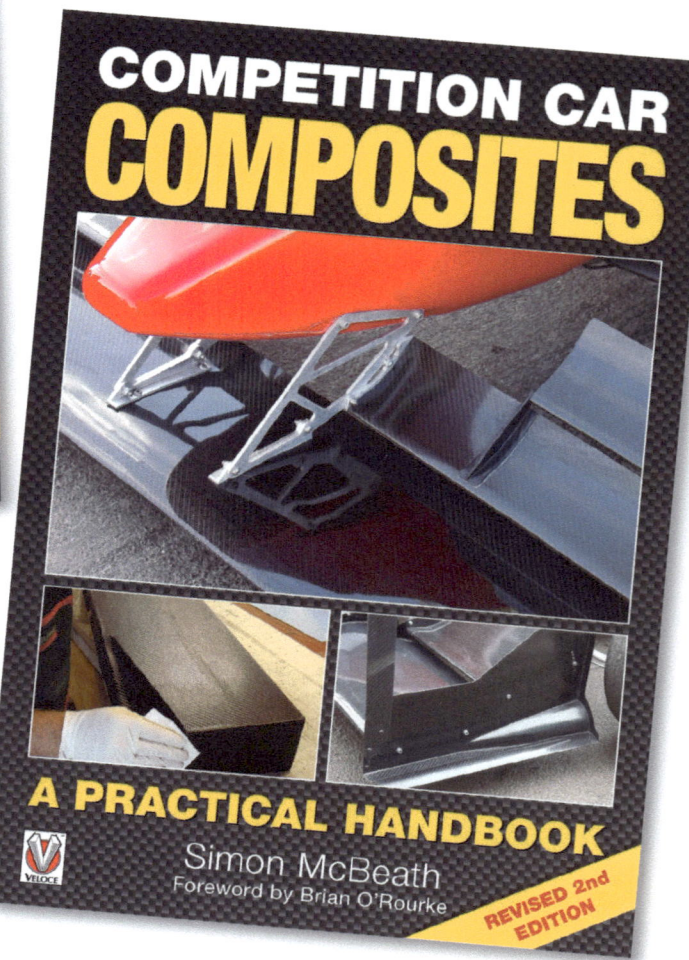

From basic methods to advanced techniques: with chapters covering materials, patterns, moulds, components and technology upgrades applicable to the home workshop, this book will help any reader – whether building, repairing, or developing competition cars or components – exploit composites technology to gain performance advantages.

ISBN: 978-1-845849-05-4
Hardback • 23.3x16.9cm • 208 pages

For more information and price details, visit the Veloce website at www.veloce.co.uk
email: info@veloce.co.uk • Tel: +44(0)1305 260068

... AND MORE FROM JULIAN EDGAR

Want to modify, restore or maintain your car at home? This book is a must-read that covers the complete setting-up and use of a home workshop. From small and humble to large and lavish – this book shows you the equipment to buy and build, the best interior workshop layouts, and how to achieve great results.

ISBN: 978-1-787112-08-7
Paperback • 25x20.7cm • 160 pages
• 250 pictures

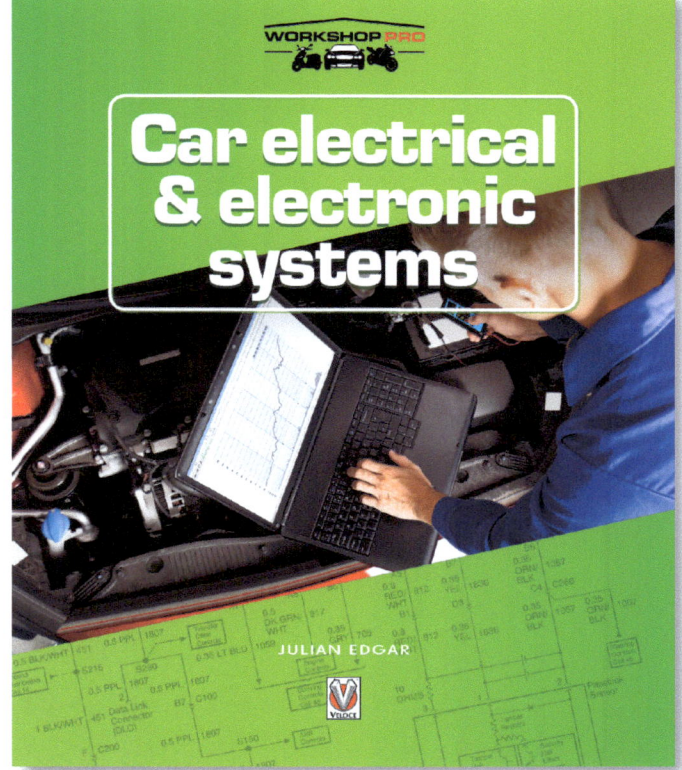

Want to restore, modify or repair your car's electrical and/or electronic systems? This handbook is a must-read that takes you from the basics of circuits right through to diagnosing and repairing complex electronic car systems.

ISBN: 978-1-787112-81-0
Paperback • 25x20.7cm • 168 pages
• 262 colour pictures

For more information and price details, visit the Veloce website at www.veloce.co.uk
email: info@veloce.co.uk • Tel: +44(0)1305 260068

SPEEDPRO SERIES

Modifying your car for increased performance? You need this book! It shows you how to easily measure on the road the gains and losses of changing air intakes, exhausts, cams and turbos. Also learn how to test suspension, brakes and car aerodynamics – accurately and at low cost.

ISBN: 978-1-787113-18-3
Paperback • 25x20.7cm • 72 pages
• 83 pictures

The first book that shows you how to fit air suspension to your car. It covers both theory and practice, and includes the step-by-step fitting of aftermarket air suspension systems and building your own with parts from other cars. If you want the best ride and handling for your road car, this is the book you need!

ISBN: 978-1-787111-79-0
Paperback • 25x20.7cm • 64 pages
• 82 colour and b&w pictures

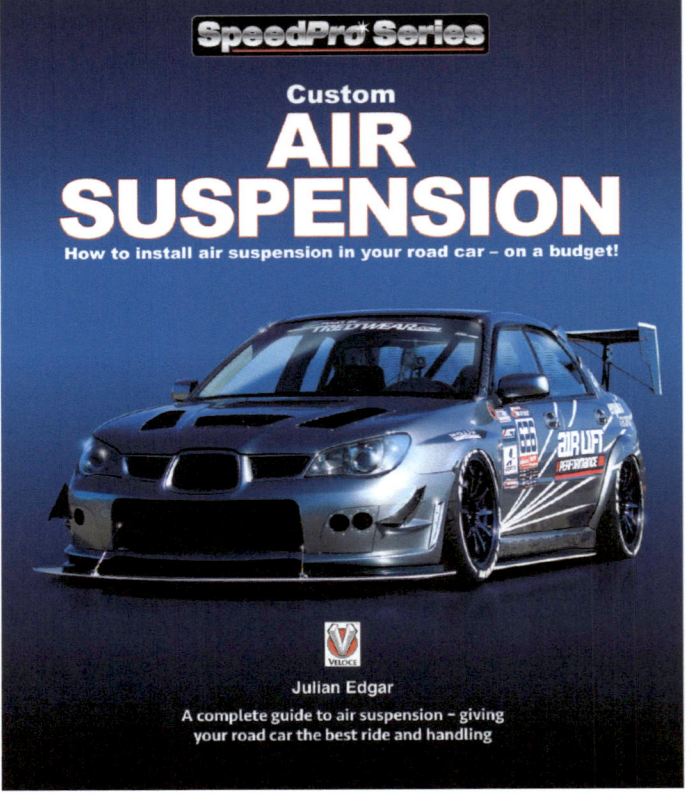

For more information and price details, visit the Veloce website at www.veloce.co.uk
email: info@veloce.co.uk • Tel: +44(0)1305 260068

Index

Active aerodynamics 204
Aerodynamic noise 233
Air brake 209
Air curtains 161
Air dam – see spoiler
Airflow speed 12, 81
Alpine 110 185
Alternator cooling duct 220
Attached flow 10, 14
Audi 80 182
Audi A2 168
Audi A4 Avant 152
Audi R8 183, 211
Audi S4 219, 232

Base 10
Belly-pan – see undertray
BMW 3.0CSL 50
Boat-tailing 127
Bonnet scoops 212, 222
Bonnet vents 217, 222
Boundary layer 12, 16
Brake duct – see alternator cooling duct

Calibra 41
Canards 194
Centre of gravity 174
    measuring 176
Centre of pressure 174
    measuring 176
Clay testing 55
Clean separation 10, 14, 137
Coast-down testing 101, 102
Coefficient of drag 20
Coefficient of lift 20
Cooling drag – reduction 120

Diffusers 182, 188
Dodge Charger 36
Downforce – creating 171
Downforce – measuring 108
Downforce – road cars 205

Drag
    form 9
    induced 17
    interference 14
    measuring 100
    reduction 115
    types 9
    viscous 11
Drag coefficient – see coefficient of drag
Ducted cooling 121, 124
Ducts – engine intake 229
Dust testing 56, 70, 133, 134

Effective backlight angle 130
Engine intakes 224

Fins 178, 202
Flow testing 52
Flow visualisation 52
Ford
    Escort 60
    Falcon 136, 232
    Galaxy 172
    Mustang 153, 161
    Probe IV 51, 160
Frontal area 20
Frontal area – reducing 115
Fuel economy testing 101, 105

General Motors
    Buick Excelle GT 141
    Cadillac ATS V 223
    Chevrolet Corvette 179
    Chevrolet Cruze 55
    Chevrolet Volt 45, 49
    EV1 43
GMC Canyon 135, 154
Grille shutters and blockers 122
Ground clearance 125

Hatch – rear angle 129
Heat exchanger flow 210

Holden Commodore 39, 196
Honda
    Accord 163
    Civic 138, 194
    Dream 51
    Insight 128, 147, 162, 188, 223, 231
    Legend 148
    NSX 210
Hood scoops – see bonnet scoops
Hood vents – see bonnet vents
HSV Group A Commodore 39
Hyundai i30 56

Intercooler airflow 212, 219

Jaguar XE 14, 18, 84, 201

Lamborghini Aventador 184
Land Rover Discovery 166
Lexus
    LC500 162
    LFA 194, 201
    LS400 62, 113
Lift 19
    measuring 108
    reducing 171

Magnehelic gauge 73, 76
Manometer
    digital 74
    fluid 72
Mazda
    Astina 67
    MX5/Miata 193, 197
    RX7 66
Mercedes Type 80 50
Mercedes-AMG GT 96
Mercedes-Benz
    CLA250 49
    CIII 175
    230 59, 16, 59, 68, 79, 80
    190E 142

SPEEDPRO SERIES

300SL 172
W126 54, 55
W211 67, 79, 80, 130
MG EX181 49
Mitsubishi
　Evo 143, 201
　Magna 67
　Mirage 129

Nissan
　Leaf 49, 164
　Maxima 212
　Micra 165
Noise –
　aerodynamic 233
　reduction 235
　testing 234
NSU
　Delphin III 175
　Ro 80 33

Opel
　Calibra 41
　Rak 2 28

Pagani Huayra 206
Panhard Dynavia 50
Pitot tube 23, 81
Plymouth Superbird 36, 200
Pop-up headlights 67
Porsche
　Boxster 207
　Carrera 195
　911 Turbo S 208
　924 13
　964 173
　993 63, 173
　996 173
　Type 64 32
Power testing 101
Pressure 12, 13

differential 211
drops in engine intake 226
dynamic 22
gradient 14
measurement 71, 213, 217, 225
static 22
Probes – pressure measuring 75, 81, 85
Projected frontal area – see frontal area

Rake 125
Rear vision drag
　measuring 106
　reduction 117
Reynolds numbers 17
Ride height – drag reduction 125
Road testing 52
Rover 820si 197
Rumpler Tropfenwagen 25

Saab
　9-3 186
　900 197
Sealing openings 223
Separated flow 10, 14
Separation bubble 16, 140
Separation edges 14
Skoda Roomster 131, 150, 227
Smoke testing 55
Speed – maximum 21
Spoilers 192, 214
Stability 173
Subaru Impreza 60, 211
Superbird 36
Suspension height sensors 109

Tail extensions 128, 170
Tatra
　T-77 29
　TT-87 29

T-97 31
Tatraplan 31
Tesla Model S 49, 92, 160
Topper 135
Toyota
　Camry 57
　Prius 16, 49, 129, 146, 154, 158, 237
　T-100 128, 154, 170
Trailer 129
Tropfenwagen 25
Turbulence 11
Turning vanes 221
Tyres – drag reduction 157

Undercar flows 19
Undertrays 151, 154, 182, 188, 213
Upwash 58

Vectors – force 15
Viscosity – air 11
Volkswagen
　1 litre 51
　Beetle 65, 239
　Transporter 32, 174
　XL-1 47, 119
Volvo XC-90 57
Vortex generators 143, 149, 197
Vortices – trailing 17, 19, 162

Wake 10
　expansion 136
　reduction 127, 130
Wheels – drag reduction 157
Wings 22, 37, 39, 198
Wool tuft testing 53, 58, 87, 92, 96, 146, 149, 195, 237

XL-1 47

Yaw 22

www.velocebooks.com / www.veloce.co.uk
Details of all current books • New book news • Special offers • Gift vouchers • Forum